Mechanical Engineering Series

Frederick F. Ling
Series Editor

Springer
New York
Berlin
Heidelberg
Barcelona
Budapest
Hong Kong
London
Milan
Paris
Santa Clara
Singapore
Tokyo

Mechanical Engineering Series

Introductory Attitude Dynamics
F.P. Rimrott

Balancing of High-Speed Machinery
M.S. Darlow

Theory of Wire Rope
G.A. Costello

Theory of Vibration: An Introduction, 2nd ed.
A.A. Shabana

Theory of Vibration: Discrete and Continuous Systems
A.A. Shabana

Laser Machining: Theory and Practice
G. Chryssolouris

Underconstrained Structural Systems
E.N. Kuznetsov

Principles of Heat Transfer in Porous Media, 2nd ed.
M. Kaviany

Mechatronics: Electromechanics and Contromechanics
D.K. Miu

Structural Analysis of Printed Circuit Board Systems
P.A. Engel

Kinematic and Dynamic Simulation of Multibody Systems: The Real-Time Challenge
J. García de Jalón and E. Bayo

High Sensitivity Moiré: Experimental Analysis for Mechanics and Materials
D. Post, B. Han, and P. Ifju

Principles of Convective Heat Transfer
M. Kaviany

(continued after index)

A.A. Shabana

Theory of Vibration

An Introduction

Second Edition

With 212 Figures

A.A. Shabana
Department of Mechanical Engineering
University of Illinois at Chicago
P.O. Box 4348
Chicago, IL 60680
USA

Series Editor
Frederick F. Ling
Ernest F. Gloyna Regents Chair in Engineering
Department of Mechanical Engineering
The University of Texas at Austin
Austin, TX 78712-1063 USA
and
William Howard Hart Professor Emeritus
Department of Mechanical Engineering,
Aeronautical Engineering and Mechanics
Rensselaer Polytechnic Institute
Troy, NY 12180-3590 USA

Library of Congress Cataloging-in-Publication Data
Shabana, Ahmed A.
Theory of vibration: an introduction / A.A. Shabana. — 2nd ed.
p. cm. — (Mechanical engineering series)
Includes bibliographical references and index.
ISBN 0-387-94524-5
1. Vibration. I. Title. II. Series: Mechanical engineering series (Berlin, Germany)
QA865.S49 1995
531'.32—dc20 95-8509

Printed on acid-free paper.

Production coordinated by Publishing Network and managed by Francine McNeill; manufacturing supervised by Joe Quatela.
Typeset by Asco Trade Typesetting Ltd., Hong Kong.
Printed and bound by Maple-Vail Book Manufacturing Group, York, PA.
Printed in the United States of America.

9 8 7 6 5 4 3 2 1

ISBN 0-387-94524-5 Springer-Verlag New York Berlin Heidelberg

Dedicated to my family

Series Preface

Mechanical engineering, an engineering discipline borne of the needs of the industrial revolution, is once again asked to do its substantial share in the call for industrial renewal. The general call is urgent as we face profound issues of productivity and competitiveness that require engineering solutions, among others. The Mechanical Engineering Series features graduate texts and research monographs intended to address the need for information in contemporary areas of mechanical engineering.

The series is conceived as a comprehensive one that covers a broad range of concentrations important to mechanical engineering graduate education and research. We are fortunate to have a distinguished roster of consulting editors on the advisory board, each an expert in one of the areas of concentration. The names of the consulting editors are listed on the next page of this volume. The areas of concentration are: applied mechanics; biomechanics; computational mechanics; dynamic systems and control; energetics; mechanics of materials; processing; thermal science; and tribology.

Professor Marshek, the consulting editor for dynamic systems and control, and I are pleased to present the second edition of *Theory of Vibration: An Introduction* by Professor Shabana. We note that this is the first of two volumes. The second deals with *discrete and continuous systems.*

Austin, Texas Frederick F. Ling

Mechanical Engineering Series

Frederick F. Ling
Series Editor

Preface

The aim of this book is to provide a presentation for the theory of vibration suitable for junior and senior undergraduate students. This book, which is based on class notes that I have used for several years, is in many ways different from existing textbooks. Basic dynamic concepts are used to develop the equations of the oscillatory motion, the assumptions used to linearize the dynamic equations are clearly stated, and the relationship between the coefficients of the differential equations and the stability of mechanical systems is discussed more thoroughly.

This text, which can be covered entirely in one semester, is intended for an introductory course on the theory of vibration. New concepts are therefore presented in simple terms and the solution procedures have been explained in detail. The material covered in the volume comprises the following chapters.

In Chapter 1, basic definitions related to the theory of vibration are presented. The elements of the vibration models, such as inertia, elastic, and damping forces, are discussed in Section 2 of this chapter. Sections 3, 4, and 5 are devoted to the use of Newton's second law and D'Alembert's principle for formulating the equations of motion of simple vibratory systems. In Section 5 the dynamic equations that describe the translational and rotational displacements of rigid bodies are presented, and it is shown that these equations can be nonlinear because of the finite rotation of the rigid bodies. The linearization of the resulting differential equations of motion is the subject of Section 6. In Section 7 methods for obtaining simple finite number of degrees of freedom models for mechanical and structural systems are discussed.

Chapter 2 describes methods for solving both homogeneous and nonhomogeneous differential equations. The effect of the coefficients in the differential equations on the stability of the vibratory systems is also examined. Even though students may have seen differential equations in other courses, I have found that presenting Chapter 2 after discussing the formulation of the equations of motion in Chapter 1 is helpful.

Chapter 3 is devoted to the free vibrations of single degree of freedom systems. Both cases of undamped and damped free vibration are considered. The stability of undamped and damped linear systems is examined, the cases of viscous, structural, Coulomb, and negative damping are discussed, and examples for oscillatory systems are presented.

Chapter 4 is concerned with the forced vibration of single degree of freedom systems. Both cases of undamped and damped forced vibration are considered, and the phenomena of resonance and beating are explained. The forced vibrations, as the result of rotating unbalance and base excitation, are discussed in Sections 5 and 6. The theoretical background required for understanding the function of vibration measuring instruments is presented in Section 7 of this chapter. Methods for the experimental evaluation of the damping coefficients are covered in Section 8.

In the analysis presented in Chapter 4, the forcing function is assumed to be harmonic. Chapter 5 provides an introduction to the vibration analysis of single degree of freedom systems subject to nonharmonic forcing functions. Periodic functions expressed in terms of Fourier series expansion are first presented. The response of the single degree of freedom system to a unit impulse is defined in Section 5. The impulse response is then used in Section 6 to obtain the response of the single degree of freedom system to an arbitrary forcing function, and a method for the frequency analysis of such an arbitrary forcing function is presented in Section 7. In Section 8, computer methods for the vibration analysis of nonlinear systems are discussed.

In Chapter 6, the linear theory of vibration of systems that have more than one degree of freedom is presented. The equations of motion are presented in a matrix form, and the case of damped and undamped free and forced vibration, as well as the theory of the vibration absorber of undamped and damped systems, are discussed.

Chapter 7 presents a brief introduction to the theory of vibration of continuous systems. The longitudinal, torsional, and transverse vibrations are discussed, and the orthogonality conditions of the mode shapes are presented and used to obtain a decoupled system of ordinary differential equations expressed in terms of the modal coordinates. A more detailed discussion on the vibration of continuous systems is presented in a second volume: *Theory of Vibration: Discrete and Continuous Systems* (Shabana, 1991).

I would like to thank many of my teachers and colleagues who contributed, directly or indirectly, to this book. I wish to acknowledge gratefully the many helpful comments and suggestions offered by my students. I would also like to thank Dr. D.C. Chen, Dr. W.H. Gau, and Mr. J.J. Jiang for their help in reviewing the manuscript and producing some of the figures. Thanks are due also to Ms. Denise Burt for the excellent job in typing the manuscript. The editorial and production staff of Springer-Verlag deserve special thanks for their cooperation and thorough professional work in producing this book. Finally, I thank my family for their patience and encouragement during the period of preparation of this book.

Chicago, Illinois — Ahmed A. Shabana

Contents

1
Introduction

The process of change of physical quantities such as displacements, velocities, accelerations, and forces may be grouped into two categories; oscillatory and nonoscillatory. The oscillatory process is characterized by alternate increases or decreases of a physical quantity. A nonoscillatory process does not have this feature. The study of oscillatory motion has a long history, extending back to more than four centuries ago. Such a study of oscillatory motion may be said to have started in 1584 with the work of Galileo (1564–1642) who examined the oscillations of a simple pendulum. Galileo was the first to discover the relationship between the frequency of the simple pendulum and its length. At the age of 26, Galileo discovered the law of falling bodies and wrote the first treatise on modern dynamics. In 1636 he disclosed the idea of the pendulum clock which was later constructed by Huygens in 1656.

An important step in the study of oscillatory motion is the formulation of the dynamic equations. Based on Galileo's work, Sir Isaac Newton (1642–1727) formulated the laws of motion in which the relationship between force, mass, and momentum is established. At the age of 45, he published his *Principle Mathematica* which is considered the most significant contribution to the field of mechanics. In particular, Newton's second law of motion has been a basic tool for formulating the dynamic equations of motion of vibratory systems. Later, the French mathematician Jean le Rond D'Alembert (1717–1783) expressed Newton's second law in a useful form, known as D'Alembert's principle, in which the inertia forces are treated in the same way as the applied forces. Based on D'Alembert's principle, Joseph Louis Lagrange (1736–1813) developed his well-known equations; Lagrange's equations, which were presented in his *Mechanique*. Unlike Newton's second law which uses vector quantities, Lagrange's equations can be used to formulate the differential equations of dynamic systems using scalar energy expressions. The Lagrangian approach, as compared to the Newtonian approach, lends itself easily to formulating the vibration equations of multidegree of freedom systems.

Another significant contribution to the theory of vibration was made by Robert Hooke (1635–1703) who was the first to announce, in 1676, the relationship between the stress and strain in elastic bodies. Hooke's law for deformable bodies states that the stress at any point on a deformable body is proportional to the strain at that point. In 1678, Hooke explained his law as "The power of any springy body is in the same proportion with extension." Based on Hooke's law of elasticity, Leonhard Euler (1707–1783) in 1744 and Daniel Bernoulli (1700–1782) in 1751 derived the differential equation that governs the vibration of beams and obtained the solution in the case of small deformation. Their work is known as Euler–Bernoulli beam theory. Daniel Bernoulli also examined the vibration of a system of n point masses and showed that such a system has n independent modes of vibration. He formulated the principle of superposition which states that the displacement of a vibrating system is given by a superposition of its modes of vibrations.

The modern theory of mechanical vibration was organized and developed by Baron William Strutt, Lord Rayleigh (1842–1919), who published his book in 1877 on the theory of sound. He also developed a method known as Rayleigh's method for finding the fundamental natural frequency of vibration using the principle of conservation of energy. Rayleigh made a correction to the technical beam theory (1894) by considering the effect of the rotary inertia of the cross section of the beam. The resulting equations are found to be more accurate in representing the propagation of elastic waves in beams. Later, in 1921, Stephen Timoshenko (1878–1972) presented an improved theory, known as Timoshenko beam theory, for the vibrations of beams. Among the contributors to the theory of vibrations is Jean Baptiste Fourier (1768–1830) who developed the well-known Fourier series which can be used to express periodic functions in terms of harmonic functions. Fourier series is widely used in the vibration analysis of discrete and continuous systems.

1.1 BASIC DEFINITIONS

In vibration theory, which is concerned with the oscillatory motion of physical systems, the motion may be harmonic, periodic, or a general motion in which the amplitude varies with time. The importance of vibration to our comfort and needs is so great that it would be pointless to try to list all the examples which come to mind. Vibration of turbine blades, chatter vibration of machine tools, electrical oscillations, sound waves, vibrations of engines, torsional vibrations of crankshafts, and vibrations of automobiles on their suspensions can all be regarded as coming within the scope of vibration theory. We shall, however, be concerned in this book with the vibrations of mechanical and structural systems.

Vibrations are encountered in many mechanical and structural applications, for example, mechanisms and machines, buildings, bridges, vehicles, and aircraft; some of these systems are shown in Fig. 1. In many of these systems, excessive vibrations produce high stress levels, which in turn may cause mechanical failure. Vibration can be classified as *free* or *forced* vibration. In free vibration, there are no external forces that act on the system, while forced

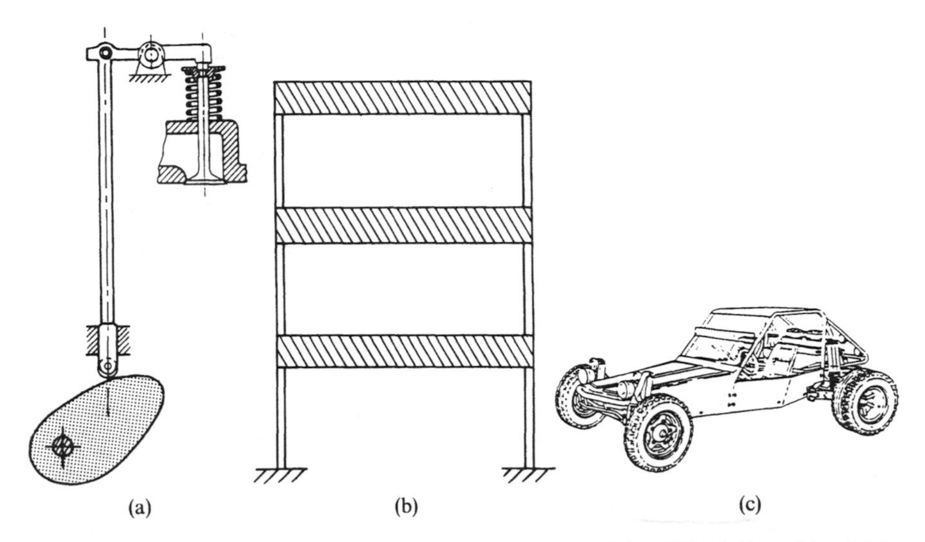

FIG. 1.1. Physical systems: (a) mechanism systems; (b) multistory buildings; (c) vehicle systems.

vibrations, are the result of external excitations. In both cases of free and forced vibration the system must be capable of producing restoring forces which tend to maintain the oscillatory motion. These restoring forces can be produced by discrete elements such as the linear and torsional springs shown, respectively, in Fig. 2(a) and (b) or by continuous structural elements such as beams and plates (Fig. 2(c), (d)).

These discrete and continuous elastic elements are commonly used in many systems, such as the suspensions and frames of vehicles, the landing gears, fuselage, and wings of aircraft, bridges, and buildings. The restoring forces produced by the elastic elements are proportional to the deflection or the elastic deformation of these elements. If the vibration is small, it is customary to assume that the force–deflection relationship is linear, that is, the force is equal to the deflection multiplied by a proportionality constant. In this case the *linear theory of vibration* can be applied. If the assumptions

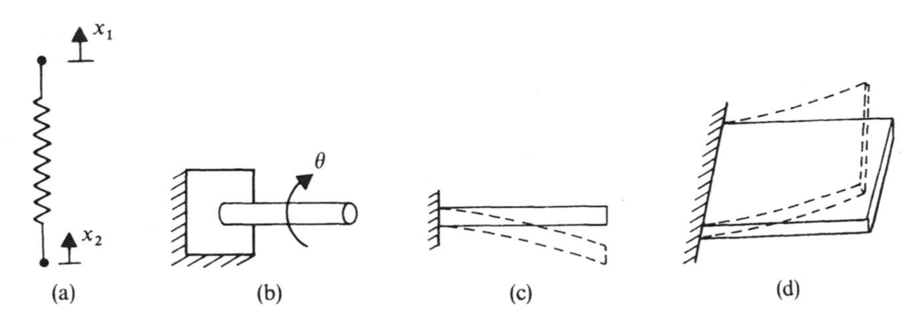

FIG. 1.2. Elastic elements.

of the linear theory of vibration are not valid, for example, if the displacement-force relationship cannot be described using linear equations, the *nonlinear theory of vibration* must be applied. Linear systems are usually easier to deal with since in many cases, where the number of equations is small, closed-form solutions can be obtained. The solution of nonlinear system equations, however, requires the use of approximation and numerical methods. Closed-form solutions are usually difficult to obtain even for simple nonlinear systems, and for this reason, linearization techniques are used in many applications in order to obtain a linear system of differential equations whose solution can be obtained in a closed form.

The level of vibration is significantly influenced by the amount of energy dissipation as a result of *dry friction* between surfaces, *viscous damping*, and/or *structural damping* of the material. The dry friction between surfaces is also called *Coulomb damping*. In many applications, energy dissipated as the result of damping can be evaluated using damping forcing functions that are velocity- dependent. In this book, we also classify vibratory systems according to the presence of damping. If the system has a damping element, it is called a *damped system*; otherwise, it is called an *undamped system*.

Mechanical systems can also be classified according to the number of degrees of freedom which is defined as the minimum number of coordinates required to define the system configuration. In textbooks on the theory of vibration, mechanical and structural systems are often classified as *single degree of freedom systems*, *two degree of freedom systems*, *multi-degree of freedom systems*, or *continuous systems* which have an infinite number of degrees of freedom. The vibration of systems which have a finite number of degrees of freedom is governed by second-order ordinary differential equations, while the vibration of continuous systems which have infinite degrees of freedom is governed by partial differential equations, which depend on time as well as on the spatial coordinates. Finite degree of freedom models, however, can be obtained for continuous systems by using approximation techniques such as the *Rayleigh–Ritz method* and the *finite element method*.

1.2 ELEMENTS OF THE VIBRATION MODELS

Vibrations are the result of the combined effects of the *inertia* and *elastic* forces. Inertia of moving parts can be expressed in terms of the masses, moments of inertia, and the time derivatives of the displacements. Elastic restoring forces, on the other hand, can be expressed in terms of the displacements and stiffness of the elastic members. While damping has a significant effect and remains as a basic element in the vibration analysis, vibration may occur without damping.

Inertia Inertia is the property of an object that causes it to resist any effort to change its motion. For a particle, the inertia force is defined as the product of the mass of the particle and the acceleration, that is,

$$\mathbf{F}_i = m\ddot{\mathbf{r}}$$

where $\mathbf{F}_i$ is the vector of the inertia forces, m is the mass of the particle, and $\ddot{\mathbf{r}}$ is the acceleration vector defined in an inertial frame of reference. Rigid bodies, on the other hand, have inertia forces and moments, and for the planar motion of a rigid body, the inertia forces and moments are given by

$$\mathbf{F}_i = m\ddot{\mathbf{r}}$$

$$M_i = I\ddot{\theta}$$

where $\mathbf{F}_i$ is the inertia forces, m is the total mass of the rigid body, $\ddot{\mathbf{r}}$ is the acceleration vector of the center of mass of the body, M_i is the inertia moment, I is the mass moment of inertia of the rigid body about its center of mass, and $\ddot{\theta}$ is the angular acceleration. The units for the inertia forces and moments are, respectively, the units of forces and moments.

Elastic Forces Components with distributed elasticity are used in mechanical and structural systems to provide flexibility, and to store or absorb energy. These elastic members produce restoring forces which depend on the stiffness of the member as well as the displacements. Consider, for example, the spring connecting the two masses shown in Fig. 3(a). If the displacement of the first mass is x_1 and the displacement of the second mass is x_2, and if we assume for the moment that x_1 is greater than x_2, the total deflection in the spring is given by

$$\Delta x = x_1 - x_2$$

where Δx is the total deflection of the spring due to the displacements of the two masses. Using Taylor's series, the spring force after the displacement Δx can be written as

$$F_s(x_0 + \Delta x) = F_s(x_0) + \left.\frac{\partial F_s}{\partial x}\right|_{x=x_0} \Delta x + \frac{1}{2!}\left.\frac{\partial^2 F_s}{\partial x^2}\right|_{x=x_0} (\Delta x)^2 + \cdots \tag{1.1}$$

where F_s is the spring force and x_0 may be defined as the pretension or

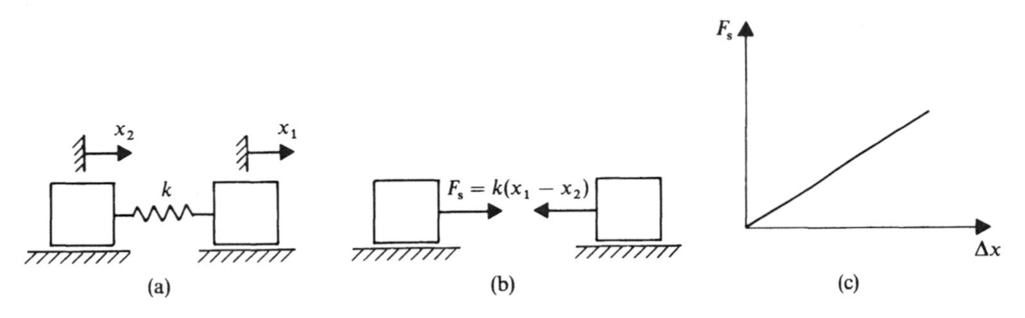

FIG. 1.3. Linear spring force.

precompression in the spring before the displacement Δx. If there is no pretension or compression in the spring, the spring force $F_s(x_0)$ is identically zero.

As a result of the displacement Δx, the spring force $F_s(x_0 + \Delta x)$ can be written as

$$F_s(x_0 + \Delta x) = F_s(x_0) + \Delta F_s$$

where ΔF_s is the change in the spring force as a result of the displacement Δx. By using Eq. 1, ΔF_s in the preceding equation can be written as

$$\Delta F_s = \left.\frac{\partial F_s}{\partial x}\right|_{x=x_0} \Delta x + \frac{1}{2}\left.\frac{\partial^2 F_s}{\partial x^2}\right|_{x=x_0} (\Delta x)^2 + \cdots$$

If the displacement Δx is assumed to be small, higher-order terms in Δx can be neglected and the spring force ΔF_s can be linearized and written as

$$\Delta F_s = \left.\frac{\partial F_s}{\partial x}\right|_{x=x_0} \Delta x$$

This equation can be written in a simpler form as

$$\begin{aligned} \Delta F_s &= k\Delta x \\ &= k(x_1 - x_2) \end{aligned} \tag{1.2}$$

where k is a proportionality constant called the *spring constant*, the *spring coefficient*, or the *stiffness coefficient*. The spring constant k is defined as

$$k = \left.\frac{\partial F_s}{\partial x}\right|_{x=x_0} \tag{1.3}$$

The effect of the spring force F_s on the two masses is shown in Fig. 3(b), and the linear relationship between the force and the displacement of the spring is shown in Fig. 3(c). Helical springs, which are widely used in many mechanical systems, as shown in Fig. 4, have a stiffness coefficient that depends on the diameter of the coil D, the diameter of the wire d, the number of coils n, and shear modulus of rigidity G. This stiffness coefficient is given by

$$k = \frac{Gd^4}{8nD^3} \tag{1.4}$$

Continuous elastic elements such as rods, beams, and shafts produce restoring elastic forces. Figure 5 shows some of these elastic elements which behave like springs. In Fig. 5(a), the rod produces a restoring elastic force that resists the longitudinal displacement in the system. If the mass of the rod is negligible compared to the mass m, one can write, from strength of materials, the following relationship

$$F = \frac{EA}{l} u \tag{1.5}$$

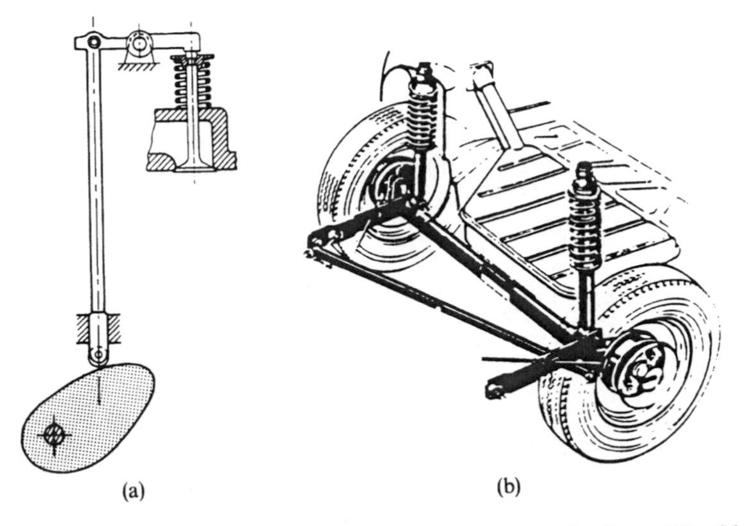

FIG. 1.4. Use of springs in mechanical systems: (a) cam mechanisms; (b) vehicle suspensions.

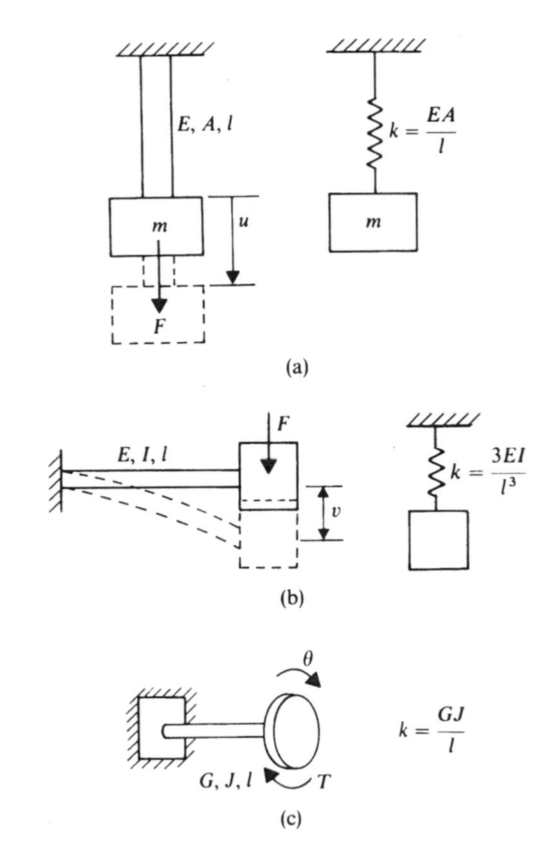

FIG. 1.5. Continuous elastic systems: (a) longitudinal vibration of rods; (b) transverse vibration of cantilever beams; (c) torsional system.

where F is the force acting at the end of the rod, u is the displacement of the end point, and l, A, and E are, respectively, the length, cross-sectional area, and modulus of elasticity of the rod. Equation 5 can be written as

$$F = ku$$

where k is the stiffness coefficient of the rod defined as

$$k = \frac{EA}{l} \tag{1.6}$$

Similarly, for the bending of the cantilever beam shown in Fig. 5(b), one can show that

$$F = \frac{3EI}{l^3} v$$

where F is the applied force, v is the transverse deflection of the end point and l, I, and E are the length, second area moment of inertia, and modulus of elasticity of the beam. In this case, we may define the beam stiffness as

$$k = \frac{3EI}{l^3} \tag{1.7}$$

From strength of materials, the relationship between the torque T and the angular torsional displacement θ of the shaft shown in Fig. 5(c) is

$$T = \frac{GJ}{l} \theta$$

where T is the torque, θ is the angular displacement of the shaft, and l, J, and G are, respectively, the length, polar moment of inertia, and modulus of rigidity. In this case, the torsional stiffness of the shaft is defined as

$$k = \frac{GJ}{l} \tag{1.8}$$

Table 1 shows the average properties of selected engineering materials. The

TABLE 1.1. Properties of Selected Engineering Materials

Materials	Density (Mg/m^3)	Modulus of Elasticity (GPa)	Modulus of Rigidity (GPa)
Wrought iron	7.70	190	—
Structural steel	7.87	200	76
Cast iron	7.20	100	—
Aluminum	2.77	71	26
Magnesium	1.83	45	16
Red brass	8.75	100	39
Bronze	8.86	100	45
Titanium	4.63	96	36

exact values of the coefficients presented in this table may vary depending on the heat treatment and the composition of the material. In this table Mg = 1000 kg, GPa = 10^9 Pa, and Pa (pascal) = N/m^2.

Damping While the effect of the inertia and elastic forces tends to maintain the oscillatory motion, the transient effect dies out because of energy dissipations. The process of energy dissipations is generally referred to as *damping*. Damping, in general, has the effect of reducing the amplitude of vibration and, therefore, it is desirable to have some amount of damping in order to achieve stability. Solid materials are not perfectly elastic, and they do exhibit damping, because of the internal friction due to the relative motion between the internal planes of the material during the deformation process. Such materials are referred to as viscoelastic solids, and the type of damping which they exhibit is known as *structural or hysteretic damping*. Another type of damping which commonly occurs as the result of the sliding contact between two surfaces is the *Coulomb or dry-friction damping*. In dry-friction damping, energy is dissipated as heat because of the friction due to the relative motion between the surfaces in contact. In this case, the damping force has a direction which is opposite to the direction of the motion. For instance, if we consider the mass sliding on the surface shown in Fig. 6, the friction force in this case is given by

$$F_f = \mu N$$

where F_f is the friction force, μ is the coefficient of *dry friction*, and N is the force normal to the contact surfaces.

The most common type of damping, however, is called *viscous damping*, in which the damping force produced is proportional to the velocity. In this case, the energy dissipating element is called *viscous damper* or *dashpot*. An example of a dashpot is the shock absorber in automobile suspensions and aircraft landing gears. Most of the actual viscous dampers consist of a piston and a cylinder filled with viscous fluid, as shown in Fig. 7. The fluid flow through holes in the piston provides the viscous resistance to the motion, and the resulting damping force is a function of the fluid viscosity, the number and size of the holes, and the dimension of the piston and cylinder. The desired damping characteristic can, therefore, be obtained by changing these parameters.

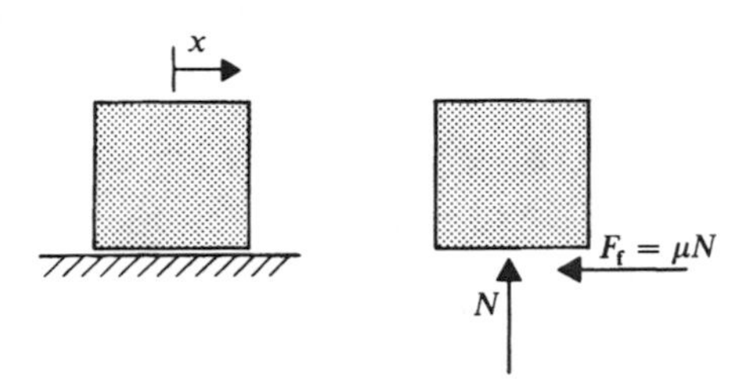

FIG. 1.6. Coulomb or dry-friction damping.

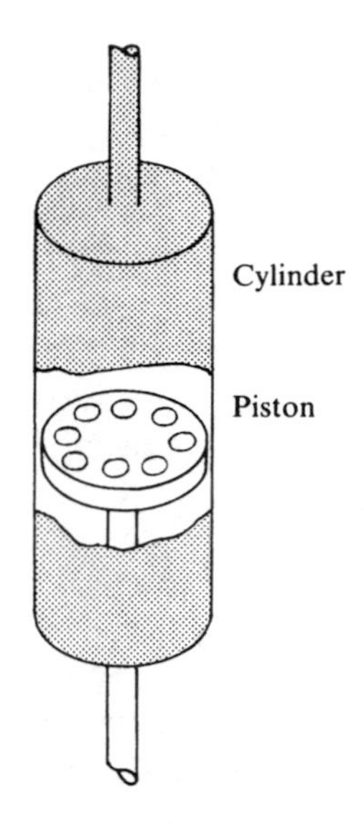

FIG. 1.7. Viscous damper.

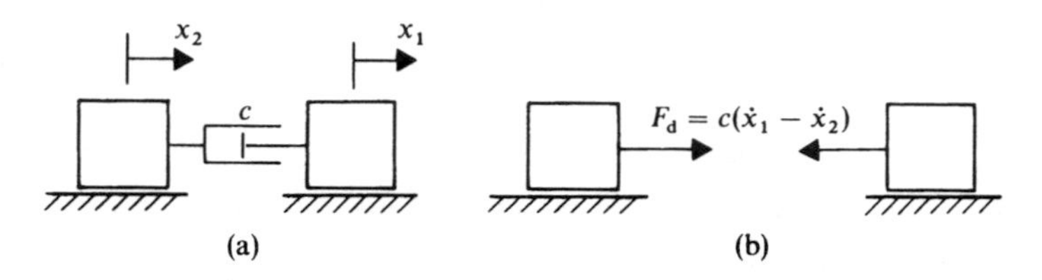

FIG. 1.8. Viscous damping force.

Figure 8 shows two masses connected by a viscous damper. The velocity of the first mass is $\dot{x}_1$, while the velocity of the second mass is $\dot{x}_2$. If $\dot{x}_1$ is assumed to be greater than $\dot{x}_2$, the resistive damping force F_d, which is proportional to the relative velocity, is given by

$$F_d \propto (\dot{x}_1 - \dot{x}_2)$$

or

$$F_d = c(\dot{x}_1 - \dot{x}_2) \tag{1.9}$$

where c is a proportionality constant called the *coefficient of viscous damping*.

1.3 PARTICLE DYNAMICS

An important step in the study of the oscillatory motion of mechanical systems is the development of the dynamic differential equations of motion. If the physical system can be modeled as a collection of lumped masses and/or rigid bodies, the equations of motion are, in general, second-order ordinary differential equations. If the system consists of continuous structural elements such as beams, plates, and shells, the governing equations are partial differential equations. The solution of the differential equations that govern the oscillatory motion of discrete and continuous systems can be used to predict the dynamic response of the system under different loading conditions. Several techniques, such as Newton's second law, can be used to formulate the dynamic differential equations of motion.

Newton's second law, which is referred to as the *law of motion*, states that the resultant force which acts on a particle is equal to the time rate of change of momentum of that particle. The particle momentum is a vector quantity defined as

$$\mathbf{p} = m\mathbf{v} \tag{1.10}$$

where $\mathbf{p}$ is the momentum of the particle, m is the mass, and $\mathbf{v}$ is the velocity vector. Newton's second law can then be expressed in a mathematical form as

$$\mathbf{F} = \dot{\mathbf{p}} \tag{1.11}$$

where $\mathbf{F}$ is the resultant force that acts on the particle and $(\dot{\ })$ denotes differentiation with respect to time. Substituting Eq. 10 into Eq. 11, and assuming that the mass of the particle remains constant, one obtains

$$\mathbf{F} = \frac{d}{dt}(m\mathbf{v}) = m\frac{d\mathbf{v}}{dt} = m\mathbf{a} \tag{1.12}$$

where $\mathbf{a}$ is the acceleration of the particle defined as

$$\mathbf{a} = \frac{d\mathbf{v}}{dt}$$

Let $\ddot{x}$, $\ddot{y}$, and $\ddot{z}$ be the components of the acceleration of the particle, and let F_x, F_y, and F_z be the components of the resultant force, then the vector equation of Eq. 12 can be written as three scalar equations given by

$$\left.\begin{aligned} m\ddot{x} &= F_x \\ m\ddot{y} &= F_y \\ m\ddot{z} &= F_z \end{aligned}\right\} \tag{1.13}$$

There are three differential equations, since the unconstrained motion of a particle in space is described by the three coordinates x, y, and z, as shown in Fig. 9.

The planar motion of a particle is described by only two coordinates x and y, as shown in Fig. 10. In this case, the dynamic differential equations of the particle reduce to two equations given by

$$\left.\begin{aligned} m\ddot{x} &= F_x \\ m\ddot{y} &= F_y \end{aligned}\right\} \tag{1.14}$$

That is, the number of independent differential equations is equal to the number of degrees of freedom of the particle.

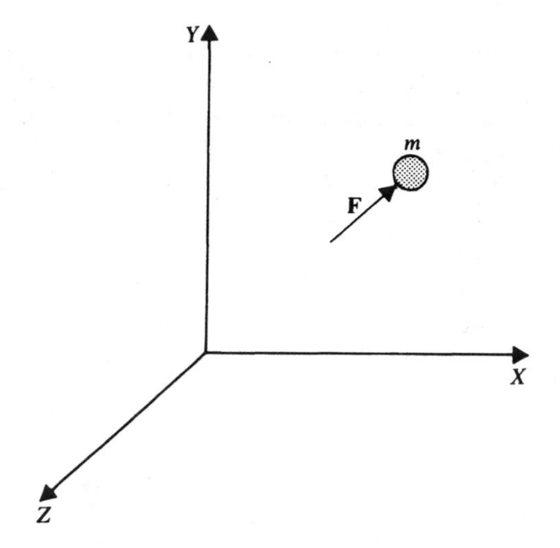

FIG. 1.9. Motion of a particle in space.

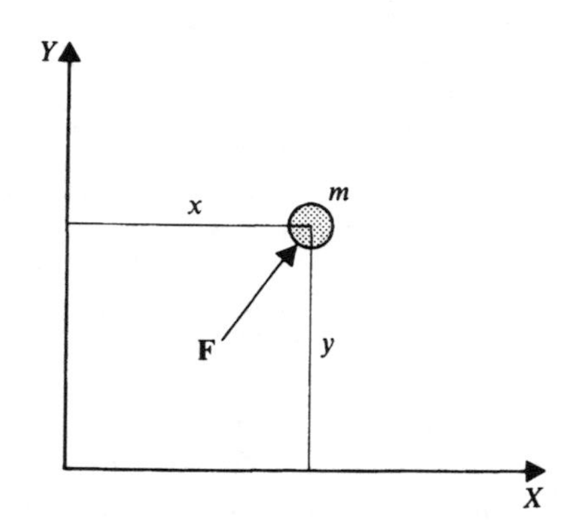

FIG. 1.10. Planar motion.

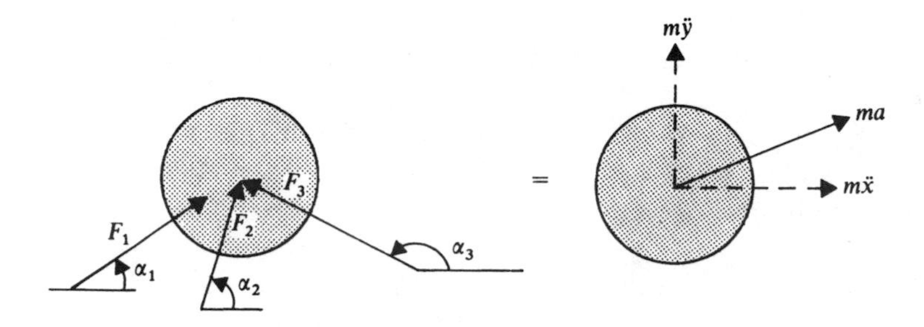

FIG. 1.11. Dynamic equilibrium.

Equation 14 can be represented by the diagram shown in Fig. 11, which indicates that the inertia force in one direction is equal to the sum of the external forces acting on the particle in that direction. For instance, if the forces F_1, F_2, and F_3 make angles α_1, α_2, and α_3 with the X-axis, Eq. 14 and the free body diagram shown in Fig. 11 imply that

$$m\ddot{x} = F_1 \cos \alpha_1 + F_2 \cos \alpha_2 + F_3 \cos \alpha_3$$
$$m\ddot{y} = F_1 \sin \alpha_1 + F_2 \sin \alpha_2 + F_3 \sin \alpha_3$$

Example 1.1

The system shown in Fig. 12(a) consists of a mass m which is allowed to move only in the horizontal direction. The mass is connected to the ground by a spring and damper. The spring has a stiffness coefficient k, and the damper has a viscous damping coefficient c. The harmonic force $F_x = F_0 \sin \omega_f t$ acts on the mass as shown in Fig. 12. Derive the differential equation of motion of this system, assuming that there is no friction force between the mass and the ground.

Solution. Since the motion is in the plane, we have the following equations

$$m\ddot{x} = F_x$$
$$m\ddot{y} = F_y$$

From the free body diagram shown in the figure, we have

$$F_x = F_0 \sin \omega_f t - c\dot{x} - kx$$
$$F_y = N - mg$$

where N is the reaction force and g is the gravitational constant. Since the motion is constrained to be only in the horizontal direction, we have

$$\ddot{y} = 0$$

That is, the two equations of motion are given by

$$m\ddot{x} = F_0 \sin \omega_f t - c\dot{x} - kx$$
$$0 = N - mg$$

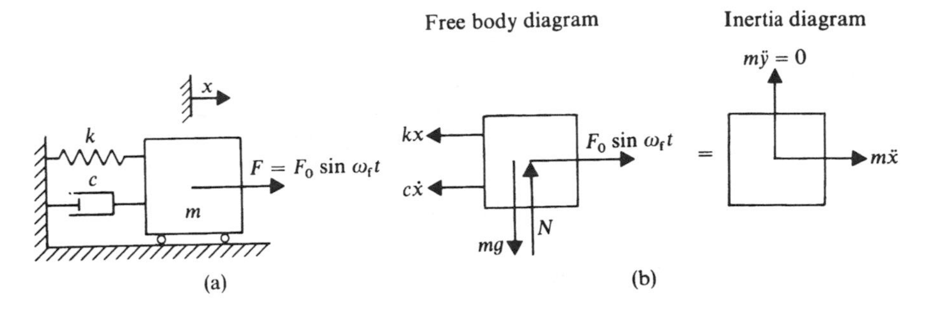

FIG. 1.12. Harmonic excitation.

which can be written as

$$m\ddot{x} + c\dot{x} + kx = F_0 \sin \omega_f t \tag{1.15}$$

$$N = mg$$

There is only one second-order differential equation since the system has only one degree of freedom. The second equation which is an algebraic equation can be used to determine the reaction force N. The differential equation of motion given by Eq. 15 is the standard equation for the *linear damped forced vibration* of single degree of freedom systems. This equation and its solution will be discussed in more detail in Chapter 4. The equation of forced vibration of undamped single degree of freedom systems can be extracted from Eq. 15 by letting the damping coefficient c equal zero. In this special case, Eq. 15 reduces to

$$m\ddot{x} + kx = F_0 \sin \omega_f t$$

If the force in the preceding equation is equal to zero, one obtains the equation of motion for the *linear free undamped vibration* of single degree of freedom systems as

$$m\ddot{x} + kx = 0$$

Integrating Eq. 11 with respect to time yields the impulse of the force $\mathbf{F}$ as

$$\int_0^t \mathbf{F}(t)\, dt = \mathbf{p}(t) - \mathbf{p}_0$$

where $\mathbf{p}_0$ is a constant. The preceding equation implies that the change in the linear momentum of the particle is equal to the impulse of the force acting on this particle, and if the force is equal to zero, the linear momentum is conserved since in this case

$$\mathbf{p}(t) = \mathbf{p}_0$$

which indicates that the momentum and velocity of the particle remain constant in this special case.

Angular Momentum It follows from Eq. 11 that

$$\mathbf{r} \times \mathbf{F} = \mathbf{r} \times \frac{d\mathbf{p}}{dt}$$

where $\mathbf{r}$ is the position vector of the particle with respect to the origin of the coordinate system. The cross product $\mathbf{r} \times \mathbf{F}$ on the left-hand side of the preceding equation is recognized as the moment of the force, while the right-hand side of this equation can be simplified using the identity

$$\frac{d}{dt}(\mathbf{r} \times \mathbf{p}) = \mathbf{v} \times \mathbf{p} + \mathbf{r} \times \frac{d\mathbf{p}}{dt}$$

The first term on the right-hand side of this equation is identically zero since

$$\mathbf{v} \times \mathbf{p} = \mathbf{v} \times m\mathbf{v} = \mathbf{0}$$

and consequently,

$$\mathbf{r} \times \mathbf{F} = \frac{d}{dt}(\mathbf{r} \times \mathbf{p}) \tag{1.16}$$

The vector $\mathbf{r} \times \mathbf{p}$ is called the *angular momentum* of the particle defined with respect to the origin of the coordinate system. Therefore, the preceding equation states that the time rate of the angular momentum of a particle is equal to the moment of the force acting on this particle.

Work and Energy Taking the dot product of both sides of Eq. 11 with the velocity vector $\mathbf{v}$, one obtains

$$\mathbf{F} \cdot \mathbf{v} = \frac{d\mathbf{p}}{dt} \cdot \mathbf{v} = \frac{d(m\mathbf{v})}{dt} \cdot \mathbf{v}$$

Using the identity

$$\frac{d(\mathbf{v} \cdot \mathbf{v})}{dt} = \frac{d\mathbf{v}}{dt} \cdot (2\mathbf{v})$$

and assuming that the mass m is constant, one gets

$$\mathbf{F} \cdot \mathbf{v} = \frac{d}{dt}\left(\frac{1}{2}m\mathbf{v} \cdot \mathbf{v}\right) = \frac{dT}{dt} \tag{1.17}$$

where T is the kinetic energy of the particle defined as

$$T = \tfrac{1}{2}m\mathbf{v} \cdot \mathbf{v} = \tfrac{1}{2}mv^2$$

Since $d\mathbf{r} = \mathbf{v}dt$, it follows from Eq. 17 that

$$\mathbf{F} \cdot \mathbf{v}dt = \mathbf{F} \cdot d\mathbf{r} = dT$$

or

$$\int \mathbf{F} \cdot d\mathbf{r} = \int dT \tag{1.18}$$

The left-hand side of this equation represents the work done by the force $\mathbf{F}$ that acts on the particle, while the right-hand side is the change in the particle kinetic energy. Therefore, the preceding equation states that the work done on the particle by the applied force is equal to the change in the kinetic energy of the particle.

A simple example of the work of a force is the work done by the spring force, which is simply given by

$$\int \mathbf{F} \cdot d\mathbf{r} = \int_0^x (-kx)\, dx = -\frac{1}{2}kx^2$$

where k is the spring stiffness and x is the spring deformation. It is clear from the preceding equation that the work done by the spring force is the negative of the strain energy $U = kx^2/2$ stored as the result of the spring deformation.

1.4 SYSTEMS OF PARTICLES

Consider the system that consists of n particles. The mass of the particle i is denoted as m_i, and its position vector is defined by the vector $\mathbf{r}_i$. The linear momentum $\mathbf{p}$ of the system of particles is defined as

$$\mathbf{p} = \sum_{i=1}^{n} \mathbf{p}_i = \sum_{i=1}^{n} m_i \mathbf{v}_i$$

where $\mathbf{p}_i$ and $\mathbf{v}_i$ are, respectively, the linear momentum and the velocity of the particle i. Let $\mathbf{F}_i$ be the resultant force vector acting on the particle i. In addition to this external force, we assume that the particle is subjected to internal forces as the result of its interaction with other particles in the system. Let $\mathbf{F}_{ij}$ be the internal force acting on particle i as the result of its interaction with particle j. With the understanding that $\mathbf{F}_{ii} = \mathbf{0}$, the equation of motion of the particle i is

$$m_i \mathbf{a}_i = \dot{\mathbf{p}}_i = \mathbf{F}_i + \sum_{j=1}^{n} \mathbf{F}_{ij}$$

where $\mathbf{a}_i$ is the absolute acceleration of the particle i. The second term on the right-hand side of the above equation represents the sum of all the internal forces acting on the particle i as the result of its interaction with other particles in the system. It follows that

$$\sum_{i=1}^{n} m_i \mathbf{a}_i = \sum_{i=1}^{n} \dot{\mathbf{p}}_i = \sum_{i=1}^{n} \mathbf{F}_i + \sum_{i=1}^{n} \sum_{j=1}^{n} \mathbf{F}_{ij}$$

From Newton's third law, the forces of interaction acting on particles i and j are equal in magnitude and opposite in direction, that is,

$$\mathbf{F}_{ij} = -\mathbf{F}_{ji}$$

and consequently,

$$\sum_{i=1}^{n} \sum_{j=1}^{n} \mathbf{F}_{ij} = \mathbf{0}$$

It follows that

$$\sum_{i=1}^{n} m_i \mathbf{a}_i = \sum_{i=1}^{n} \dot{\mathbf{p}}_i = \sum_{i=1}^{n} \mathbf{F}_i$$

Since the position vector of the center of mass of the system of particles $\mathbf{r}_c$ must satisfy

$$\sum_{i=1}^{n} m_i \mathbf{r}_i = m \mathbf{r}_c$$

where

$$m = \sum_{i=1}^{n} m_i$$

is the total mass of the system of particles, one has

$$\sum_{i=1}^{n} m_i \mathbf{a}_i = m\mathbf{a}_c$$

where $\mathbf{a}_c$ is the absolute acceleration of the center of mass of the system of particles. It follows that

$$m\mathbf{a}_c = \dot{\mathbf{p}} = \sum_{i=1}^{n} \mathbf{F}_i \tag{1.19}$$

which implies that the product of the total mass of the system of particles and the absolute acceleration of the center of mass of this system is equal to the sum of all the external forces acting on the system of particles. Note that if the sum of the external forces acting on a system of particles is equal to zero, then

$$\dot{\mathbf{p}} = \mathbf{0}$$

and we have the principle of conservation of momentum.

Example 1.2

The system shown in Fig. 13(a) consists of the two masses m_1 and m_2 which move in the horizontal direction on a friction-free surface. The two masses are

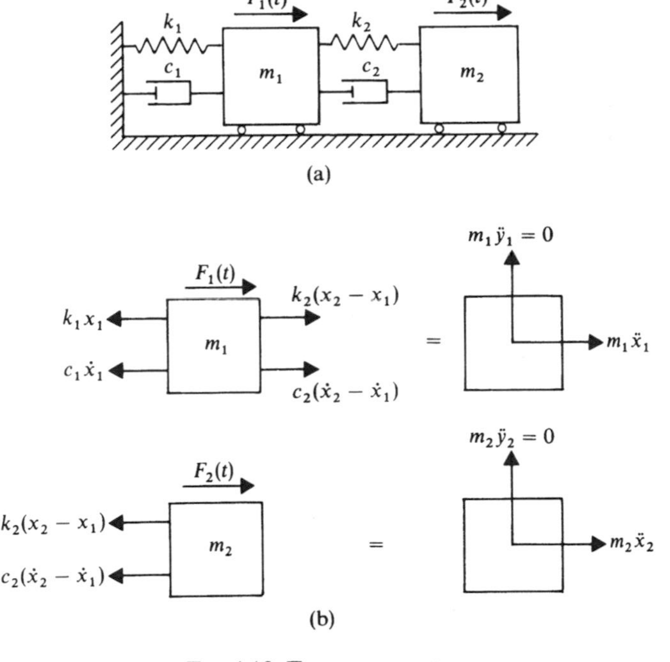

FIG. 1.13. Two-mass system.

connected to each other and to the surface by springs and dampers, as shown in the figure. The external forces F_1 and F_2 act, respectively, on the masses m_1 and m_2. Obtain the differential equations of motion of this system.

Solution. The system shown in Fig. 13 has two degrees of freedom which can be represented by two independent coordinates x_1 and x_2. Without any loss of generality, we assume that x_2 is greater than x_1, and $\dot{x}_2$ is greater than $\dot{x}_1$. By using the free body diagram shown in the figure, the differential equation of motion for the mass m_1 is given by

$$\begin{aligned} m_1\ddot{x}_1 &= F_{x_1} \\ &= F_1(t) - k_1x_1 - c_1\dot{x}_1 + k_2(x_2 - x_1) + c_2(\dot{x}_2 - \dot{x}_1) \end{aligned}$$

which can be written as

$$m_1\ddot{x}_1 + (c_1 + c_2)\dot{x}_1 - c_2\dot{x}_2 + (k_1 + k_2)x_1 - k_2x_2 = F_1(t) \tag{1.20}$$

Similarly, for the second mass, we have

$$\begin{aligned} m_2\ddot{x}_2 &= F_{x_2} \\ &= F_2(t) - k_2(x_2 - x_1) - c_2(\dot{x}_2 - \dot{x}_1) \end{aligned}$$

or

$$m_2\ddot{x}_2 + c_2\dot{x}_2 - c_2\dot{x}_1 + k_2x_2 - k_2x_1 = F_2(t) \tag{1.21}$$

Observe that there are two second-order differential equations of motion (Eqs. 20 and 21) since the system has two degrees of freedom. Adding the equations of motion of the two particles leads to

$$m_1\ddot{x}_1 + m_2\ddot{x}_2 = F_1(t) + F_2(t) - c_1\dot{x}_1 - k_1x_1$$

If the external forces are equal to zero, which is the case of free vibration, one has

$$m_1\ddot{x}_1 + m_2\ddot{x}_2 = -c_1\dot{x}_1 - k_1x_1$$

Angular Momentum The angular momentum $\mathbf{L}$ of a system of particles is defined as the sum of the angular momenta of the individual particles, namely,

$$\mathbf{L} = \sum_{i=1}^{n} (\mathbf{r}_i \times m_i\mathbf{v}_i) \tag{1.22}$$

Taking the time derivative of the angular momentum, one gets

$$\frac{d\mathbf{L}}{dt} = \sum_{i=1}^{n} (\mathbf{v}_i \times m_i\mathbf{v}_i) + \sum_{i=1}^{n} (\mathbf{r}_i \times m_i\mathbf{a}_i)$$

Since $\mathbf{v}_i \times m_i\mathbf{v}_i = \mathbf{0}$, the above equation reduces to

$$\frac{d\mathbf{L}}{dt} = \sum_{i=1}^{n} (\mathbf{r}_i \times m_i\mathbf{a}_i) \tag{1.23}$$

The inertia force $m_i\mathbf{a}_i$ of the particle i is equal to the resultant of the applied forces acting on this particle. Therefore, we can write

$$\frac{d\mathbf{L}}{dt} = \sum_{i=1}^{n}\left[\mathbf{r}_i \times \left(\mathbf{F}_i + \sum_{j=1}^{n}\mathbf{F}_{ij}\right)\right]$$

$$= \sum_{i=1}^{n}\mathbf{r}_i \times \mathbf{F}_i + \sum_{i=1}^{n}\sum_{j=1}^{n}\mathbf{r}_i \times \mathbf{F}_{ij}$$

Since the internal forces of interaction acting on two particles are equal in magnitude, opposite in direction, and act along the line connecting the two particles, one has

$$\mathbf{r}_i \times \mathbf{F}_{ij} + \mathbf{r}_j \times \mathbf{F}_{ji} = (\mathbf{r}_i - \mathbf{r}_j) \times \mathbf{F}_{ij} = \mathbf{0}$$

and consequently,

$$\sum_{i=1}^{n}\sum_{j=1}^{n}\mathbf{r}_i \times \mathbf{F}_{ij} = \mathbf{0}$$

Therefore, the rate of change of the angular momentum can be written as

$$\frac{d\mathbf{L}}{dt} = \sum_{i=1}^{n}\mathbf{r}_i \times \mathbf{F}_i \tag{1.24}$$

That is, the rate of change of the angular momentum of a system of particles is equal to the moment of all the external forces acting on the system. If the resultant moment is equal to zero, we obtain the *principle of conservation of angular momentum*, which can be stated as

$$\dot{\mathbf{L}} = \mathbf{0} \tag{1.25}$$

which implies that $\mathbf{L}$ = constant.

The angular momentum of the system of particles can be expressed in terms of the velocity of the center of mass $\mathbf{v}_i$. It can be shown that

$$\mathbf{L} = \mathbf{r}_c \times m\mathbf{v}_c + \sum_{i=1}^{n}\mathbf{r}_{ic} \times m_i\mathbf{v}_{ic} \tag{1.26}$$

where $\mathbf{r}_{ic}$ and $\mathbf{v}_{ic}$ are, respectively, the relative position and velocity vectors of the particle i with respect to the center of mass of the system of particles, and m is total mass of the system of particles. It also can be shown that the kinetic energy of the system of particles can be written as

$$T = \frac{1}{2}\sum_{i=1}^{n} m_i(\mathbf{v}_i \cdot \mathbf{v}_i) = \frac{1}{2}\sum_{i=1}^{n} m_i v_i^2$$

$$= \frac{1}{2}mv_c^2 + \sum_{i=1}^{n}\frac{1}{2}m_i v_{ic}^2 \tag{1.27}$$

1.5 DYNAMICS OF RIGID BODIES

In particle kinematics, the objects are assumed to be so small that they can be represented by points in the three-dimensional space, and as a consequence,

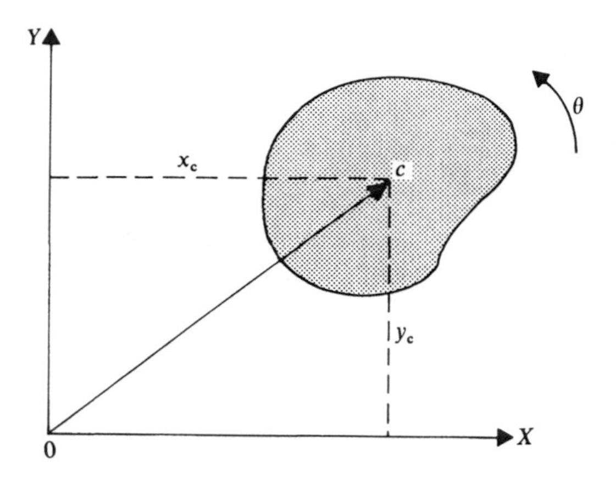

FIG. 1.14. Unconstrained motion of rigid bodies.

the position of the particle can be defined by the coordinates of a point. That is, three coordinates are required for the spatial motion and only two coordinates are required for the planar motion. The configuration of a rigid body in space, however, can be determined using six coordinates, three coordinates define the translation of the body center of mass and three coordinates define the orientation of the body in a fixed coordinate system. The unconstrained planar motion of a rigid body can be described in terms of three coordinates, as shown in Fig. 14, where two coordinates define the translation of the center of mass and one coordinate θ defines the orientation of the rigid body in the XY plane. Therefore, there are three differential equations that govern the unconstrained planar motion of a rigid body. These three equations are

$$\left.\begin{aligned} m\ddot{x}_c &= F_x \\ m\ddot{y}_c &= F_y \\ I_c\ddot{\theta} &= M \end{aligned}\right\} \tag{1.28}$$

where m is the mass of the rigid body, $\ddot{x}_c$ and $\ddot{y}_c$ are the accelerations of the center of mass of the rigid body, F_x and F_y are the sum of the forces that act at the center of mass in the x and y directions, respectively, M is the sum of the moments, and I_c is the mass moment of inertia with respect to a perpendicular axis through the center of mass. The left-hand sides of the first two equations in Eq. 28 are called the *inertia forces* or *effective forces*, and the left-hand side of the third equation is called the *inertia moment* or *effective moment*.

The first two equations in Eq. 28 imply that the inertia forces or effective forces are equal to the external forces, while the third equation implies that the inertia moment or effective moment is equal to the external moments that act on the rigid body. Equation 28 can be represented by the diagram shown in Fig. 15. For instance, if the forces F_1, F_2, and F_3 make, respectively, angles

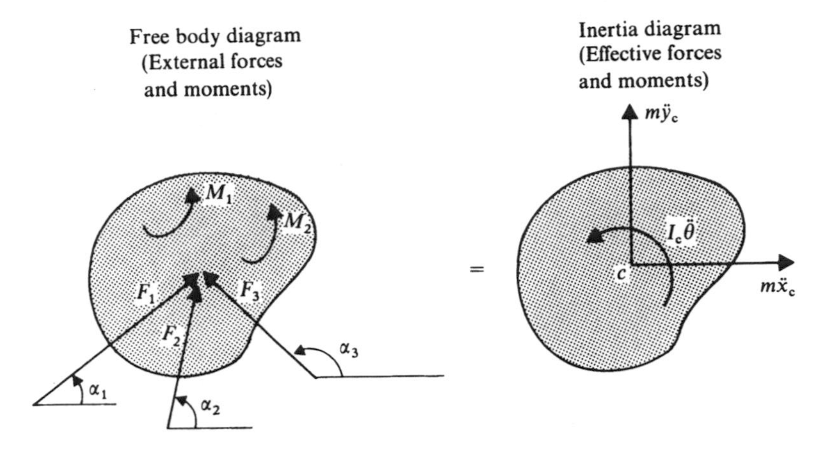

FIG. 1.15. Dynamic equilibrium of rigid bodies.

α_1, α_2, and α_3 with the horizontal axis, Eq. 28 implies that

$$m\ddot{x}_c = F_1 \cos \alpha_1 + F_2 \cos \alpha_2 + F_3 \cos \alpha_3$$

$$m\ddot{y}_c = F_1 \sin \alpha_1 + F_2 \sin \alpha_2 + F_3 \sin \alpha_3$$

$$I_c\ddot{\theta} = M_1 + M_2$$

Note that if the rigid body undergoes pure planar translation, only the first two equations of Eq. 28 are required. If the rigid body undergoes pure rotation, only the third equation is sufficient. That is, the number of independent differential equations is equal to the number of degrees of freedom of the rigid body.

It is also important to emphasize that the two systems of forces (external and effective) shown in Fig. 15, are equal in the sense that the inertia (effective) forces and moments can be treated in the same manner that the external forces and moments are treated. As a consequence, one can take the moment of the inertia forces and moments about any point and equate this with the moment of the external forces and moments about the same point. This can be stated mathematically as

$$M_a = M_{eff} \tag{1.29}$$

where M_a is the moment of the externally applied moments and forces and M_{eff} is the moment of the inertia forces and moments about the same point. Equation 29 is useful in many applications and its use is demonstrated by the following example.

Example 1.3

Figure 16 depicts a rigid rod of length l, mass m, and mass moment of inertia about its mass center, I_c. One end of the rod is connected to the ground by a pin joint, and

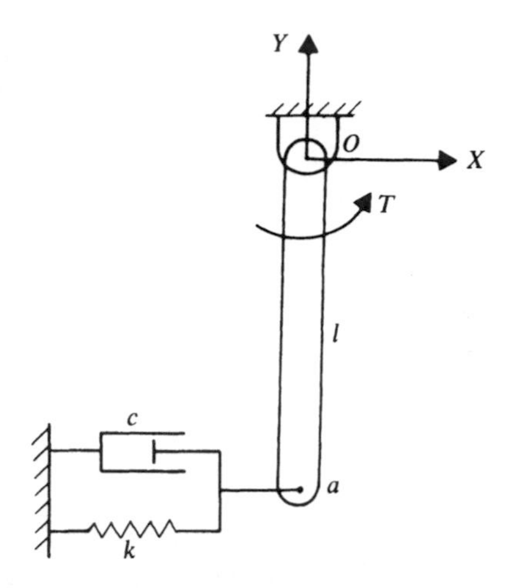

FIG. 1.16. Pendulum vibration.

the other end is connected to the ground by spring and damper. The spring stiffness is k and the viscous damping coefficient is c. If T is an externally applied moment that acts on the rod, derive the system differential equations of motion.

Solution. For a given angular orientation θ, the free body diagram of the rod is shown in Fig. 17. In this figure, R_x and R_y denote the reaction forces of the pin joint. The pendulum system shown in Fig. 16 has only one degree of freedom which is described by the angular rotation θ, and therefore, the equations of motion for this system can be described by only one differential equation. In order to avoid the unknown reaction forces R_x and R_y, we use Eq. 29 and take the moment about point O. From the free body diagram shown in Fig. 17, the moment of the applied forces about O is given by

$$M_a = T - mg\frac{l}{2}\sin\theta - (c\dot{x}_a + kx_a)l\cos\theta \tag{1.30}$$

where x_a and $\dot{x}_a$ are the displacement and velocity of point a expressed in terms of the angle θ as

$$x_a = l\sin\theta$$
$$\dot{x}_a = l\dot{\theta}\cos\theta$$

Substituting these two equations into Eq. 30, one obtains

$$M_a = T - mg\frac{l}{2}\sin\theta - (c\dot{\theta}\cos\theta + k\sin\theta)l^2\cos\theta \tag{1.31}$$

The moments of the effective forces and moments about O are

$$M_{eff} = I_c\ddot{\theta} + m\ddot{x}_c\frac{l}{2}\cos\theta + m\ddot{y}_c\frac{l}{2}\sin\theta \tag{1.32}$$

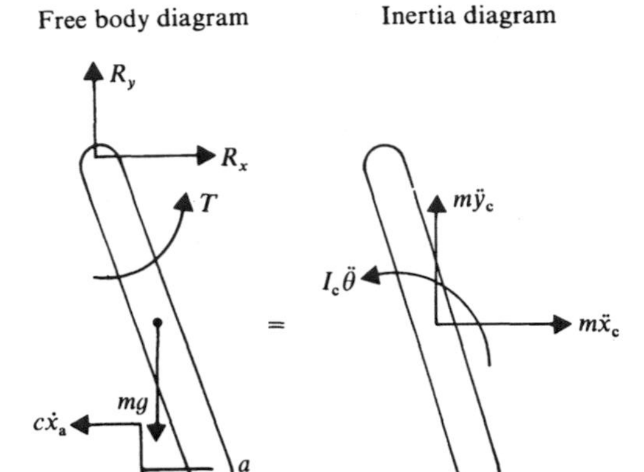

FIG. 1.17. Dynamic equilibrium.

where the coordinates of the center of mass x_c and y_c are given by

$$x_c = \frac{l}{2} \sin \theta$$

$$y_c = -\frac{l}{2} \cos \theta$$

The velocity and acceleration of the center of mass can be obtained by direct differentation of the above equations, that is,

$$\dot{x}_c = \frac{l}{2}\dot{\theta} \cos \theta, \qquad \ddot{x}_c = \frac{l}{2}\ddot{\theta} \cos \theta - \frac{l}{2}\dot{\theta}^2 \sin \theta$$

$$\dot{y}_c = \frac{l}{2}\dot{\theta} \sin \theta, \qquad \ddot{y}_c = \frac{l}{2}\ddot{\theta} \sin \theta + \frac{l}{2}\dot{\theta}^2 \cos \theta$$

Substituting the expressions of the accelerations into the moment equation, Eq. 32, yields

$$M_{\text{eff}} = I_c\ddot{\theta} + m\left[\frac{l}{2}\ddot{\theta} \cos \theta - \frac{l}{2}\dot{\theta}^2 \sin \theta\right]\frac{l}{2} \cos \theta + m\left[\frac{l}{2}\ddot{\theta} \sin \theta + \frac{l}{2}\dot{\theta}^2 \cos \theta\right]\frac{l}{2} \sin \theta$$

$$= I_c\ddot{\theta} + m\left(\frac{l}{2}\right)^2 \ddot{\theta}\,[\cos^2 \theta + \sin^2 \theta]$$

Using the trigonometric identity

$$\cos^2 \theta + \sin^2 \theta = 1,$$

the effective moment M_{eff} reduces to

$$M_{\text{eff}} = I_c\ddot{\theta} + \frac{ml^2}{4}\ddot{\theta} = \left(I_c + \frac{ml^2}{4}\right)\ddot{\theta}$$

$$= I_O\ddot{\theta}$$

where I_O is the mass moment of inertia of the rod about point O and is defined as

$$I_O = I_c + \frac{ml^2}{4}$$

Using Eq. 29, we have

$$M_a = M_{eff}$$

that is,

$$T - mg\frac{l}{2}\sin\theta - (c\dot{\theta}\cos\theta + k\sin\theta)l^2\cos\theta = I_O\ddot{\theta}$$

which can be written as

$$I_O\ddot{\theta} + cl^2\dot{\theta}\cos^2\theta + kl^2\sin\theta\cos\theta + mg\frac{l}{2}\sin\theta = T \tag{1.33}$$

Principle of Work and Energy For rigid body systems, the principle of work and energy can be conveniently used to obtain the equations of motion in many applications. In order to derive this principle for rigid bodies, we note that upon the use of the relation $dx = \dot{x}\,dt$, the accelerations of the center of mass of the rigid body can be written as

$$\ddot{x}_c = \dot{x}_c\frac{d\dot{x}_c}{dx_c}, \qquad \ddot{y}_c = \dot{y}_c\frac{d\dot{y}_c}{dy_c}, \qquad \ddot{\theta} = \dot{\theta}\frac{d\dot{\theta}}{d\theta}$$

Substituting these equations into Eq. 28, and integrating, one obtains

$$\int m\dot{x}_c\,d\dot{x}_c = \int F_x\,dx_c, \qquad \int m\dot{y}_c\,d\dot{y}_c = \int F_y\,dy_c,$$

$$\int I_c\dot{\theta}\,d\dot{\theta} = \int M\,d\theta$$

which yield

$$\frac{1}{2}m\dot{x}_c^2 - c_x = \int F_x\,dx_c, \qquad \frac{1}{2}m\dot{y}_c^2 - c_y = \int F_y\,dy_c,$$

$$\frac{1}{2}I_c\dot{\theta}^2 - c_\theta = \int M\,d\theta$$

where c_x, c_y, and c_θ are the constants of integration which define the kinetic energy of the rigid body at the initial configuration as $T_0 = c_x + c_y + c_\theta$. Adding the preceding equations, one obtains

$$\frac{1}{2}m(\dot{x}_c^2 + \dot{y}_c^2) + \frac{1}{2}I_c\dot{\theta}^2 - T_0 = \int F_x\,dx_c + \int F_y\,dy_c + \int M\,d\theta \tag{1.34}$$

which can be written as

$$T = W \tag{1.35}$$

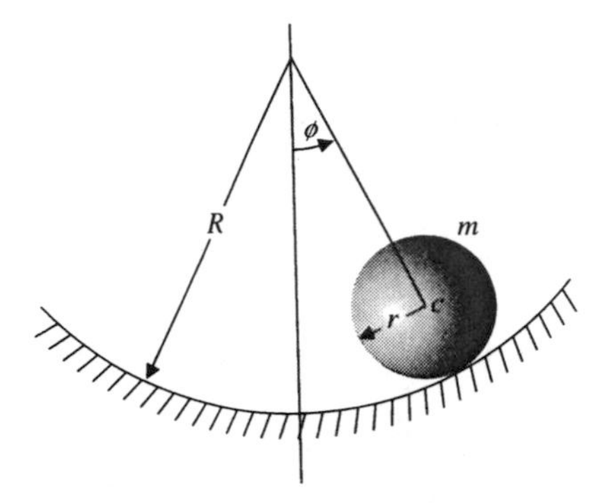

FIG. 1.18. Principle of work and energy.

where T is the change in the kinetic energy of the rigid body, and W is the work of the forces, both defined as

$$T = \tfrac{1}{2}m(\dot{x}_c^2 + \dot{y}_c^2) + \tfrac{1}{2}I_c\dot{\theta}^2 - T_0 = \tfrac{1}{2}mv_c^2 + \tfrac{1}{2}I_c\dot{\theta}^2 - T_0$$

$$W = \int F_x\,dx_c + \int F_y\,dy_c + \int M\,d\theta$$

where v_c is the absolute velocity of the center of mass of the rigid body.

Equation 34 or 35 is a statement of the *principle of work and energy* for rigid bodies. According to this principle, the change in the kinetic energy of the rigid body is equal to the work of the applied forces and moments. In order to demonstrate the application of this principle in the case of rigid body dynamics, we consider the system shown in Fig. 18, which depicts a homogeneous circular cylinder of radius r, mass m, and mass moment of inertia I_c about the center of mass; $I_c = mr^2/2$. The cylinder rolls without slipping on a curved surface of radius R. The change in the kinetic energy and the work done by the gravity force of the cylinder are

$$T = \tfrac{1}{2}mv_c^2 + \tfrac{1}{2}I_c\dot{\theta}^2 - T_0$$

$$W = -mg(R - r)(1 - \cos\phi)$$

where v_c is the absolute velocity of the center of mass, and $\dot{\theta}$ is the angular velocity of the cylinder, both defined as

$$v_c = (R - r)\dot{\phi}$$

$$\dot{\theta} = \frac{v_c}{r} = \frac{(R - r)\dot{\phi}}{r}$$

The change in the kinetic energy of the cylinder can then be written as

$$T = \tfrac{3}{4}m(R - r)^2\dot{\phi}^2 - T_0$$

and the principle of work and energy as stated by Eq. 35 leads to

$$T - W = 0$$

or

$$\tfrac{3}{4}m(R-r)^2\dot{\phi}^2 + mg(R-r)(1-\cos\phi) - T_0 = 0$$

Differentiating this equation with respect to time and dividing by $\dot{\phi}$, one obtains the equation of motion of the cylinder as

$$\tfrac{3}{2}(R-r)\ddot{\phi} + g\sin\phi = 0$$

1.6 LINEARIZATION OF THE DIFFERENTIAL EQUATIONS

In many cases the dynamics of physical systems is governed by nonlinear differential equations, such as Eq. 33 of the preceding example. It is difficult, however, to obtain a closed-form solution to many of the resulting nonlinear differential equations. If the assumptions of small oscillations are made, linear second-order ordinary differential equations can be obtained, and a standard procedure can then be used to obtain the solution of these linear equations in a closed form. For instance, if the angular oscillation in the preceding example is assumed to be small ($\theta \leq 10°$),

$$\sin\theta \approx \tan\theta \approx \theta$$

$$\cos\theta \approx 1$$

and Eq. 33 can be simplified and written as

$$I_O\ddot{\theta} + cl^2\dot{\theta} + \left(kl^2 + mg\frac{l}{2}\right)\theta = T \tag{1.36}$$

which is a linear second-order differential equation with constant coefficients.

The preceding equation also can be obtained by carrying out the linearization at an early stage, that is, by linearizing the kinematic relationships presented in the preceding example. In this case, we have

$$x_a = l\theta$$

$$x_c = \frac{l}{2}\theta$$

$$y_c = -\frac{l}{2}$$

It follows that

$$\dot{x}_a = l\dot{\theta}, \qquad \ddot{x}_c = \frac{l}{2}\ddot{\theta}, \qquad \text{and} \qquad \ddot{y}_c = 0$$

Substituting these equations into Eqs. 30 and 32 and using the assumption of small angular oscillations, one obtains the following simplified expressions for the applied and effective moments about point O

$$M_a = T - mg\frac{l}{2}\theta - (c\dot{\theta} + k\theta)l^2$$

$$M_{eff} = I_c\ddot{\theta} + m\frac{l^2}{4}\ddot{\theta} = \left(I_c + \frac{ml^2}{4}\right)\ddot{\theta} = I_O\ddot{\theta}$$

If we put

$$M_a = M_{eff}$$

we obtain

$$I_O\ddot{\theta} = T - mg\frac{l}{2}\theta - (c\dot{\theta} + k\theta)l^2$$

which is the same as Eq. 36 previously obtained by linearizing the nonlinear differential equation given by Eq. 33. It is important, however, to point out that early linearization in some problems could lead to the loss of some important terms. Nonetheless, in the following chapters, for the sake of simplicity and to avoid unnecessary and laborious calculations, we prefer to linearize the kinematic equations at an early stage, provided that an early linearization leads to the same equations obtained by linearizing the nonlinear differential equations.

Example 1.4

Find the nonlinear and linear differential equations of motion of the system shown in Fig. 19, assuming that $z(t)$ is a specified known displacement.

Solution. Since the displacement $z(t)$ of the slider block is specified, the number of degrees of freedom of the system reduces to one. From the free body diagram shown

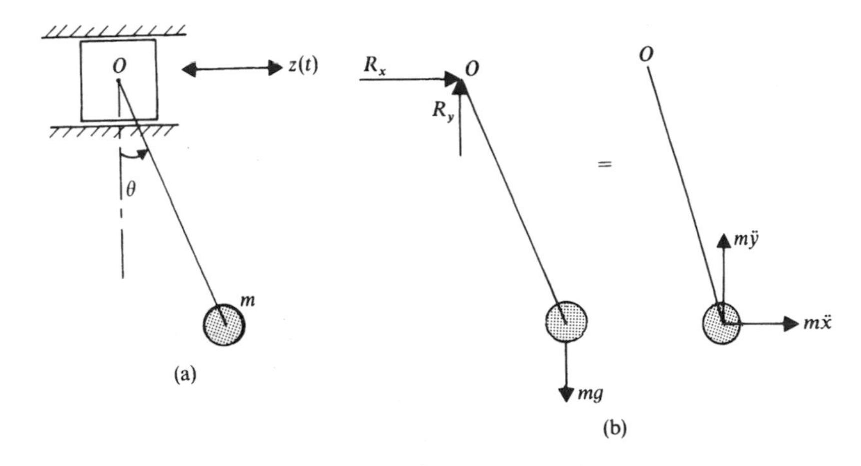

FIG. 1.19. Support motion.

in the figure and by taking the moment about point O, we have

$$M_a = -mgl \sin \theta \tag{1.37}$$

$$M_{eff} = m\ddot{x}l \cos \theta + m\ddot{y}l \sin \theta \tag{1.38}$$

The mass m is assumed to be a point mass or a particle and therefore it has no moment of inertia about its center of mass. The coordinates x and y of this mass are given by

$$x = z + l \sin \theta \tag{1.39}$$

$$y = -l \cos \theta \tag{1.40}$$

Differentiating these coordinates with respect to time yields

$$\dot{x} = \dot{z} + \dot{\theta} l \cos \theta$$

and

$$\dot{y} = \dot{\theta} l \sin \theta$$

By differentiating these velocities with respect to time, one obtains

$$\ddot{x} = \ddot{z} + \ddot{\theta} l \cos \theta - \dot{\theta}^2 l \sin \theta$$

$$\ddot{y} = \ddot{\theta} l \sin \theta + \dot{\theta}^2 l \cos \theta$$

Substituting these equations into Eq. 38 yields

$$\begin{aligned} M_{eff} &= m(\ddot{z} + \ddot{\theta} l \cos \theta - \dot{\theta}^2 l \sin \theta) l \cos \theta + m(\ddot{\theta} l \sin \theta + \dot{\theta}^2 l \cos \theta) l \sin \theta \\ &= m\ddot{z}l \cos \theta + ml^2 \ddot{\theta} \end{aligned} \tag{1.41}$$

Equating Eqs. 37 and 41, the nonlinear differential equation of motion can be obtained as

$$ml^2 \ddot{\theta} + mgl \sin \theta = -m\ddot{z}l \cos \theta \tag{1.42}$$

In order to obtain the linear differential equation, we assume small oscillations and use early linearization by linearizing the kinematic relationships. In this case, Eqs. 39 and 40 are

$$x \approx z + l\theta$$

$$y \approx -l$$

which yields

$$\dot{x} = \dot{z} + l\dot{\theta}, \qquad \ddot{x} = \ddot{z} + l\ddot{\theta}$$

$$\dot{y} = \ddot{y} = 0$$

The applied and effective moment equations can be written as

$$M_a = -mgl\theta$$

$$M_{eff} = m\ddot{x}l = m\ddot{z}l + ml^2\ddot{\theta}$$

Equating these two equations, we obtain the following linear differential equation of motion

$$ml^2\ddot{\theta} + mgl\theta = -m\ddot{z}l \tag{1.43}$$

Clearly, this equation can be obtained from the nonlinear equation given by Eq. 42 by using the assumption of small oscillations.

1.7 IDEALIZATION OF MECHANICAL AND STRUCTURAL SYSTEMS

Modern mechanical and structural systems may consist of several components connected together by different types of joints. In order to develop a mathematical model for a system, the actual system may be represented by a simplified model which has equivalent inertia, damping, and stiffness characteristics. Several assumptions such as neglecting small effects, replacing the distributed characteristics by lumped characteristics, and neglecting uncertainities and noise may have to be made in order to obtain a simplified model which is more amenable to analytical studies. Adopting a very complex model may be considered just as poor a judgement as adopting an oversimplified model because of the waste of energy and time required to study complex models. A reasonably simplified model makes the analysis much simpler as the result of reducing the number of variables and the complexity of the resulting dynamic equations.

Mechanism Systems Figure 20(a) shows the overhead value arrangement in an automotive engine. The cam is driven by the camshaft which rotates with a certain angular velocity, and the cam-follower train consists of the pushrod, the rocker arm, and the valve stem. If the speed of operation of the system is low, the system may be analyzed as consisting of rigid parts, and in this case, the rigid body analysis produces satisfactory results. In many

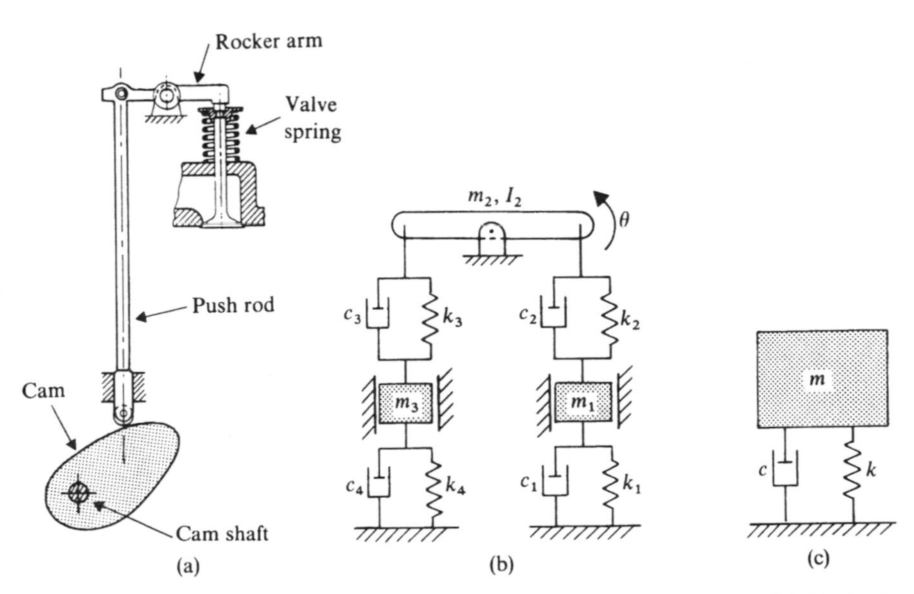

FIG. 1.20. Cam system: (a) cam mechanism; (b) three degree of freedom model; (c) single degree of freedom model.

applications, the speed of operation is very high such that elastic-body analysis must be used, and therefore, the simplified model must account for the elasticity of the system components. Several models with different numbers of degrees of freedom may be developed. The validity of each model, however, in representing the actual system will depend on the inertia, damping, and stiffness characteristics of the system as well as the operating speed. Figure 20(b), for example, shows a three degree of freedom model for the elastic-body cam system. In this model, the masses m_1 and m_2 are the lumped mass characteristics of the follower train. The mass m_3 is the equivalent mass of the cam and a portion of the camshaft. The spring coefficient k_1 represents the stiffness of the follower retaining spring, and the coefficients k_2 and k_3 represent the stiffness characteristics of the follower train. The coefficient k_4 is the bending stiffness of the camshaft. The damping coefficient c_1, c_2, c_3, and c_4 are introduced in this model to account for the dissipation of energy as the result of friction. In this model, the distributed inertia and stiffness characteristics of the true model are replaced by lumped characteristics. If the cam, camshaft, rocker arm, and the valve spring are relatively stiff compared to the pushrod, the simple model shown in Fig. 20(c) can be used where k is the equivalent stiffness of the pushrod and m is the equivalent mass of the cam-follower train.

Structural Systems Another example which can be used to demonstrate the idealization of mechanical and structural systems is the multistory building shown in Fig. 21(a). If we assume that the mass of the frame is small compared to the mass of the floor, we may represent this system by the multidegree of freedom system shown in Fig. 21(b), where m_i is the equivalent mass of the ith floor and k_i is the equivalent stiffness of the ith frame.

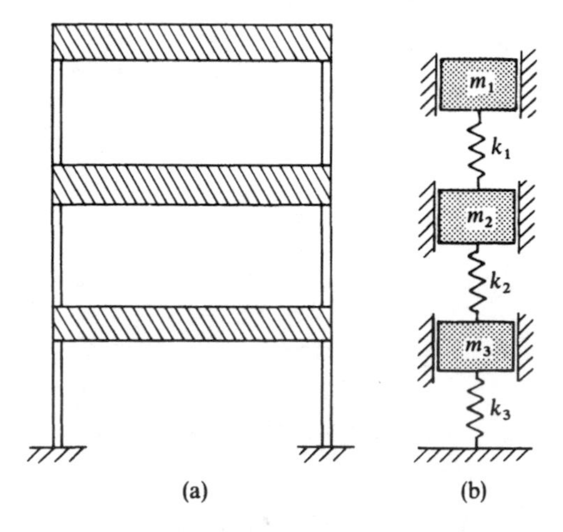

FIG. 1.21. Multistory building.

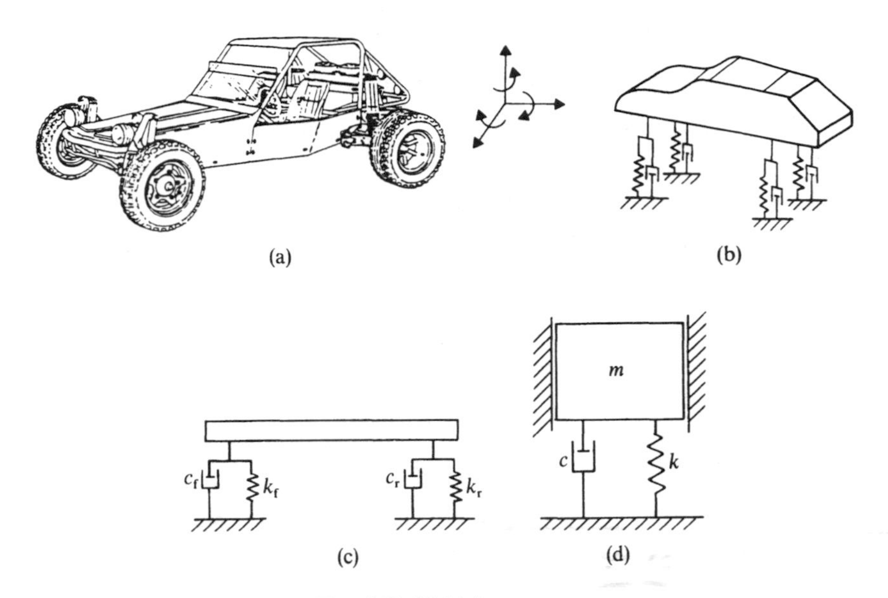

FIG. 1.22. Vehicle system.

Vehicle Systems A third example is the vehicle system shown in Fig. 22(a). The chassis of an actual vehicle consists of beams, plates and other elastic components which may deform. An accurate mathematical model that describes the chassis deformation may require the use of approximation and numerical techniques such as the finite element method. One, however, may assume that the chassis deformation has no significant effect on the gross motion of the vehicle. In order to examine the effect of the suspension characteristics on the gross displacement of the chassis we may assume, for the moment, that the chassis can be treated as a rigid body. We can, therefore, obtain the model shown in Fig. 22(b) which has six degrees of freedom, three degrees of freedom describe the translation of the center of mass of the chassis and three degrees of freedom define the chassis orientation. If the interest is focused only on the planar vibration of the system, a simpler model, such as the one shown in Fig. 22(c), can be used, where in this model k_f and c_f are, respectively, the equivalent stiffness and damping coefficients of the front suspension and k_r and c_r are, respectivly, the equivalent stiffness and damping coefficients of the rear suspension. In general, this system has three degrees of freedom; two degrees of freedom represent the translation of the chassis and one degree of freedom describes the chassis rotation. If we are only concerned with the vibration of the vehicle in the vertical direction, the three degree of freedom model can be further simplified in order to obtain the single degree of freedom model shown in Fig. 22(d), where k and c are, respectively, equivalent stiffness and damping coeffieients that represent the flexibility and damping of the front and rear suspension of the vehicle.

Problems

1.1. What are the basic elements of the vibration model?

1.2. Give examples of systems that have one degree, two degrees, and three degrees of freedom.

1.3. Explain why continuous systems have infinite numbers of degrees of freedom.

1.4. Derive the stiffness coefficient of the helical spring defined by Eq. 4.

1.5. Explain the difference between the static and sliding coefficients of friction.

1.6. For the mass spring system of Example 1, use the equation of motion in the case of undamped free vibration to show that the work done by the spring force is equal to the kinetic energy of the mass.

1.7. For the two degree of freedom system of Example 2, obtain the equation of motion of the forced undamped vibration.

1.8. For the two degree of freedom system of Example 2, obtain the equations of motion of the undamped free vibration.

1.9. Show that the angular momentum of a system of particles can be written as

$$\mathbf{L} = \mathbf{r}_c \times m\mathbf{v}_c + \sum_{i=1}^{n} \mathbf{r}_{ic} \times m_i \mathbf{v}_{ic}$$

where $\mathbf{r}_{ic}$ and $\mathbf{v}_{ic}$ are, respectively, the relative position and velocity vectors of the particle i with respect to the center of mass of the system of particles, m_i is the mass of the particle i, and m is total mass of the system of particles.

1.10. Show that the kinetic energy of the system of particles can be written as

$$T = \frac{1}{2} m v_c^2 + \sum_{i=1}^{n} \frac{1}{2} m_i v_{ic}^2$$

where m is the total mass of the system of particles, v_c is the velocity of the center of mass, m_i is the mass of the particle i, and v_{ic} is the velocity of the particle i with respect to the center of mass.

1.11. For the two degree of freedom system of Example 2, determine the acceleration of the center of mass using Eq. 19. Verify the results you obtain by adding the two equations of motion of the system presented in the example.

1.12. For the two degree of freedom system of Example 2, show that the linear momentum of the system is conserved if $k_1 = c_1 = F_1 = F_2 = 0$.

1.13. For the pendulum system discussed in Section 6, calculate the errors in the displacements, velocities, and accelerations of the center of mass as the result of using the linearized equations. Assume that the rod has length 1 m and it rotates with a constant angular velocity 10 rad/s. Determine the errors when $\theta = 5, 10, 20, 30°$.

1.14. Repeat the preceding problem if the angular velocity of the rod is 100 rad/s.

1.15. Develop several possible simplified vibration models for an airplane and discuss the degrees of freedom of each model and the basic assumptions used to develop such a model.

2
Solution of the Vibration Equations

It was shown in the preceding chapter that the application of Newton's second law to study the motion of physical systems leads to second-order ordinary differential equations. The coefficients of the accelerations, velocities, and displacements in these differential equations represent physical parameters such as inertia, damping, and restoring elastic forces. These coefficients not only have a significant effect on the response of the mechanical and structural systems, but they also affect the stability as well as the speed of response of the system to a given excitation. Changes in these coefficients may result in a stable or unstable system, and/or an oscillatory or nonoscillatory system.

In order to examine, understand, and analyze the behavior of physical systems, we will attempt first to solve the differential equations that govern the vibration of these systems. In this chapter the interest will be focused on solving the following differential equation

$$a_1\ddot{x} + a_2\dot{x} + a_3x = f(t) \tag{2.1}$$

In the linear theory of vibration, a_1, a_2, and a_3 are constant coefficients that represent, respectively, the inertia, damping, and stiffness coefficients. The variable x represents the displacement, $\dot{x}$ is the velocity, and $\ddot{x}$ is the acceleration. The first and second time derivatives of the variable x are given by

$$\dot{x} = \frac{dx}{dt}, \qquad \ddot{x} = \frac{d^2x}{dt^2} = \frac{d\dot{x}}{dt} \tag{2.2}$$

The right-hand side of Eq. 1, denoted as $f(t)$, represents a forcing function which may depend on time t. If the forcing function $f(t)$ is not equal to zero, Eq. 1 is described as a *linear, nonhomogeneous, second-order ordinary differential equation with constant coefficients*. The *homogeneous* differential equations, in which $f(t) = 0$, correspond to the case of *free vibration*. In the case of *undamped vibration*, the coefficient a_2 of the velocity $\dot{x}$ is identically zero. In the following sections we present methods for obtaining solutions for both homogeneous and nonhomogeneous differential equations.

2.1 HOMOGENEOUS DIFFERENTIAL EQUATIONS

In this section, techniques for solving linear, homogeneous, second-order differential equations with constant coefficients are discussed. Whenever the right-hand side of Eq. 1 is identically zero, that is,

$$f(t) = 0, \tag{2.3}$$

the equation is called a homogeneous differential equation. In this case, Eq. 1 reduces to

$$a_1\ddot{x} + a_2\dot{x} + a_3x = 0 \tag{2.4}$$

By a solution to Eq. 4 we mean a function $x(t)$ which, with its derivatives, satisfies the differential equation. A solution to Eq. 4 can be obtained by trial and error. A trial solution is to assume the function $x(t)$ in the following form

$$x(t) = Ae^{pt} \tag{2.5}$$

where A and p are constants to be determined. Differentiating Eq. 5 with respect to time yields

$$\dot{x}(t) = pAe^{pt} \tag{2.6}$$

$$\ddot{x}(t) = p^2Ae^{pt} \tag{2.7}$$

Substituting Eqs. 5–7 into Eq. 4 leads to

$$a_1p^2Ae^{pt} + a_2pAe^{pt} + a_3Ae^{pt} = 0$$

that is,

$$(a_1p^2 + a_2p + a_3)Ae^{pt} = 0 \tag{2.8}$$

Note that e^{pt} is not equal to zero for all values of time t. Also, if the constant A is equal to zero, this implies, from Eq. 5, that $x(t)$ is equal to zero, which is the case of a trivial solution. Therefore, for Eq. 8 to be satisfied for a nontrivial $x(t)$, one must have

$$a_1p^2 + a_2p + a_3 = 0 \tag{2.9}$$

This is called the *characteristic equation* of the second-order differential equation. Equation 9 has two roots p_1 and p_2 which can be determined from the quadratic formula as

$$p_1 = \frac{-a_2 + \sqrt{a_2^2 - 4a_1a_3}}{2a_1} \tag{2.10}$$

$$p_2 = \frac{-a_2 - \sqrt{a_2^2 - 4a_1a_3}}{2a_1} \tag{2.11}$$

Accordingly, we have the following two independent solutions

$$x_1(t) = A_1e^{p_1t} \tag{2.12}$$

and

$$x_2(t) = A_2e^{p_2t} \tag{2.13}$$

The general solution of the differential equation can then be written as the sum of these two independent solutions, provided that the roots of the characteristic equation are not equal, that is,

$$x(t) = x_1(t) + x_2(t) = A_1 e^{p_1 t} + A_2 e^{p_2 t} \tag{2.14}$$

The complete solution of the second-order ordinary differential equations contains two arbitrary constants A_1 and A_2. These arbitrary constants can be determined from the initial conditions, as discussed in later sections.

Clearly, the solution of the differential equation depends on the roots p_1 and p_2 of the characteristic equation. There are three different cases for the roots p_1 and p_2. In the first case, in which p_1 and p_2 are real numbers and $p_1 \neq p_2$, one has $a_2^2 > 4a_1 a_3$. In vibration systems, this case corresponds to the case in which the damping coefficient is relatively high, and for this reason, the system is said to be *overdamped*. If $a_2^2 = 4a_1 a_3$, the roots p_1 and p_2 are real numbers and $p_1 = p_2$, and the system is said to be *critically damped*. In the third case, $a_2^2 < 4a_1 a_3$, and the roots p_1 and p_2 are complex conjugates. This is the case in which the damping coefficient is relatively small. In this case, the system is said to be *underdamped*. In the following, these three different cases are discussed in more detail.

Real Distinct Roots This is the case in which the following inequality is satisfied

$$a_2^2 > 4a_1 a_3 \tag{2.15}$$

In this case, the quantity $\sqrt{a_2^2 - 4a_1 a_3}$ is real and the roots p_1 and p_2 are distinct, that is,

$$p_1 \neq p_2 \tag{2.16}$$

Therefore, we have two independent solutions and the complete solution is the sum of two exponential functions and is given by

$$x(t) = A_1 e^{p_1 t} + A_2 e^{p_2 t} \tag{2.17}$$

If both p_1 and p_2 are positive, the solution $x(t)$ will be exponentially increasing with time. If both roots are negative, the solution $x(t)$ will be exponentially decreasing with time, and the rate of decay and growth will depend on the magnitude of p_1 and p_2. Another possibility is that one root, say p_1, is positive and the other root p_2 is negative.

Example 2.1

Find the solution of the following homogeneous second-order ordinary differential equation

$$\ddot{x} - 4\dot{x} + 3x = 0$$

Solution. Assume a solution in the form

$$x(t) = Ae^{pt}$$

Substituting this solution into the differential equation yields

$$p^2 A e^{pt} - 4pAe^{pt} + 3Ae^{pt} = 0$$

That is,

$$(p^2 - 4p + 3)Ae^{pt} = 0$$

The characteristic equation can then be defined as

$$p^2 - 4p + 3 = 0$$

That is,

$$(p - 1)(p - 3) = 0$$

The roots p_1 and p_2 are then given by

$$p_1 = 1 \qquad \text{and} \qquad p_2 = 3$$

There are two independent solutions $x_1(t)$ and $x_2(t)$ given by

$$x_1(t) = A_1 e^{p_1 t} = A_1 e^t$$

$$x_2(t) = A_2 e^{p_2 t} = A_2 e^{3t}$$

The solutions $x_1(t)$ and $x_2(t)$ are shown in Fig. 1. The complete solution is the sum of the two solutions, that is,

$$x(t) = x_1(t) + x_2(t) = A_1 e^t + A_2 e^{3t}$$

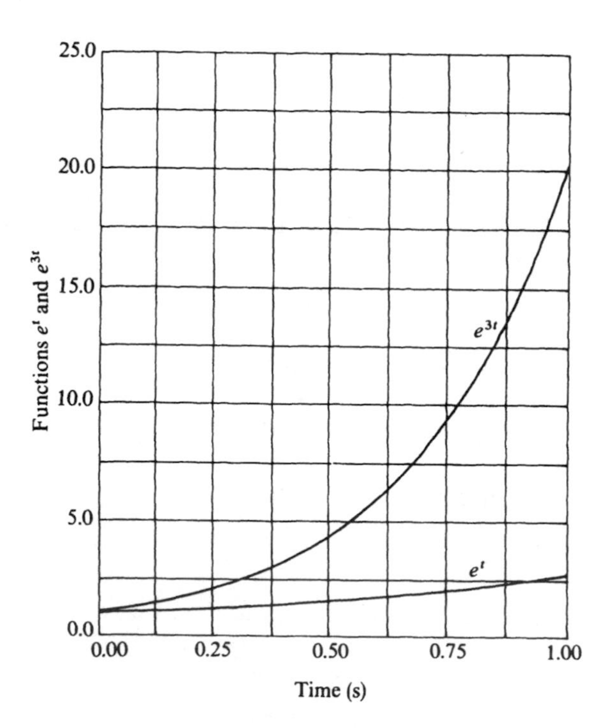

FIG. 2.1. Independent solutions.

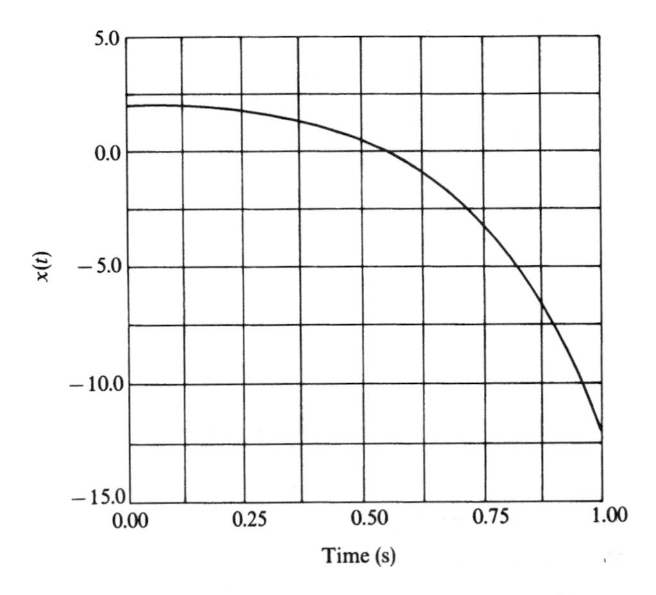

FIG. 2.2. Complete solution ($A_1 = 3$, $A_2 = -1$).

The complete solution depends on the constants A_1 and A_2 which can be determined using the initial conditions. This solution is shown is Fig. 2 for the case in which $A_1 = 3$ and $A_2 = -1$.

Example 2.2

Find the solution of the following second-order differential equation

$$\ddot{x} + \dot{x} - 6x = 0$$

Solution. We assume a solution in the follwing exponential form

$$x(t) = Ae^{pt}$$

Substituting this solution into the differential equation leads to

$$p^2Ae^{pt} + pAe^{pt} - 6Ae^{pt} = 0$$

which can be written as

$$(p^2 + p - 6)Ae^{pt} = 0$$

The characteristic equation can then be defined as

$$p^2 + p - 6 = 0$$

or

$$(p - 2)(p + 3) = 0$$

That is,

$$p_1 = 2, \qquad p_2 = -3$$

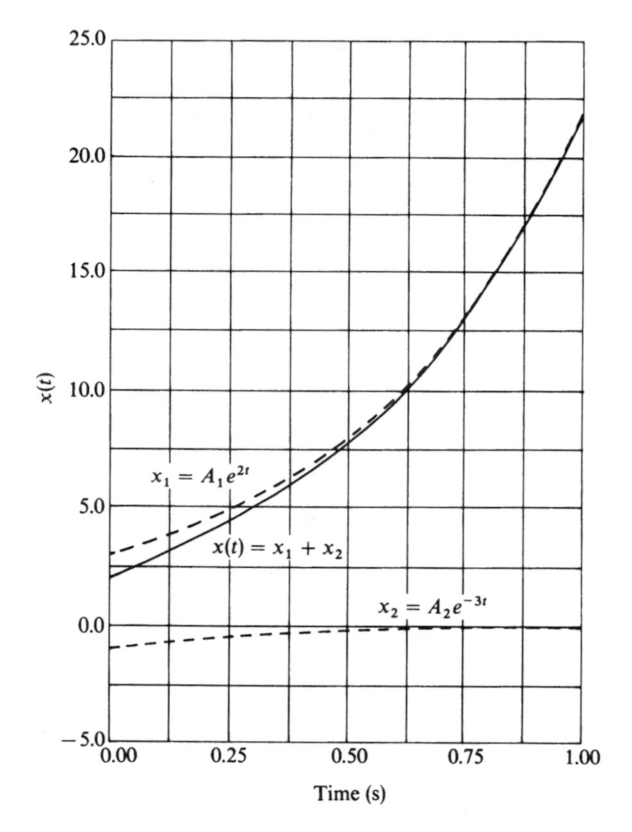

FIG. 2.3. Distinct real roots ($A_1 = 3$, $A_2 = -1$).

The two independent solutions are then given by

$$x_1(t) = A_1 e^{2t}$$

$$x_2(t) = A_2 e^{-3t}$$

and the complete solution is

$$x(t) = x_1(t) + x_2(t) = A_1 e^{2t} + A_2 e^{-3t}$$

The independent solutions $x_1(t)$ and $x_2(t)$, as well as the complete solution $x(t)$, are shown in Fig. 3 in the case in which $A_1 = 3$ and $A_2 = -1$.

We observe from this and the preceding example that if the roots p_1 and p_2 are distinct and real, the solution is not oscillatory, but it can be represented as the sum of exponential functions which increase or decrease with time.

Repeated Roots If the roots of the characteristic equation are real and equal,

$$p_1 = p_2 \tag{2.18}$$

This condition will be satisfied if the coefficients of the differential equation satisfy the following equality

$$a_2^2 = 4a_1a_3 \tag{2.19}$$

In this case, we have only one solution

$$x_1(t) = A_1e^{p_1t} \tag{2.20}$$

From the theory of differential equations, it is shown that the complete solution $x(t)$ can be assumed in the following form

$$x(t) = x_1(t)u(t) \tag{2.21}$$

where $u(t)$ is a function that can be determined by substituting $x(t)$ into the original differential equation

$$a_1\ddot{x} + a_2\dot{x} + a_3x = 0 \tag{2.22}$$

Differentiating Eq. 21 with respect to time yields

$$\begin{aligned}\dot{x}(t) &= \dot{x}_1(t)u(t) + x_1(t)\dot{u}(t) = p_1A_1e^{p_1t}u(t) + A_1e^{p_1t}\dot{u}(t)\\ &= [p_1u + \dot{u}]A_1e^{p_1t}\end{aligned} \tag{2.23}$$

$$\begin{aligned}\ddot{x}(t) &= [p_1\dot{u} + \ddot{u}]A_1e^{p_1t} + p_1[p_1u + \dot{u}]A_1e^{p_1t}\\ &= [p_1^2u + 2p_1\dot{u} + \ddot{u}]A_1e^{p_1t}\end{aligned} \tag{2.24}$$

Substituting Eqs. 21, 23, and 24 into Eq. 22 yields

$$a_1(p_1^2u + 2p_1\dot{u} + \ddot{u})A_1e^{p_1t} + a_2(p_1u + \dot{u})A_1e^{p_1t} + a_3uA_1e^{p_1t} = 0$$

That is,

$$a_1(p_1^2u + 2p_1\dot{u} + \ddot{u}) + a_2(p_1u + \dot{u}) + a_3u = 0$$

or

$$a_1\ddot{u} + (2a_1p_1 + a_2)\dot{u} + (a_1p_1^2 + a_2p_1 + a_3)u = 0 \tag{2.25}$$

Using Eqs. 10 and 19, one can verify that the root p_1 is given by

$$p_1 = -\frac{a_2}{2a_1} \tag{2.26}$$

Substituting this equation into Eq. 25 and using the identity of Eq. 19 one gets

$$\ddot{u} = 0$$

which, upon integration, yields

$$u = A_2 + A_3t \tag{2.27}$$

where A_2 and A_3 are constants. The complete solution in the case of repeated roots then can be written as

$$x(t) = x(t)u(t) = (A_2 + A_3t)A_1e^{p_1t}$$

which can be rewritten as

$$x(t) = (c_1 + c_2t)e^{p_1t} \tag{2.28}$$

where the two arbitrary constants c_1 and c_2 can be determined from the initial conditions.

Example 2.3

Find the complete solution of the following second-order ordinary differential equation

$$\ddot{x} + 6\dot{x} + 9x = 0$$

Solution. We assume a solution in the form

$$x(t) = Ae^{pt}$$

Substituting this solution into the ordinary differential equation we obtain

$$p^2 Ae^{pt} + 6pAe^{pt} + 9Ae^{pt} = 0$$

or

$$(p^2 + 6p + 9)Ae^{pt} = 0$$

The characteristic equation can then be defined as

$$p^2 + 6p + 9 = 0$$

which can be written as

$$(p + 3)(p + 3) = 0$$

that is,

$$p_1 = p_2 = -3$$

which is the case of repeated roots. In this case, the complete solution can be written as

$$x(t) = (c_1 + c_2 t)e^{-3t}$$

This solution is shown in Fig. 4. As in the case of real distinct roots, the solution is not of oscillatory nature.

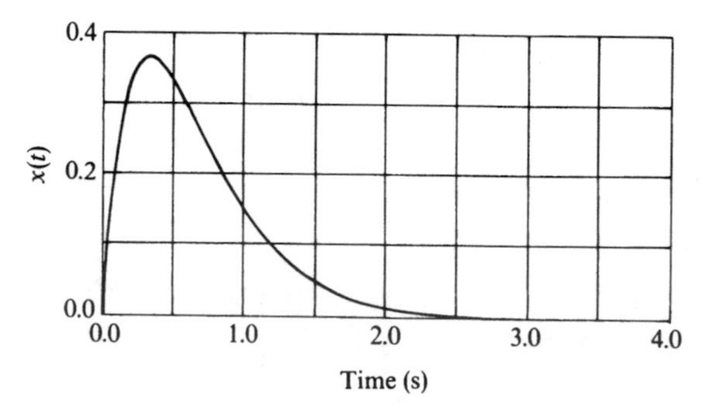

FIG. 2.4. Repeated roots ($c_1 = 0$, $c_2 > 0$).

Complex Conjugate Roots This is the case in which the coefficients of the differential equation satisfy the inequality

$$a_2^2 < 4a_1a_3 \tag{2.29}$$

In this case, we can write

$$\sqrt{a_2^2 - 4a_1a_3} = \sqrt{-(4a_1a_3 - a_2^2)} = i\sqrt{4a_1a_3 - a_2^2}$$

where i is the imaginary operator defined as

$$i = \sqrt{-1}$$

Let

$$\alpha = -\frac{a_2}{2a_1}, \qquad \beta = \frac{1}{2a_1}\sqrt{4a_1a_3 - a_2^2} \tag{2.30}$$

Then the roots of Eqs. 10 and 11 can be written as

$$p_1 = -\frac{a_2}{2a_1} + \frac{1}{2a_1}\sqrt{a_2^2 - 4a_1a_3} = \alpha + i\beta \tag{2.31}$$

$$p_2 = -\frac{a_2}{2a_1} - \frac{1}{2a_1}\sqrt{a_2^2 - 4a_1a_3} = \alpha - i\beta \tag{2.32}$$

That is, p_1 and p_2 are complex conjugates. Since p_1 is not equal to p_2, the complete solution is the sum of two independent solutions and can be expressed as follows

$$x(t) = A_1e^{p_1t} + A_2e^{p_2t} \tag{2.33}$$

where A_1 and A_2 are constants.

Substituting Eqs. 31 and 32 into Eq. 33, one gets

$$\begin{aligned} x(t) &= A_1e^{(\alpha+i\beta)t} + A_2e^{(\alpha-i\beta)t} \\ &= e^{\alpha t}(A_1e^{i\beta t} + A_2e^{-i\beta t}) \end{aligned} \tag{2.34}$$

The complex exponential functions $e^{i\beta t}$ and $e^{-i\beta t}$ can be written in terms of trigonometric functions using *Euler's formulas* which are given by

$$e^{i\theta} = \cos\theta + i\sin\theta \tag{2.35}$$

$$e^{-i\theta} = \cos\theta - i\sin\theta \tag{2.36}$$

Using these identities with Eq. 34, one obtains

$$\begin{aligned} x(t) &= e^{\alpha t}[A_1(\cos\beta t + i\sin\beta t) + A_2(\cos\beta t - i\sin\beta t)] \\ &= e^{\alpha t}[(A_1 + A_2)\cos\beta t + i(A_1 - A_2)\sin\beta t] \end{aligned} \tag{2.37}$$

Since the displacement $x(t)$ must be real, the coefficients of the sine and cosine functions in the above equations must be real. This will be the case if and only if A_1 and A_2 are complex conjugates. In this case,

$$A_1 + A_2 = c_1$$
$$i(A_1 - A_2) = c_2$$

where c_1 and c_2 are constants. The complete solution of Eq. 37 can then be written as

$$x(t) = e^{\alpha t}[c_1 \cos \beta t + c_2 \sin \beta t] \tag{2.38}$$

The constants c_1 and c_2 can be determined from the initial conditions. Note that the solution $x(t)$ is of an oscillatory nature, since it is the product of the exponential function $e^{\alpha t}$ and the harmonic functions $\cos \beta t$ and $\sin \beta t$.

The solution $x(t)$ given by Eq. 38 can also be expressed in another simple form. To this end, let

$$X = \sqrt{c_1^2 + c_2^2}$$

One can write $x(t)$ as

$$x(t) = Xe^{\alpha t}\left[\frac{c_1}{X} \cos \beta t + \frac{c_2}{X} \sin \beta t\right] \tag{2.39}$$

As shown in Fig. 5, we define the angle ϕ such that

$$\phi = \tan^{-1} \frac{c_1}{c_2} \tag{2.40}$$

It follows that

$$\sin \phi = \frac{c_1}{X}, \qquad \cos \phi = \frac{c_2}{X} \tag{2.41}$$

Substituting Eq. 41 into Eq. 39 yields

$$x(t) = Xe^{\alpha t}[\sin \phi \cos \beta t + \cos \phi \sin \beta t] \tag{2.42}$$

Recall the following trigonometric identity

$$\sin \phi \cos \beta t + \cos \phi \sin \beta t = \sin(\beta t + \phi) \tag{2.43}$$

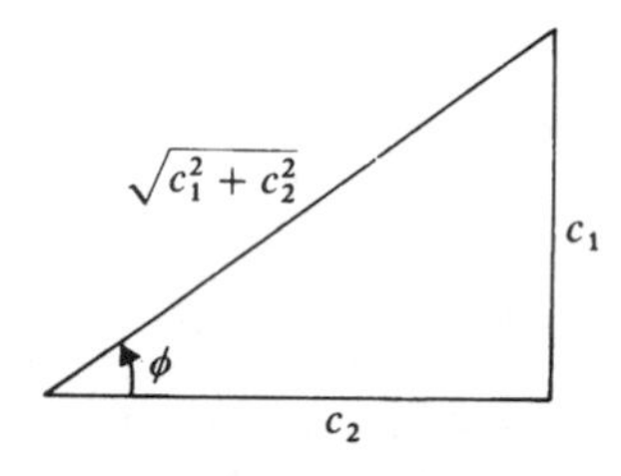

FIG. 2.5. Definition of the phase angle ϕ.

By using this identity, Eq. 42 can be written as

$$x(t) = Xe^{\alpha t} \sin(\beta t + \phi) \tag{2.44}$$

This form of the solution is useful in the analysis of the free vibration of mechanical systems, and as in the preceding two cases, there are two arbitrary constants, X and ϕ, which can be determined from the initial conditions. The constant X is called the *amplitude of displacement* and the constant ϕ is called the *phase angle*. Note that α and β are known since they are functions of the coefficients a_1, a_2, and a_3 of the differential equation (see Eqs. 31 and 32).

Example 2.4

Find the complete solution of the following second-order ordinary differential equation

$$5\ddot{x} + 2\dot{x} + 50x = 0$$

Solution. We assume a solution in the form

$$x = Ae^{pt}$$

Substituting this solution into the differential equation we obtain

$$(5p^2 + 2p + 50)Ae^{pt} = 0$$

The characteristic equation is then given by

$$5p^2 + 2p + 50 = 0$$

which has the following roots

$$p_1 = \frac{-a_2 + \sqrt{a_2^2 - 4a_1a_3}}{2a_1} = \frac{-2 + \sqrt{4 - 4(5)(50)}}{2(5)} = -0.2 + 3.156i$$

$$p_2 = \frac{-a_2 - \sqrt{a_2^2 - 4a_1a_3}}{2a_1} = \frac{-2 - \sqrt{4 - 4(5)(50)}}{2(5)} = -0.2 - 3.156i$$

The two roots, p_1 and p_2, are complex conjugates, and the constants α and β can be recognized as

$$\alpha = -0.2, \qquad \beta = 3.156$$

The complete solution is then given by Eq. 38 as

$$x(t) = e^{-0.2t}[c_1 \cos 3.156t + c_2 \sin 3.156t]$$

or, equivalently, by Eq. 44 as

$$x(t) = Xe^{-0.2t} \sin(3.156t + \phi)$$

where the constants c_1 and c_2 or X and ϕ can be determined from the initial conditions. As shown in Fig. 6, the solution is of oscillatory nature with an amplitude that decreases with time. A system with this type of solution is said to be a *stable system*.

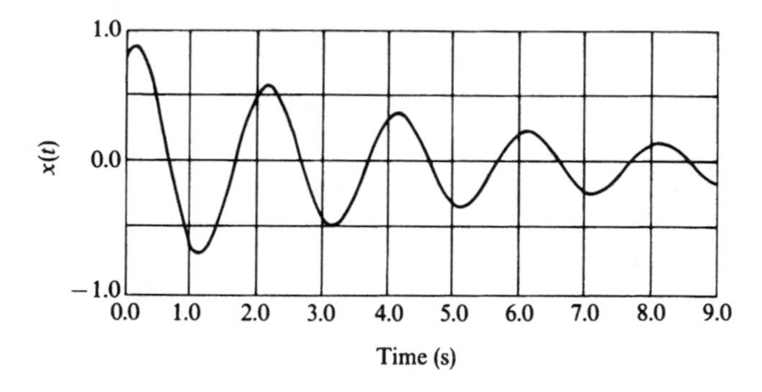

FIG. 2.6. Complex conjugate roots with negative real part.

Example 2.5

Find the complete solution of the following second-order differential equation

$$5\ddot{x} - 2\dot{x} + 50x = 0$$

Solution. This differential equation is the same as the one given in the preceding example, except the coefficient a_2 of $\dot{x}$ becomes negative. In this case, the characteristic equation is

$$5p^2 - 2p + 50 = 0$$

The roots of this characteristic equation are

$$p_1 = \frac{-a_2 + \sqrt{a_2^2 - 4a_1a_3}}{2a_1} = \frac{2 + \sqrt{4 - 4(5)(50)}}{2(5)} = 0.2 + 3.156i$$

$$p_2 = \frac{-a_2 - \sqrt{a_2^2 - 4a_1a_3}}{2a_1} = \frac{2 - \sqrt{4 - 4(5)(50)}}{2(5)} = 0.2 - 3.156i$$

The constants α and β of Eqs. 31 and 32 are, respectively, given by

$$\alpha = 0.2 \qquad \text{and} \qquad \beta = 3.156$$

The complete solution is then given by

$$x(t) = e^{0.2t}[c_1 \cos 3.156t + c_2 \sin 3.156t]$$

or, equivalently, as

$$x(t) = Xe^{0.2t} \sin(3.156t + \phi)$$

The solution $x(t)$ is shown in Fig. 7. The solution is of oscillatory nature with increasing amplitude. A physical system with this type of solution is said to be an *unstable system*.

Example 2.6

Find the complete solution of the following second-order differential equation

$$5\ddot{x} + 50x = 0$$

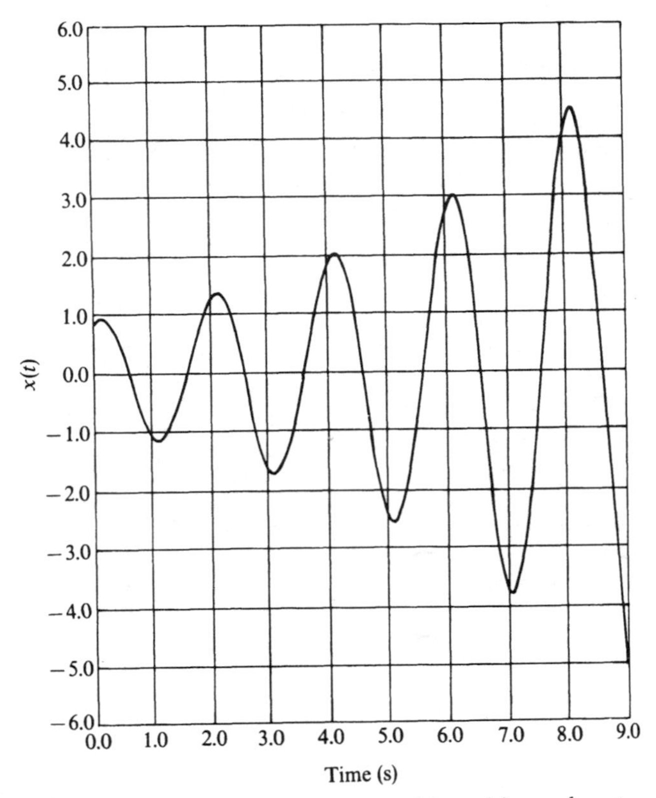

FIG. 2.7. Complex conjugate roots with positive real part.

Solution. This is, the same equation as in the preceding example, except that the coefficient a_2 of $\dot{x}$ is equal to zero. In this case, the characteristic equation is given by

$$5p^2 + 50 = 0$$

or

$$p^2 + 10 = 0$$

the roots of this equation are

$$p_1 = 3.162i, \qquad p_2 = -3.162i$$

The roots p_1 and p_2 are complex conjugates with the real parts equal to zero. In this case.

$$\alpha = 0 \qquad \text{and} \qquad \beta = 3.162$$

and the solution can be written as

$$x(t) = c_1 \cos 3.162t + c_2 \sin 3.162t$$

or, equivalently,

$$x(t) = X \sin(3.162t + \phi)$$

This solution is a harmonic function which has a constant amplitude, as shown in

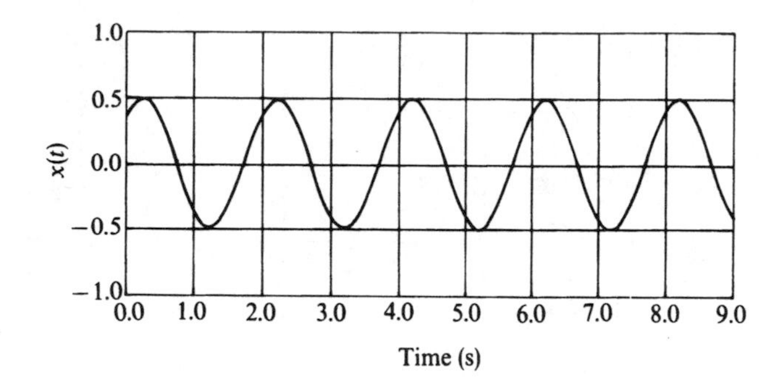

FIG. 2.8. Complex conjugate roots with zero real parts.

Fig. 8. If the response of a physical system has a similar nature, the system is said to be *critically stable* or to have *sustained oscillation*. This case of free vibration in which the damping coefficient is equal to zero will be studied in more detail in Chapter 3.

2.2 INITIAL CONDITIONS

It was shown in the preceding section that the solution of the homogeneous second-order ordinary differential equations contains two arbitrary constants which can be obtained by imposing two initial conditions. This leads to two algebraic equations which can be solved for the two constants. In the following, we discuss the determination of these constants in the three different cases of real distinct roots, repeated roots, and complex conjugate roots.

Real Distinct Roots If $p_1 \neq p_2$ and both p_1 and p_2 are real, the solution is given by

$$x(t) = A_1 e^{p_1 t} + A_2 e^{p_2 t} \tag{2.45}$$

where p_1 and p_2 are defined by Eqs. 10 and 11, and A_1 and A_2 are two arbitrary constants to be determined from the initial conditions.

Differentiating Eq. 45 with respect to time yields

$$\dot{x}(t) = p_1 A_1 e^{p_1 t} + p_2 A_2 e^{p_2 t} \tag{2.46}$$

If any two specific conditions on the displacement and/or the velocity are given, Eqs. 45 and 46 can be used to determine the constants A_1 and A_2. For instance, let x_0 and $\dot{x}_0$ be the initial values for the displacement and velocity such that

$$x_0 = x(t = 0), \qquad \dot{x}_0 = \dot{x}(t = 0) \tag{2.47}$$

Substituting Eq. 47 into Eqs. 45 and 46 yields

$$x_0 = A_1 + A_2 \tag{2.48}$$

$$\dot{x}_0 = p_1 A_1 + p_2 A_2 \tag{2.49}$$

These are two algebraic equations which can be used to determine the constants A_1 and A_2. One can verify that, in this case, A_1 and A_2 are given by

$$A_1 = \frac{x_0 p_2 - \dot{x}_0}{p_2 - p_1}, \qquad A_2 = \frac{\dot{x}_0 - p_1 x_0}{p_2 - p_1} \tag{2.50}$$

Example 2.7

Find the complete solution of the following second-order ordinary differential equation

$$\ddot{x} - 4\dot{x} + 3x = 0$$

subject to the initial conditions

$$x_0 = 2, \qquad \dot{x}_0 = 0$$

Solution. It was shown in Example 1, that the solution of this differential equation is

$$x(t) = A_1 e^t + A_2 e^{3t}$$

It follows that

$$\dot{x}(t) = A_1 e^t + 3A_2 e^{3t}$$

Using the initial conditions, we have the following two algebraic equations

$$2 = A_1 + A_2$$

$$0 = A_1 + 3A_2$$

The solution of these two algebraic equations yields

$$A_1 = 3, \qquad A_2 = -1$$

and the complete solution can be written as

$$x(t) = 3e^t - e^{3t}$$

Repeated Roots If p_1 and p_2 are real and equal, the solution, as demonstrated in the preceding section, is given by

$$x(t) = (c_1 + c_2 t)e^{p_1 t} \tag{2.51}$$

which yields

$$\begin{aligned} \dot{x}(t) &= c_2 e^{p_1 t} + (c_1 + c_2 t) p_1 e^{p_1 t} \\ &= [c_2 + p_1(c_1 + c_2 t)]e^{p_1 t} \end{aligned} \tag{2.52}$$

where c_1 and c_2 are arbitrary constants to be determined from the initial conditions. Let x_0 and $\dot{x}_0$ be, respectively, the initial displacement and velocity at time $t = 0$, it follows that

$$x_0 = c_1 \tag{2.53}$$

$$\dot{x}_0 = c_2 + p_1 c_1 \tag{2.54}$$

These two algebraic equations can be solved for the constants c_1 and c_2 as

$$c_1 = x_0 \tag{2.55}$$

$$c_2 = \dot{x}_0 - p_1 x_0 \tag{2.56}$$

Example 2.8

Find the complete solution of the following second-order ordinary differential equation

$$\ddot{x} + 6\dot{x} + 9x = 0$$

subject to the initial conditions

$$x_0 = 0, \qquad \dot{x}_0 = 3$$

Solution. It was shown in Example 3 that the solution of this differential equation is given by

$$x(t) = (c_1 + c_2 t)e^{-3t}$$

Differentiating this equation with respect to time yields

$$\dot{x}(t) = [c_2 - 3(c_1 + c_2 t)]e^{-3t}$$

By using the initial conditions, we obtain the following two algebraic equations

$$0 = c_1$$

$$3 = c_2 - 3c_1$$

or

$$c_1 = 0$$

$$c_2 = 3$$

The complete solution can then be written as

$$x(t) = 3te^{-3t}$$

Complex Conjugate Roots If the roots of the characteristic equations p_1 and p_2 can be written in the following form

$$p_1 = \alpha + i\beta \tag{2.57}$$

$$p_2 = \alpha - i\beta \tag{2.58}$$

where α and β are constant real numbers, then the solution of the homogeneous differential equation can be written as

$$x(t) = e^{\alpha t}[c_1 \cos \beta t + c_2 \sin \beta t] \tag{2.59}$$

or, equivalently,

$$x(t) = Xe^{\alpha t} \sin(\beta t + \phi) \tag{2.60}$$

where c_1 and c_2 or X and ϕ are constants to be determined using the initial conditions. Differentiating Eq. 60 with respect to time yields

$$\begin{aligned}\dot{x}(t) &= \alpha X e^{\alpha t} \sin(\beta t + \phi) + \beta X e^{\alpha t} \cos(\beta t + \phi) \\ &= X e^{\alpha t}[\alpha \sin(\beta t + \phi) + \beta \cos(\beta t + \phi)] \qquad (2.61)\end{aligned}$$

If x_0 and $\dot{x}_0$ are, respectively, the initial displacement and velocity at time $t = 0$, substituting these initial conditions into Eqs. 60 and 61 leads to the following algebraic equations

$$x_0 = X \sin \phi \qquad (2.62)$$

$$\dot{x}_0 = X(\alpha \sin \phi + \beta \cos \phi) \qquad (2.63)$$

The solution of these two equations defines X and ϕ as

$$X = \sqrt{x_0^2 + \left(\frac{\dot{x}_0 - \alpha x_0}{\beta}\right)^2} \qquad (2.64)$$

$$\phi = \tan^{-1} \frac{\beta x_0}{(\dot{x}_0 - \alpha x_0)} \qquad (2.65)$$

Example 2.9

Obtain the complete solution of the following second-order ordinary differential equation

$$5\ddot{x} + 2\dot{x} + 50x = 0$$

subject to the initial conditions

$$x_0 = 0.01, \qquad \dot{x}_0 = 3$$

Solution. It was shown in Example 4, that the solution of this equation is given by

$$x(t) = X e^{-0.2t} \sin(3.156t + \phi)$$

and the derivative is

$$\dot{x}(t) = X e^{-0.2t}[-0.2 \sin(3.156t + \phi) + 3.156 \cos(3.156t + \phi)]$$

Using the initial conditions we obtain the following algebraic equations

$$0.01 = X \sin \phi$$

$$3 = -0.2X \sin \phi + 3.156X \cos \phi$$

from which we have

$$X = \sqrt{(0.01)^2 + \left(\frac{(3 + 0.2(0.01)}{3.156}\right)^2} = 0.9512$$

$$\phi = \tan^{-1} \frac{(3.156)(0.01)}{3 - (0.01)(-0.2)} = 0.6023^\circ$$

The complete solution is then given by

$$x(t) = 0.9512e^{-0.2t} \sin(3.156t + 0.6023^\circ)$$

2.3 SOLUTION OF NONHOMOGENEOUS EQUATIONS WITH CONSTANT COEFFICIENTS

In the preceding sections, methods for obtaining the solutions of homogeneous second-order ordinary differential equations with constant coefficients were discussed, and it was shown that the solution of such equations contains two constants which can be determined using the initial conditions. In this section, we will learn how to solve nonhomogeneous differential equations with constant coefficients. These equations can be written in the following general form

$$a_1\ddot{x} + a_2\dot{x} + a_3x = f(t) \tag{2.66}$$

This equation, which represents the mathematical model of many vibratory systems, is in a similar form to the standard equation used to study the forced vibration of damped single degree of freedom systems, where $f(t)$ represents the forcing function. If $f(t) = 0$, Eq. 66 reduces to the case of homogeneous equations which are used to study the free vibration of single degree of freedom systems. Thus, the solution of Eq. 66 consists of two parts First, the solution x_h of Eq. 66 where the right-hand side is equal to zero, that is, $f(t) = 0$; this part of the solution is called the *complementary function.* Methods for obtaining the complementary functions were discussed in the preceding sections. The second part of the solution is a solution x_p which satisfies Eq. 66; this part of the solution is called the *particular solution.* That is, the complete solution of Eq. 66 can be written as

$$x = \text{complementary function} + \text{particular solution}$$

or

$$x = x_h + x_p \tag{2.67}$$

where x_h is the solution of the equation

$$a_1\ddot{x}_h + a_2\dot{x}_h + a_3x_h = 0 \tag{2.68}$$

and x_p is the solution of the equation

$$a_1\ddot{x}_p + a_2\dot{x}_p + a_3x_p = f(t) \tag{2.69}$$

The particular solution x_p can be found by the method of *undetermined coefficients.* In this method, we determine the linearly independent functions which result from repeated differentiation of the function $f(t)$. Then we assume x_p as a linear combination of all the independent functions that appear in the function $f(t)$ and its independent derivatives. The particular solution x_p is then substituted into the differential equation, and the independent constants can be found by equating the coefficients of the independent functions in both sides of the differential equation.

In most of the applications in this book the function $f(t)$ possesses only a finite number of independent derivatives. For example, if the function $f(t)$ is constant, such that

$$f(t) = b$$

where b is a given constant, we have

$$\frac{df(t)}{dt} = \frac{d^2f(t)}{dt^2} = \cdots = \frac{d^nf(t)}{dt^n} = 0$$

Since all the derivatives of $f(t)$ are zeros in this case, we assume the particular solution x_p in the following form

$$x_p = k_1 f(t) = k_1 b$$

where k_1 is a constant. It follows that $\dot{x}_p = \ddot{x}_p = 0$. Substituting x_p, $\dot{x}_p$, and $\ddot{x}_p$ into Eq. 69, and keeping in mind that $f(t) = b$, we obtain

$$0 + 0 + a_3 k_1 b = f(t) = b$$

from which $k_1 = 1/a_3$ and $x_p = b/a_3$.

Examples of other functions which have a finite number of independent derivatives are t^n, e^{bt}, $\cos bt$, and $\sin bt$, where n is an integer and b is an arbitrary constant. If the function $f(t)$ possesses an infinite number of independent derivatives, such as the functions $1/t$ or $1/t^n$ where n is a positive integer, the particular solution x_p can be assumed as an infinite series whose terms are the derivatives of $f(t)$ multiplied by constants. In this case the convergence of the solution must be checked. In general, the method for solving the linear nonhomogeneous second-order ordinary differential equation (Eq. 66) can be summarized in the following steps:

1. The complementary function x_h, which is the solution of the homogeneous equation given by Eq. 68, is first obtained by using the method described in the preceding sections.
2. The independent functions that appear in the function $f(t)$ and its derivatives are then determined by repeated differentiation of $f(t)$. Let these independent functions be denoted as $f_1(t), f_2(t), \ldots, f_n(t)$ where n is a positive integer.
3. The particular solution x_p is assumed in the form

$$x_p = k_1 f_1(t) + k_2 f_2(t) + \cdots + k_n f_n(t) \tag{2.70}$$

   where $k_1, k_2, \ldots, k_n$ are n constants to be determined.
4. Substituting this assumed solution into the differential equation defined by Eq. 69, and equating the coefficients of the independent functions in both sides of the equations, we obtain n algebraic equations which can be solved for the unknowns $k_1, k_2, \ldots, k_n$.
5. The complete solution x is then defined as $x = x_h + x_p$. This complete solution contains only two arbitrary constants which appear in the complementary function x_h. At this stage, the particular solution contains no unknown constants, and the initial conditions $x_0 = x(t = 0)$ and $\dot{x} = \dot{x}(t = 0)$ can be used to determine the arbitrary constants in the complementary function. It is important, however, to emphasize that the initial conditions must be imposed on the complete solution x and not just the complementary function x_h.

The above procedure for solving linear, nonhomogeneous, second-order ordinary differential equations with constant coefficients is demonstrated by the following example.

Example 2.10

Find the solution of the differential equation

$$\ddot{x} + 4x = te^{3t}$$

subject to the initial conditions

$$x_0 = 0.01, \qquad \dot{x}_0 = 0$$

Solution. First, we will solve the homogeneous equation

$$\ddot{x}_h + 4x_h = 0$$

by assuming a complementary function in the form

$$x_h = Ae^{pt}$$

Upon substituting this solution into the homogeneous differential equation, the following characteristic equation is obtained

$$p^2 + 4 = 0$$

which has the roots

$$p_1 = 2i, \qquad p_2 = -2i$$

and, accordingly, the complementary function x_h is defined as

$$x_h = X \sin(2t + \phi)$$

where X and ϕ are constants to be determined later, using the initial conditions. The function $f(t)$, in this example, is given by

$$f(t) = te^{3t}$$

By repeated differentiation of this function with respect to time, we have

$$\frac{df}{dt} = 3te^{3t} + e^{3t}, \qquad \frac{d^2f}{dt^2} = 9te^{3t} + 6e^{3t}, \ldots$$

Clearly, by repeated differentiation, only the independent functions te^{3t} and e^{3t} appear and, as such, the function $f(t)$ and its derivatives contain the following independent functions

$$f_1(t) = te^{3t}, \qquad f_2(t) = e^{3t}$$

Therefore, we assume a particular solution in the form

$$x_p = k_1 f_1(t) + k_2 f_2(t) = k_1 te^{3t} + k_2 e^{3t}$$

Differentiating this equation with respect to time yields

$$\dot{x}_p = k_1(3te^{3t} + e^{3t}) + 3k_2 e^{3t}$$

$$= (k_1 + 3k_2 + 3k_1 t)e^{3t}$$

$$\ddot{x}_p = (6k_1 + 9k_2 + 9k_1 t)e^{3t}$$

Substituting x_p, $\dot{x}_p$, and $\ddot{x}_p$ into the differential equation

$$\ddot{x}_p + 4x_p = te^{3t}$$

yields

$$(6k_1 + 9k_2 + 9k_1 t)e^{3t} + 4(k_2 + k_1 t)e^{3t} = te^{3t}$$

or

$$(6k_1 + 13k_2)e^{3t} + 13k_1 te^{3t} = te^{3t}$$

Equating the coefficients of the independent functions on both sides yields the following two algebraic equations

$$13k_1 = 1$$

$$6k_1 + 13k_2 = 0$$

From which k_1 and k_2 can be determined as

$$k_1 = \frac{1}{13}, \qquad k_2 = -\frac{6}{13}\left(\frac{1}{13}\right) = -\frac{6}{169}$$

The particular solution x_p can then be written as

$$x_p = \frac{1}{13}\left(te^{3t} - \frac{6}{13}e^{3t}\right)$$

It is clear that the particular solution, at this stage, contains no unknown coefficients. The complete solution is given by summing the complementary function and the particular solution as

$$x = x_h + x_p = X\sin(2t + \phi) + \frac{1}{13}\left(te^{3t} - \frac{6}{13}e^{3t}\right)$$

The constants X and ϕ can now be determined using the initial conditions

$$x_0 = 0.01 = X\sin\phi - \frac{6}{169}$$

$$\dot{x}_0 = 0 = 2X\cos\phi - \frac{5}{169}$$

or

$$X\sin\phi = 0.0455$$

$$X\cos\phi = 0.01479$$

By dividing the first equation by the second and by squaring the two equations and adding, we obtain, respectively, the following

$$\phi = \tan^{-1}\frac{0.0455}{0.01479} = \tan^{-1}(3.0764) = 71.993°$$

$$X = \sqrt{(0.0455)^2 + (0.01479)^2} = 0.04784$$

that is,

$$x = 0.04784\sin(2t + 71.993°) + \frac{1}{13}\left(te^{3t} - \frac{6}{13}e^{3t}\right)$$

Special Case In the method presented in this chapter for finding the solution of the differential equations, the particular solution is assumed as a linear combination of a set of independent functions. If one of these functions, however, is the same as one of the functions that appear in the homogeneous solution, this function in the particular solution must be altered in order to make it independent of the homogeneous solution. This can be achieved by multiplying this function by the independent variable t. In order to demonstrate this procedure, consider the differential equation

$$\ddot{x} - 9x = 5t + e^{-3t}$$

The homogeneous solution of this equation is given by

$$x_h = A_1 e^{3t} + A_2 e^{-3t}$$

where A_1 and A_2 are arbitrary constants. By using the method described in this chapter, the particular solution can be assumed in the following form

$$x_p = k_1 t + k_2 + k_3 e^{-3t}$$

where k_1, k_2, and k_3 are constants to be determined by substituting this solution into the differential equation. One, however, may observe that the last function used in the particular solution e^{-3t} is the same as one of the functions that appear in the solution of the homogeneous equation. In this case, the particular solution must be modified by multiplying this function by the independent variable t. The particular solution then must be assumed as

$$x_p = k_1 t + k_2 + k_3 t e^{-3t}$$

By so doing, all the functions that appear in the assumed particular solution become independent of those that appear in the homogeneous solution.

The special case discussed in this section has practical significance in the vibration analysis, because the solution procedure discussed in this special case is used to obtain the solution of the vibration equation in an important case known as *resonance*. This resonance situation will be examined in Chapter 4.

Principle of Superposition One important advantage in dealing with linear differential equations is the fact that the *principle of superposition* can be applied. According to this principle, if the right-hand side of the differential equation can be written as the sum of several functions, the particular solution of the differential equation can be expressed as the sum of the particular solutions as the result of application of each of these functions separately. Consider the differential equation

$$a_1 \ddot{x} + a_2 \dot{x} + a_3 x = F_1(t) + F_2(t) + \cdots + F_n(t) \tag{2.71}$$

Let x_{pi}, $i = 1, 2, \ldots, n$, be the particular solution of the differential equation

$$a_1 \ddot{x} + a_2 \dot{x} + a_3 x = F_i(t)$$

then, according to the principle of superposition which is applicable only to linear systems, the particular solution x_p of the differential equation defined by Eq. 71 can be written as

$$x_p = x_{p1} + x_{p2} + \cdots + x_{pn} = \sum_{i=1}^{n} x_{pi} \tag{2.72}$$

2.4 STABILITY OF MOTION

It has been shown in the preceding sections that the solution of the vibration equation consists of two parts; the complementary function and the particular solution. In the theory of vibration, the particular solution depends on the external excitation. In the special case of free vibration, on the other hand, one has to determine only the complementary function which contains two arbitrary constants that can be determined using the initial conditions. As was shown in the preceding section, the complementary function depends on the roots of the characteristic equations, and these roots depend on the coefficients of the displacement and its time derivatives in the differential equation. Consequently, these coefficients, which represent inertia, damping, and stiffness coefficients, affect the form of the complementary function and the complete solution of the vibration problem. In fact, as was shown in the examples presented in this chapter, the stability of the system depends on the roots of the characteristic equations. In terms of these roots, the general form of the complementary function is given by

$$x(t) = A_1 e^{p_1 t} + A_2 e^{p_2 t}$$

where A_1 and A_2 are arbitrary constants and p_1 and p_2 are the roots of the characteristic equation. In the following, we examine the effect of the roots on the stability of the system and summarize some of the important results obtained in the preceding sections.

Negative Real Roots If both the roots p_1 and p_2 are real and negative, the solution $x(t)$ (as shown in Fig. 9(a)) approaches zero as time t becomes large. In this case, the solution is bounded and nonoscillatory; and the rate at which the solution decreases as time increases depends on the magnitude of the roots p_1 and p_2. It is quite usual to use the complex plane to examine the stability of the system. As shown in Fig. 9(b), in the case of negative real roots, the roots of the characteristic equations p_1 and p_2 lie on the negative portion of the real axis. The system can be made more stable by increasing the magnitude of the negative roots or, equivalently, by changing the inertia, damping, and/or stiffness parameters of the system.

Positive Real Roots If one of the roots or both become positive, the solution is nonoscillatory and grows without bound as time increases; an

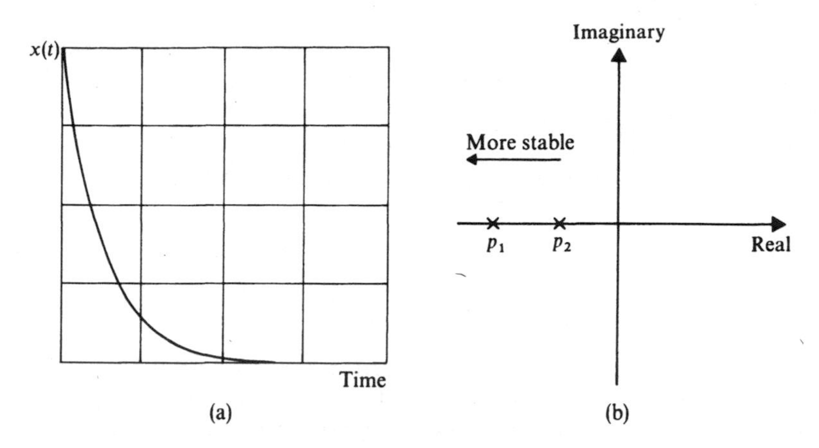

FIG. 2.9. Negative real roots.

example of this case is shown in Fig. 10(a). The rate at which the solution increases with time depends on the magnitude of the positive roots, and as this magnitude increases, the solution grows more rapidly. This case of instability, which can be encountered in some engineering applications, is the result of the exponential form of the solution, and that instability occurs when at least one of the roots is positive. Different combinations that lead to instability are shown in the complex planes shown in Fig. 10(b)–(d).

Complex Roots In this case, the roots p_1 and p_2 appear as complex conjugates which can be written as

$$p_1 = \alpha + i\beta$$

$$p_2 = \alpha - i\beta$$

The solution, in this case, is oscillatory and can be expressed as

$$x(t) = Xe^{\alpha t} \sin(\beta t + \phi)$$

where X and ϕ are arbitrary constants which can be determined from the initial conditions. As shown in the examples presented in the preceding sections, there are three possibilities which may be encountered. First, the real part of the root α is negative, such that the roots p_1 and p_2 are located in the left-hand side of the complex plane, as shown in Fig. 11(a). As shown in Fig. 11(b), the solution of the vibration equation, in this case, is oscillatory and bounded, with an amplitude which decreases with time. The frequency of oscillation depends on β while the rate of decay depends on the magnitude of the real part α. This is the case of a stable system.

The second possibility occurs when the real part α is identically zero. In this case, the roots p_1 and p_2 consist of only imaginary parts, as shown in Fig. 12(a). In this case, the solution is a harmonic function which has a frequency

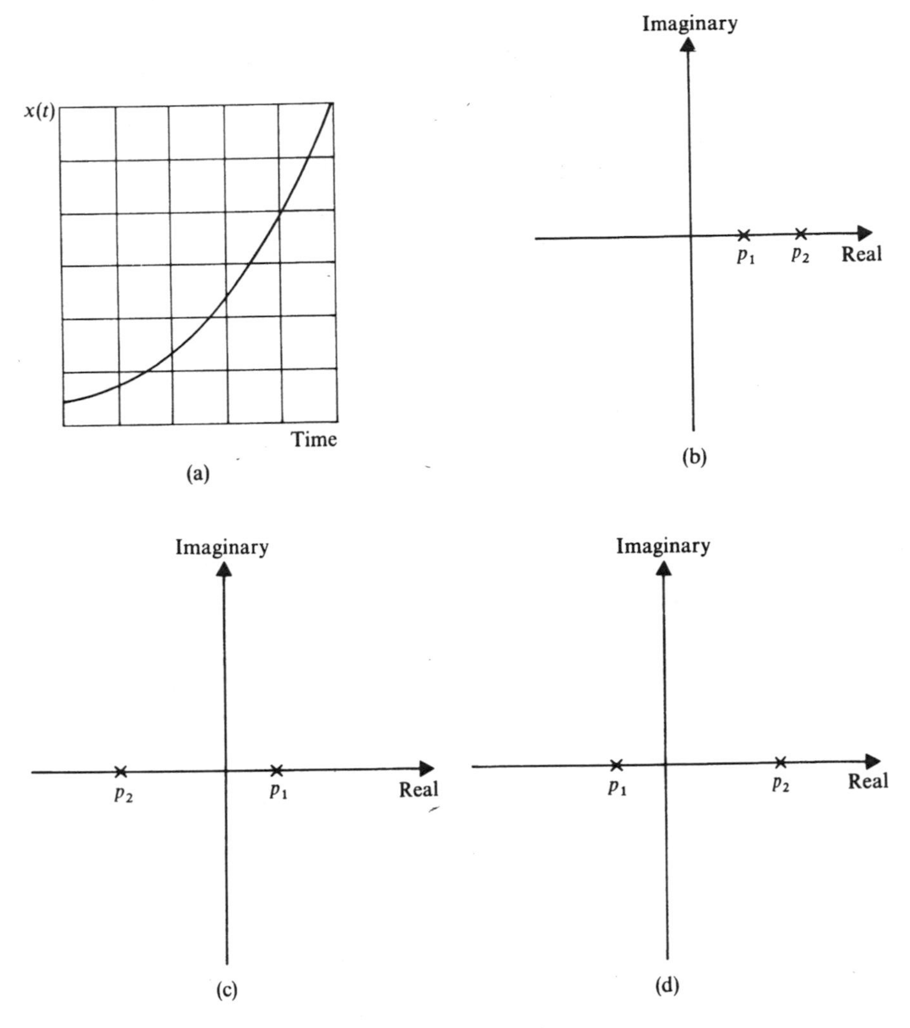

FIG. 2.10. Positive real roots.

equal to β, as shown in Fig. 12(b). In this case, the system is called *critically stable* since the solution is bounded, but the amplitude does not increase or decrease as time t increases.

The third possibility occurs when the real part α is positive, and as a consequence, the roots p_1 and p_2 are located in the right-hand side of the complex plane, as shown in Fig. 13(a). The solution, in this case, is a product of an exponential function which grows with time and a harmonic function which has a constant amplitude. The result is an oscillatory solution with an amplitude that increases as time increases, as shown in Fig. 13(b). This is the case of an unstable system.

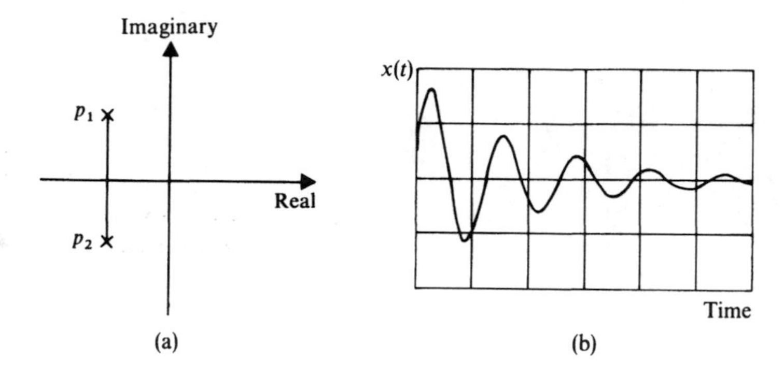

FIG. 2.11. Stable oscillatory motion.

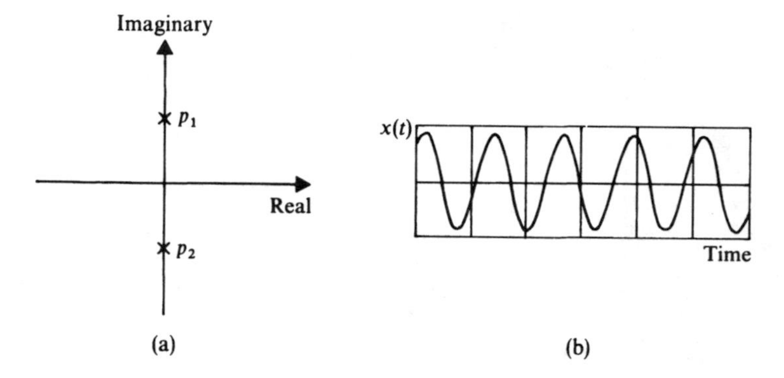

FIG. 2.12. Critically stable system.

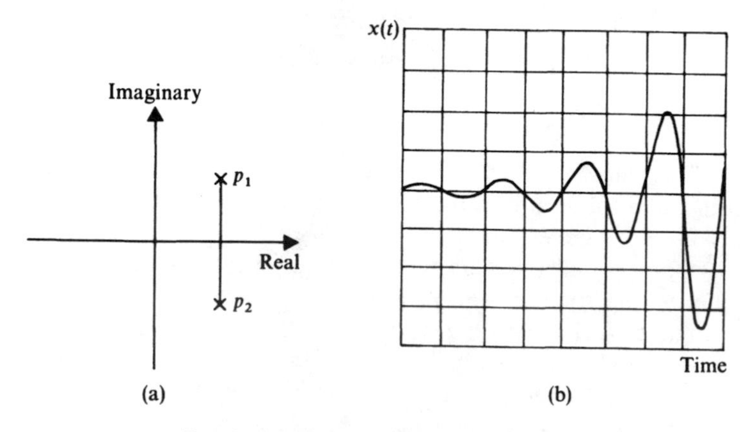

FIG. 2.13. Unstable oscillatory motion.

It is clear from the analysis presented in this section, that stability of motion is achieved if the roots p_1 and p_2 of the characteristic equation lie in the left-hand side of the complex plane. If any of the roots lie in the right-hand side of the complex plane, the system becomes unstable, and the degree of stability or instability depends on the magnitude of the roots. The locations of the roots of the characteristic equations can be altered by changing the physical parameters of the system, such as the inertia, damping, and stiffness coefficients.

Problems

Find the solution of each of the following second-order differential equations:

2.1. $\ddot{x} + 9x = 0$.

2.2. $5\ddot{x} + 16x = 0$.

2.3. $\ddot{x} - 9x = 0$.

2.4. $2\ddot{x} + 3\dot{x} + 7x = 0$.

2.5. $2\ddot{x} - 3\dot{x} + 7x = 0$.

2.6. $2\ddot{x} - 3\dot{x} - 7x = 0$.

Find the solution of each of the following second-order ordinary differential equations:

2.7. $\ddot{x} + 9x = 6e^t$.

2.8. $\ddot{x} + 16x = 10 + 3e^t$.

2.9. $3\ddot{x} + 25x = t^2e^{2t} + t$.

2.10. $3\ddot{x} + 25x = 5\cos 2t$.

2.11. $2\ddot{x} + 3\dot{x} + 7x = 2\cos 2t + 3\sin t$.

2.12. $\ddot{x} + 3\dot{x} + 2x = 10e^{3t} + 4t^2$.

2.13. $2\ddot{x} - 3\dot{x} + 7x = 3t\sin t + t^2$.

2.14. $\ddot{x} + 5\dot{x} + 6x = 3e^{-2t} + e^{3t}$.

Find the complete solution of the following differential equations:

2.15. $\ddot{x} + 9x = 3\cos t;\ x_0 = 1,\ \dot{x}_0 = 0$.

2.16. $\ddot{x} + 9x = 3\cos t;\ x_0 = 0,\ \dot{x}_0 = 3$.

2.17. $5\ddot{x} + 16x = 0;\ x_0 = 2,\ \dot{x}_0 = 0$.

2.18. $5\ddot{x} + 16x = 0;\ x_0 = 0,\ \dot{x}_0 = 3$.

2.19. $5\ddot{x} + 16x = 0;\ x_0 = 2,\ \dot{x}_0 = 3$.

2.20. $\ddot{x} + 3\dot{x} + 2x = 10t(1 + \sin 2t);\ x_0 = 1,\ \dot{x}_0 = 0$.

2.21. $\ddot{x} - 2\dot{x} + 5x = e^{2t}(1 - t\sin 3t);\ x_0 = 1,\ \dot{x}_0 = 2$.

2.22. $\ddot{x} + 15\dot{x} + 6x = 3e^{-2t} + te^{3t}$; $x_0 = 1$, $\dot{x}_0 = 5$.

2.23. Discuss the stability of the dynamic system whose differential equation is given in Problem 1.

2.24. Compare the stability results obtained in Problem 23 with the stability of the system whose dynamics is defined by the differential equation given in Problem 3.

2.25. Examine the stability of the solutions of the two differential equations given in Problems 4 and 5.

3
Free Vibration of Single Degree of Freedom Systems

The term *free vibration* is used to indicate that there is no external force causing the motion, and that the motion is primarily the result of initial conditions, such as an initial displacement of the mass element of the system from an equilibrium position and/or an initial velocity. The free vibration is said to be *undamped free vibration* if there is no loss of energy throughout the motion of the system. This is the case of the simplest vibratory system, which consists of an inertia element and an elastic member which produces a restoring force which tends to restore the inertia element to its equilibrium position. Dissipation of energy may be caused by friction or if the system contains elements such as dampers which remove energy from the system. In such cases, the oscillation is said to be *free damped vibration*. The mathematical models that govern the free vibration of single degree of freedom systems can be described in terms of homogeneous second-order ordinary differential equations that contain displacement, velocity, and acceleration terms. The displacement coefficients describe the stiffness of the elastic members or the restoring forces. The velocity coefficients define the damping constants and determine the amount of energy dissipated, and the acceleration coefficients define the inertia of the system.

3.1 FREE UNDAMPED VIBRATION

In this section, the standard form of the differential equation that governs the linear free undamped vibration of single degree of freedom systems is derived, and the solution of this equation is obtained and is used to introduce several basic definitions that will be frequently used in this text. For this purpose, a single degree of freedom mass–spring system is first used, and it is shown that the linear free undamped vibration of other single degree of freedom systems is governed by a mathematical model that resembles the model obtained for the mass–spring system.

Static Equilibrium Configuration Figure 1(a) shows an elastic element represented by a spring, which has undeformed length denoted as l_0.

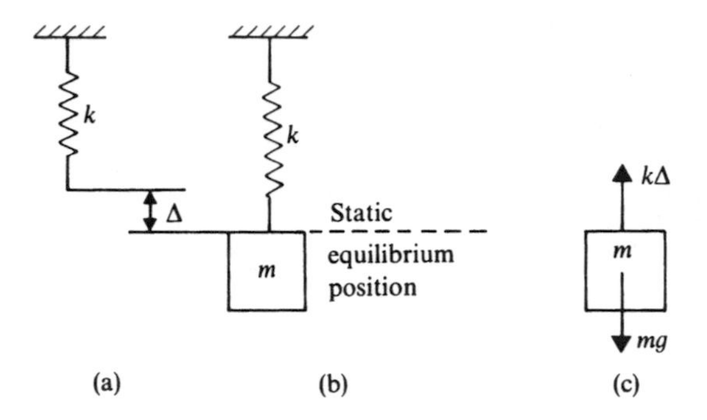

FIG. 3.1. Static equilibrium position.

If the mass m is attached to this spring, there will be an elongation Δ in the spring due to the weight of the mass as a static load, as shown in Fig. 1(b). In this configuration, the mass is in a position called the *static equilibrium position*. The free body diagram shown in Fig. 1(c) reveals that the spring force must be equal in magnitude and opposite in direction to the weight of the mass, that is,

$$k\Delta = mg$$

or

$$k\Delta - mg = 0 \tag{3.1}$$

where k is the spring stiffness, Δ is the static deflection, and g is the gravitational constant. Equation 1 is called the *static equilibrium condition.*

Natural Frequency Suppose now that the system is set in motion from the static equilibrium position due to initial displacement and/or initial velocity. In order to derive the differential equation of motion that governs the free vibration of this system, we consider the mass at an arbitrary position x from the static equilibrium position. The extension in the spring at this position will be the displacement x plus the static deflection Δ. The spring force denoted as F_s can then be written as

$$F_s = -k(\Delta + x) \tag{3.2}$$

The negative sign indicates that the spring force is a restoring force which is in an opposite direction to the direction of motion. Applying Netwon's second law and using the free body diagram shown in Fig. 2, we have

$$m\ddot{x} = mg - k(x + \Delta)$$

or

$$m\ddot{x} = mg - kx - k\Delta \tag{3.3}$$

Using the static equilibrium condition given by Eq. 1, Eq. 3 can be written as

$$m\ddot{x} = -kx$$

or

$$m\ddot{x} + kx = 0 \tag{3.4}$$

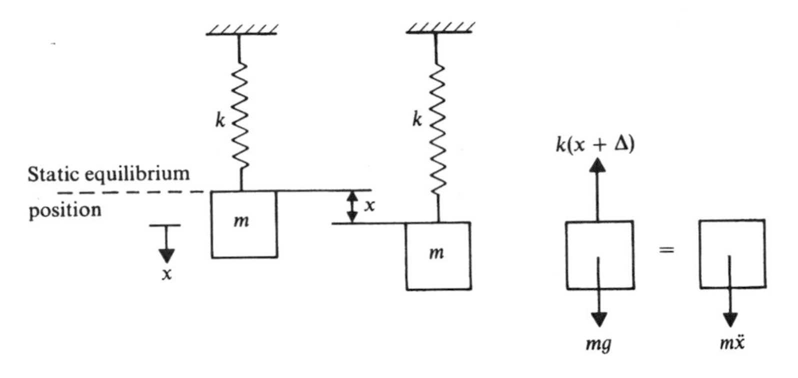

FIG. 3.2. Free undamped vibration.

This is the standard form of the equation of motion that governs the linear free vibration of single degree of freedom systems. Equation 4 can be rewritten as

$$\ddot{x} + \frac{k}{m}x = 0$$

or

$$\ddot{x} + \omega^2 x = 0 \tag{3.5}$$

where ω is a constant that depends on the inertia and stiffness characteristics of the system and is defined as

$$\omega = \sqrt{\frac{k}{m}} \tag{3.6}$$

The constant ω is called the *circular* or the *natural frequency* of the system. That is, the natural frequency ω is defined to be the square root of the coefficient of x divided by the coefficient of $\ddot{x}$. The unit of the natural frequency ω is radians/second or simply rad/s.

Harmonic Motion As discussed in the preceding chapter, the complete solution of Eq. 5 can be assumed in the following form

$$x = Ae^{pt} \tag{3.7}$$

Substituting this solution into the differential equation (Eq. 5) leads to

$$(p^2 + \omega^2)Ae^{pt} = 0$$

The characteristic equation of the system is then defined as

$$(p^2 + \omega^2) = 0$$

which has the roots

$$p_1 = i\omega \qquad \text{and} \qquad p_2 = -i\omega$$

These roots are complex and conjugate, and as a consequence the solution can be assumed in the following form

$$x = A \cos \omega t + B \sin \omega t \tag{3.8}$$

where A and B are arbitrary constants which can be determined by using the initial conditions. Equation 8 can be expressed in another form as

$$x = X \sin(\omega t + \phi) \tag{3.9}$$

where

$$X = \sqrt{A^2 + B^2} \tag{3.10}$$

$$\phi = \tan^{-1}\left(\frac{A}{B}\right) \tag{3.11}$$

The constants X and ϕ which are called, respectively, the *amplitude of vibration* and the *phase angle*, can be determined if the constants A and B are known.

Equation 9 represents a harmonic oscillation, and the maximum or minimum displacement x occurs when

$$|\sin(\omega t + \phi)| = 1$$

That is,

$$\omega t + \phi = \frac{2n + 1}{2}\pi, \qquad n = 0, 1, 2, \ldots$$

which implies that the solution has an infinite number of peaks, and the time at which these peaks occur depends on the phase angle ϕ.

Applications Equation 4 or 5 and its oscillatory solution as defined by Eq. 8 or 9 describe the vibration of many single degree of freedom systems, and their use is not only restricted to the simple mass–spring system. In order to illustrate this, we consider the pulley system shown in Fig. 3, where the spring has a stiffness coefficient k and the pulley, which is assumed to have a frictionless surface, has a negligible mass. In this system, the displacement of the mass is twice the displacement of the spring. Using the static equilibrium configuration, one has

$$mg = F_r = \tfrac{1}{2}F_s$$

where the spring force $F_s = k\Delta$, Δ is the static deflection of the spring, and F_r is the tension in the rope. The static equilibrium condition then can be written as

$$mg - \tfrac{1}{2}k\Delta = 0$$

If the mass oscillates as the result of an initial displacement or an initial velocity, the equation of motion of the mass is given by

$$m\ddot{x} = mg - F_r = mg - \frac{1}{2}k\left(\frac{x}{2} + \Delta\right)$$

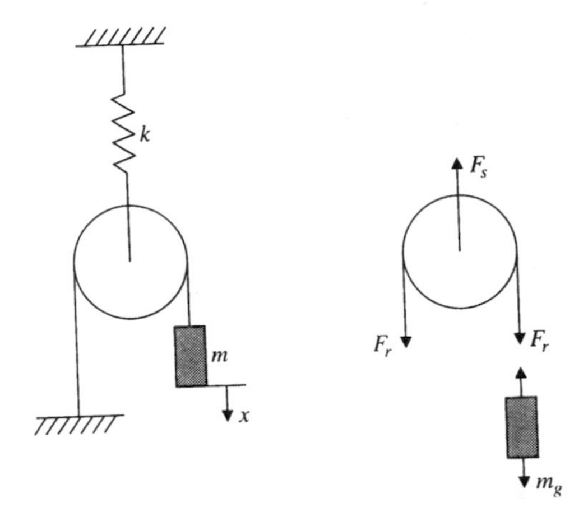

FIG. 3.3. Pulley system.

where x is the displacement of the mass from the static equilibrium position. The preceding equation, upon the use of the static equilibrium condition, leads to the equation of motion of the system shown in Fig. 3 as

$$m\ddot{x} + \frac{k}{4}x = 0$$

which defines the natural frequency of the system as

$$\omega = \sqrt{\frac{k}{4m}}$$

As another example, we use the U- *tube manometer*, shown in Fig. 4, which can be used in pressure measurements. The liquid mercury in the tube has length l and mass density ρ, and the cross section of the tube is A and it is assumed to be uniform. The total mass of the mercury in the tube is $m = \rho A l$, and the pressure on the cross section that acts in a direction opposite to the direction of the acceleration is $p_u = -2\rho Axg$. Using Newton's second law, one has

$$m\ddot{x} = p_u$$

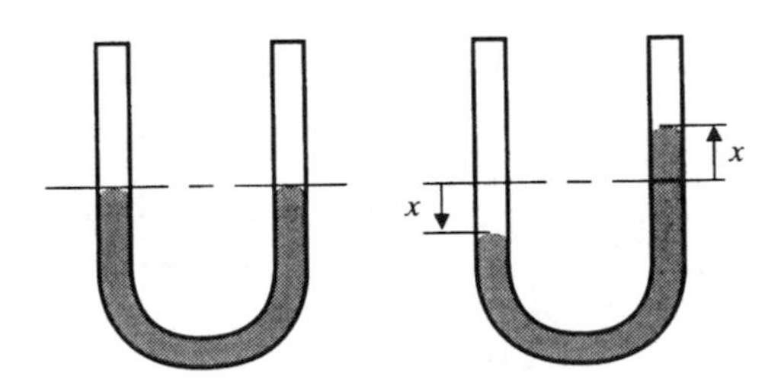

FIG. 3.4. U-tube manometer.

which leads to the linear second-order ordinary differential equation

$$\rho A l \ddot{x} = -2\rho A g x$$

This equation can be written in the form of Eq. 5 as

$$\ddot{x} + \frac{2g}{l} x = 0$$

which defines the natural frequency of oscillation as

$$\omega = \sqrt{\frac{2g}{l}}$$

3.2 ANALYSIS OF THE OSCILLATORY MOTION

For the mass–spring system discussed in the preceding section, let us consider the case in which the mass m has both initial displacement and velocity which can be expressed as

$$x(t = 0) = x_0 \tag{3.12}$$

$$\dot{x}(t = 0) = \dot{x}_0 \tag{3.13}$$

Substituting these initial conditions into Eq. 8 and its time derivative, one can show that the constants A and B can be written as

$$A = x_0, \qquad B = \frac{\dot{x}_0}{\omega} \tag{3.14}$$

Therefore, Eq. 8 can be expressed in terms of the initial conditions as

$$x = x_0 \cos \omega t + \frac{\dot{x}_0}{\omega} \sin \omega t \tag{3.15}$$

and the velocity $\dot{x}$ is

$$\dot{x} = -\omega x_0 \sin \omega t + \dot{x}_0 \cos \omega t \tag{3.16}$$

Using Eq. 14, the amplitude X of Eq. 10 and the phase angle ϕ of Eq. 11 are defined as

$$X = \sqrt{x_0^2 + \left(\frac{\dot{x}_0}{\omega}\right)^2} \tag{3.17}$$

$$\phi = \tan^{-1}\left(\frac{x_0}{\dot{x}_0/\omega}\right) \tag{3.18}$$

If the initial displacement x_0 is equal to zero, the phase angle ϕ is equal to zero and, accordingly, the displacement is equal to $X \sin \omega t$. On the other hand, if the initial velocity is equal to zero, the phase angle ϕ is equal to $\pi/2$ and the displacement, in this case, is equal to $X \cos \omega t$.

The velocity of the mass can be obtained by differentiating Eq. 9 with respect to time. This yields

$$\dot{x} = \omega X \cos(\omega t + \phi) \tag{3.19}$$

Using the following trigonometric identity

$$\cos \theta = \sin\left(\theta + \frac{\pi}{2}\right), \tag{3.20}$$

Eq. 19 can also be written as

$$\dot{x} = \omega X \sin\left(\omega t + \phi + \frac{\pi}{2}\right) \tag{3.21}$$

which indicates that the velocity $\dot{x}$ has a phase lead angle of $\pi/2$ compared to the displacement x given by Eq. 9.

The acceleration of the mass can be obtained by differentiating Eq. 19 with respect to time. This yields

$$\ddot{x} = -\omega^2 X \sin(\omega t + \phi) \tag{3.22}$$

The acceleration $\ddot{x}$ can also be obtained by differentiating Eq. 21 with respect to time, that is,

$$\ddot{x} = \omega^2 X \cos\left(\omega t + \phi + \frac{\pi}{2}\right)$$

Using the trigonometric identity of Eq. 20, the acceleration $\ddot{x}$ can be expressed as

$$\ddot{x} = \omega^2 X \sin(\omega t + \phi + \pi) \tag{3.23}$$

which is the same as Eq. 22, since $\sin \theta = -\sin(\theta + \pi)$. Equation 23, however, shows that the phase difference between the displacement x and the acceleration $\ddot{x}$ is 180°, and comparing Eqs. 21 and 23 shows that the phase difference between the velocity $\dot{x}$ and the acceleration $\ddot{x}$ is 90°. The displacement x of Eq. 9, the velocity $\dot{x}$ of Eq. 19, and the acceleration $\ddot{x}$ of Eq. 22 are shown graphically in Fig. 5.

Natural Period of Oscillation The natural period of oscillation is denoted as τ and defined from the equation

$$\omega\tau = 2\pi$$

that is,

$$\tau = \frac{2\pi}{\omega} = 2\pi\sqrt{\frac{m}{k}} \tag{3.24}$$

The system natural frequency can be expressed as the number of cycles per second as

$$f = \frac{1}{\tau} = \frac{\omega}{2\pi} = \frac{1}{2\pi}\sqrt{\frac{k}{m}} \tag{3.25}$$

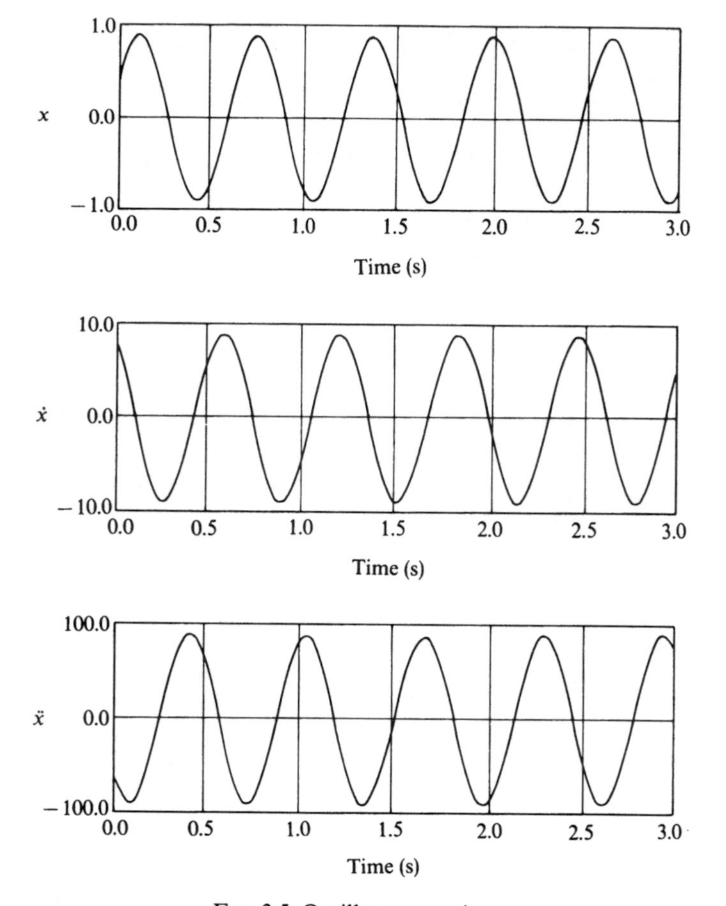

FIG. 3.5. Oscillatory motion.

The frequency of the system can also be written in terms of the static deflection Δ. To this end, we may rewrite Eq. 1 in the following form

$$\frac{k}{m} = \frac{g}{\Delta}$$

Substituting this equation into Eq. 25 yields

$$f = \frac{1}{2\pi}\sqrt{\frac{g}{\Delta}} \tag{3.26}$$

which implies that the natural frequency of this particular single degree of freedom system can be obtained once the static deflection Δ is known. The unit used for the frequency f is Hertz or simply Hz.

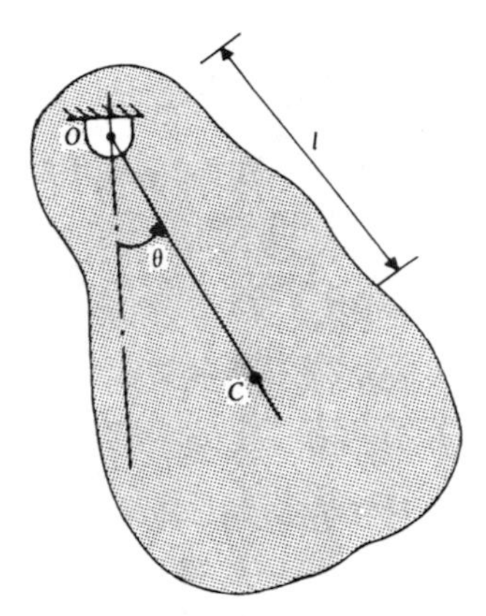

FIG. 3.6. Experimental determination of the mass moment of inertia.

Measuring the natural frequency of oscillations provides a convenient method for the experimental evaluation of the mass moment of inertia of a body with complex geometry, about a fixed axis. If the mass of the body as well as the location of its center of gravity can be determined, a simple analysis can be used in order to determine the mass moment of inertia of the body. To this end, the body can be suspended like a pendulum and caused to oscillate freely about an axis passing through a fixed point O, as shown in Fig. 6. The linear equation of free vibration of this body is given by

$$I_O\ddot{\theta} + mgl\theta = 0$$

where I_O is the mass moment of inertia of the body about an axis passing through O, m is the mass of the body, g is the gravity constant, l is the distance of the center of mass from point O, and θ is the angular oscillation. The frequency of the free oscillations is then given by

$$\omega = \sqrt{\frac{mgl}{I_O}}$$

Therefore, by measuring the number of oscillations in a certain period of time, the natural frequency ω can be determined and the mass moment of inertia I_O can be calculated by using the preceding equation. The mass moment of inertia of the body about its center of mass can be determined by using the parallel axis theory

$$I_O = I + ml^2$$

where I is the mass moment of inertia defined with respect to the center of mass of the body.

Conservation of Energy Two forms of energy exist as the result of the free vibration of the undamped single degree of freedom system shown in Fig. 2. The first form is the kinetic energy T as the result of the motion of the mass, while the second form is the strain energy U resulting from the deformation of the spring from the static equilibrium position. The kinetic energy T and the strain energy U are given, for the single degree of freedom mass–spring system discussed in this section at an arbitrary position x, by

$$T = \tfrac{1}{2}m\dot{x}^2$$

$$U = \tfrac{1}{2}kx^2$$

The system total energy E is defined as the sum of the kinetic and strain energies, that is,

$$E = \tfrac{1}{2}m\dot{x}^2 + \tfrac{1}{2}kx^2$$

Using the definition of x and $\dot{x}$ given, respectively, by Eqs. 9 and 19, the total energy E can be written as

$$E = \tfrac{1}{2}m\omega^2X^2\cos^2(\omega t + \phi) + \tfrac{1}{2}kX^2\sin^2(\omega t + \phi)$$

Since $k = \omega^2 m$ (Eq. 6), the preceding equation reduces to

$$E = \tfrac{1}{2}kX^2[\cos^2(\omega t + \phi) + \sin^2(\omega t + \phi)]$$

By using the trigonometric identity

$$\cos^2(\omega t + \phi) + \sin^2(\omega t + \phi) = 1,$$

the total energy of the system at any arbitrary position x reduces to

$$E = \tfrac{1}{2}kX^2 = \text{constant}$$

Therefore, the total energy of the single degree of freedom system at any instant in time is constant. One may observe that the total energy E given by the preceding equation is equal to the strain energy, as the result of the deformation of the spring when the displacement is maximum. At this position the velocity is equal to zero. Since the total energy is equal to the sum of the kinetic and strain energies, one may expect that the total energy at any instant in time is also equal to the kinetic energy when the strain energy is equal to zero or, equivalently, when the deformation of the spring is equal to zero. This is indeed the case, and can be demonstrated using the fact that $k = \omega^2 m$ which, upon substitution in the preceding equation for E, leads to

$$E = \tfrac{1}{2}kX^2 = \tfrac{1}{2}\omega^2 mX^2 = \text{constant}$$

observe that ωX is equal to the maximum velocity which occurs when the displacement is equal to zero. Therefore, the total energy of the system at any instant in time is equal to the maximum kinetic energy. This implies that in the free vibration of the system under consideration, the maximum kinetic energy is equal to the maximum strain energy which is equal to the total energy. Between the points at which the displacement is zero (maximum kinetic energy) and the points at which the velocity is zero (maximum strain energy), both forms of energy (kinetic and strain) exist and the total energy E is the sum of both and remains constant. In this case, we say energy is conserved and the undamped single degree of freedom system is said to be a *conservative system.*

Phase Plane It is clear from the analysis presented in this section and the preceding section that both the displacement x and the velocity $\dot{x}$ are represented by harmonic functions. By using the velocity expression of Eq. 19, one may define the following new variable

$$y = \frac{\dot{x}}{\omega} = X \cos(\omega t + \phi)$$

By using this equation and the expression for the displacement given by Eq. 9, one has

$$\begin{aligned} x^2 + y^2 &= X^2 \sin^2(\omega t + \phi) + X^2 \cos^2(\omega t + \phi) \\ &= X^2[\sin^2(\omega t + \phi) + \cos^2(\omega t + \phi)] \\ &= X^2 \end{aligned}$$

The preceding relationship between x and y describes a circle of radius X which is the amplitude of the free oscillation. This relationship, as shown in *the phase plane* of Fig. 7(a), can be represented by a vector $\mathbf{x}$ rotating with a constant angular velocity ω, and the position of the vector is located by the angle $\omega t + \phi$. Each point in the graphical representation of the phase plane corresponds to a unique displacement and velocity that satisfy the equation of motion of the single degree of freedom system. Therefore, the motion

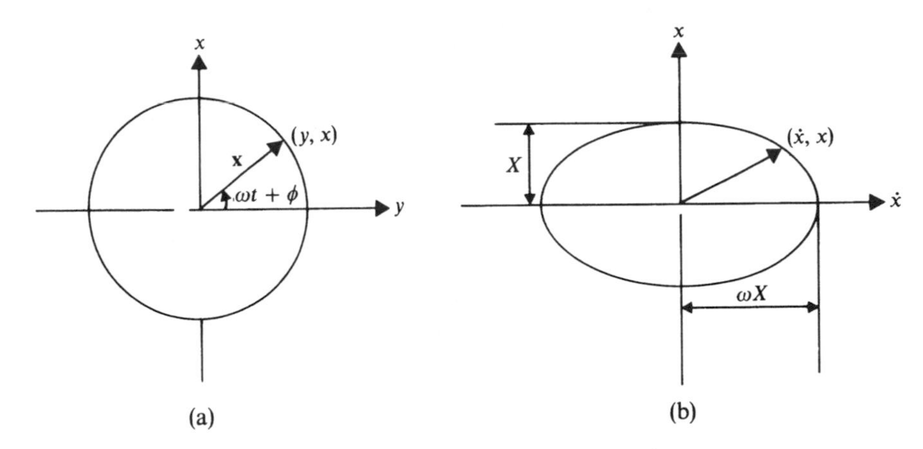

FIG. 3.7. Analysis of the oscillatory motion.

resulting from the free vibration of the single degree of freedom system is such that x and $\dot{x}/\omega$ are the coordinates of a point on the circle of the phase plane, that is,

$$x^2 + \left(\frac{\dot{x}}{\omega}\right)^2 = X^2$$

which can also be written as

$$\left(\frac{x}{X}\right)^2 + \left(\frac{\dot{x}}{\omega X}\right)^2 = 1$$

which represents an ellipse on which the coordinates of every point corresponds to a unique displacement x and velocity $\dot{x}$, as shown in Fig. 7(b).

Example 3.1

A mass of 0.5 kg is suspended in a vertical plane by a spring having a stiffness coefficient of 300 N/m. If the mass is displaced downward from its static equilibrium position through a distance 0.01 m determine:

(a) the differential equation of motion;
(b) the natural frequency of the system;
(c) the response of the system as a function of time;
(d) the system total energy.

Solution. The differential equation of motion is given by

$$m\ddot{x} + kx = 0$$

Since $m = 0.5$ kg and $k = 300$ N/m, the above differential equation can be written as

$$0.5\ddot{x} + 300x = 0$$

The natural frequency ω is given by

$$\omega = \sqrt{\frac{k}{m}} = \sqrt{\frac{300}{0.5}} = 24.495 \text{ rad/s}$$

The frequency f is given by

$$f = \frac{\omega}{2\pi} = \frac{24.495}{2\pi} = 3.898 \text{ Hz}$$

Using Eq. 9, the response of the system is given by

$$x = X \sin(\omega t + \phi)$$

where

$$X = \sqrt{x_0^2 + \left(\frac{\dot{x}_0}{\omega}\right)^2} = \sqrt{(0.01)^2 + 0} = 0.01$$

$$\phi = \tan^{-1}\left(\frac{x_0}{\dot{x}_0/\omega}\right) = \tan^{-1}(\infty) = \frac{\pi}{2}$$

that is,

$$x = 0.01 \cos 24.495t$$

For this simple conservative system, the system total energy is equal to the maximum kinetic energy or the maximum strain energy. One therefore has

$$E = \tfrac{1}{2}kX^2 = \tfrac{1}{2}(300)(0.01)^2 = 0.015 \text{ N}\cdot\text{m}$$

This is the same as the maximum kinetic energy given by

$$E = \tfrac{1}{2}m(\omega X)^2 = \tfrac{1}{2}(0.5)(24.495 \times 0.01)^2 = 0.015 \text{ N}\cdot\text{m}$$

Example 3.2

Figure 8 depicts a uniform slender bar of mass m and length l. The bar, which is connected to the ground by a pin joint at O, is supported by a spring which has a stiffness coefficient k, as shown in the figure. The undeformed length of the spring is such that the bar is in static equilibrium when it is in the horizontal position. Assuming small angular oscillations, find the differential equation of motion and the natural frequency.

Solution. Let R_x and R_y be the reaction forces at the joint at O. We first consider the static equilibrium condition of the bar. From the free body diagram shown in Fig. 8(b), and by taking the moment about O, we have

$$k\Delta a - mg\frac{l}{2} = 0$$

where Δ is the static deflection of the spring.

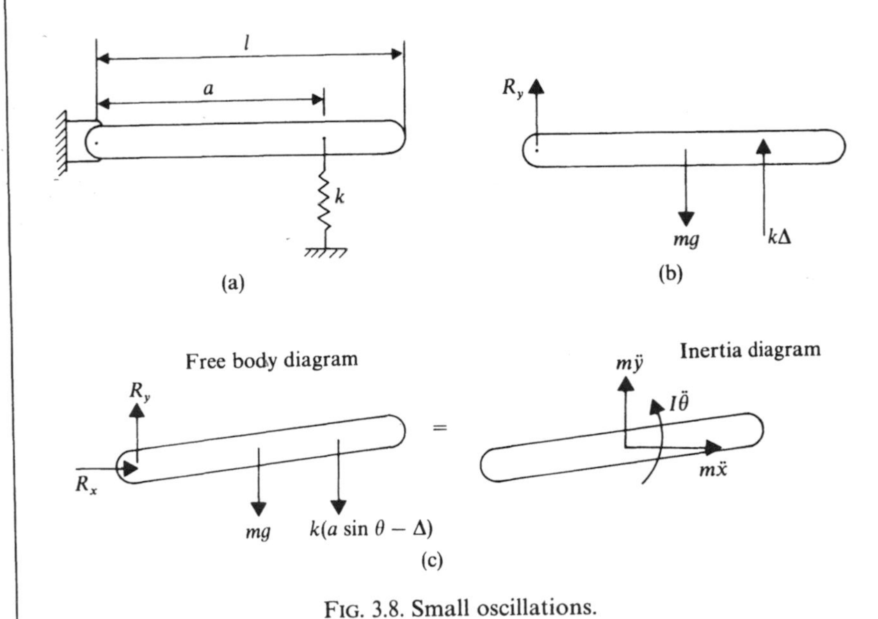

FIG. 3.8. Small oscillations.

Now we consider the case in which the bar oscillates. In order to develop the differential equation of motion of the bar, we consider the bar at an arbitrary angular position θ. In this case, the condition of dynamic equilibrium is defined by

$$M_a = M_{eff}$$

where M_a is the applied external moment and M_{eff} is the moment of the inertia forces. If we consider the moment equation at point O, we get

$$M_a = -mg\frac{l}{2}\cos\theta - k(a\sin\theta - \Delta)a\cos\theta$$

By using the static equilibrium condition, this equation reduces to

$$M_a = -ka^2\sin\theta\cos\theta$$

Series expansions of $\sin\theta$ and $\cos\theta$ can be expressed, respectively, as

$$\sin\theta = \theta - \frac{\theta^3}{3!} + \frac{\theta^5}{5!} - \frac{\theta^7}{7!} + \cdots$$

$$\cos\theta = 1 - \frac{\theta^2}{2!} + \frac{\theta^4}{4!} - \frac{\theta^6}{6!} + \cdots$$

For small oscillations, we can neglect higher-order terms and write

$$\sin\theta \approx \theta, \qquad \cos\theta \approx 1$$

Therefore, the applied moment M_a can be written as

$$M_a = -ka^2\theta$$

The moment of the inertia forces M_{eff} is given by

$$M_{eff} = I\ddot{\theta} - m\ddot{x}\frac{l}{2}\sin\theta + m\ddot{y}\frac{l}{2}\cos\theta$$

where I is the mass moment of inertia about the center of mass of the bar given by

$$I = \frac{ml^2}{12}$$

and $\ddot{x}$ and $\ddot{y}$ are the accelerations of the mass center, which can be obtained using the kinematic relationships

$$x = \frac{l}{2}\cos\theta$$

$$y = \frac{l}{2}\sin\theta$$

Differentiating these two equations yields

$$\ddot{x} = -(\ddot{\theta}\sin\theta + \dot{\theta}^2\cos\theta)\frac{l}{2}$$

$$\ddot{y} = (\ddot{\theta}\cos\theta - \dot{\theta}^2\sin\theta)\frac{l}{2}$$

Therefore, M_{eff} can be written as

$$M_{\text{eff}} = I\ddot{\theta} + m(\ddot{\theta}\sin\theta + \dot{\theta}^2\cos\theta)\left(\frac{l}{2}\right)^2 \sin\theta + m(\ddot{\theta}\cos\theta - \dot{\theta}^2\sin\theta)\left(\frac{l}{2}\right)^2 \cos\theta$$

which upon simplifying and utilizing the following trigonometric identity

$$\cos^2\theta + \sin^2\theta = 1$$

leads to

$$M_{\text{eff}} = I\ddot{\theta} + m\frac{l^2}{4}\ddot{\theta} = \left(I + \frac{ml^2}{4}\right)\ddot{\theta} = \left(\frac{ml^2}{12} + \frac{ml^2}{4}\right)\ddot{\theta} = \frac{ml^2}{3}\ddot{\theta}$$

Therefore, the differential equation of motion can be written as

$$-ka^2\theta = \frac{ml^2}{3}\ddot{\theta}$$

or

$$\frac{ml^2}{3}\ddot{\theta} + ka^2\theta = 0$$

The natural frequency ω is then defined as

$$\omega = \sqrt{\frac{ka^2}{ml^2/3}} \quad \text{rad/s}$$

This frequency can be defined in Hertz as

$$f = \frac{\omega}{2\pi} = \frac{1}{2\pi}\sqrt{\frac{ka^2}{ml^2/3}} \quad \text{Hz}$$

3.3 STABILITY OF UNDAMPED LINEAR SYSTEMS

As was demonstrated in the preceding chapter, the mass and stiffness coefficients affect the roots of the characteristic equation which define the response of the system and, as such, changes in these coefficients lead to changes in the system dynamic behavior. It is shown, in this section, that by a proper selection of the stiffness and inertia coefficients, instability of the motion of the undamped single degree of freedom systems can be avoided. In later sections, the effect of damping on the stability of motion will be discussed.

Illustrative Example The inverted pendulum shown in Fig. 9 consists of a mass m and a massless rod of length l. The mass is supported by a spring which has a stiffness coefficient k. Let R_x and R_y be the reaction forces at the pin joint. Since this is a one degree of freedom system, we can obtain the differential equation of motion by taking the moment about point O and applying the formula

$$M_a = M_{\text{eff}}$$

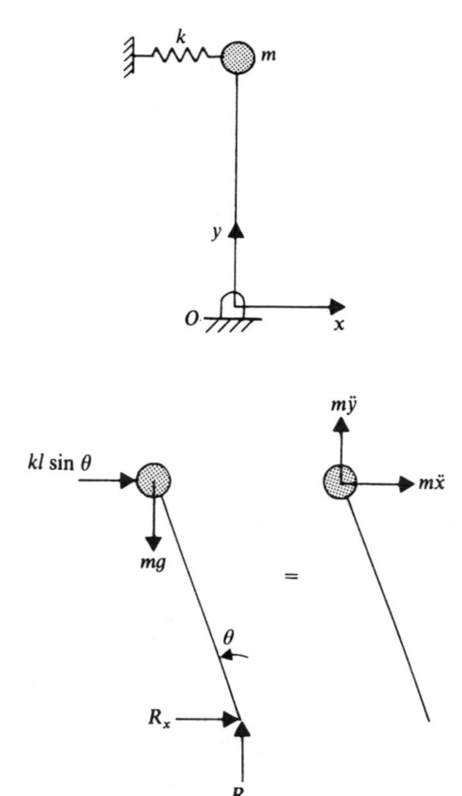

FIG. 3.9. Inverted pendulum.

In this case, the applied external moment M_a about point O is given by

$$M_a = mgl \sin \theta - kl \sin \theta(l \cos \theta)$$

The moment of the inertia forces is given by

$$M_{eff} = -m\ddot{x}l \cos \theta - m\ddot{y}l \sin \theta$$

where $\ddot{x}$ and $\ddot{y}$ are the accelerations of the concentrated mass m. Using the Cartesian coordinate system shown in Fig. 9, the acceleration $\ddot{x}$ and $\ddot{y}$ can be obtained by differentiating the following kinematic relationships

$$x = -l \sin \theta$$

$$y = l \cos \theta$$

that is,

$$\ddot{x} = -l\ddot{\theta} \cos \theta + l\dot{\theta}^2 \sin \theta$$

$$\ddot{y} = -l\ddot{\theta} \sin \theta - l\dot{\theta}^2 \cos \theta$$

Therefore, the moment of the inertia forces can be written as

$$M_{\text{eff}} = m(l\ddot{\theta}\cos\theta - l\dot{\theta}^2\sin\theta)l\cos\theta + m(l\ddot{\theta}\sin\theta + l\dot{\theta}^2\cos\theta)l\sin\theta = ml^2\ddot{\theta}(\cos^2\theta + \sin^2\theta) = ml^2\ddot{\theta}$$

Thus, the second-order differential equation of motion is given by

$$mgl\sin\theta - kl\sin\theta(l\cos\theta) = ml^2\ddot{\theta}$$

or

$$ml^2\ddot{\theta} + kl^2\sin\theta\cos\theta - mgl\sin\theta = 0$$

If we use the assumption of small oscillations, that is,

$$\sin\theta \approx \theta, \qquad \cos\theta \approx 1,$$

the linear differential equation of motion can be obtained as

$$ml^2\ddot{\theta} + (kl - mg)l\theta = 0$$

Let

$$kl - mg = b$$

The differential equation of the system in terms of the parameter b can be written as

$$ml^2\ddot{\theta} + bl\theta = 0$$

There are three different cases which lead to different solutions. These cases are:

(1) the constant b is positive;
(2) the constant b is zero;
(3) the constant b is negative.

Critically Stable System If the constant b is positive, we have

$$kl > mg$$

In this case, the natural frequency of the system is defined as

$$\omega = \sqrt{\frac{bl}{ml^2}} = \sqrt{\frac{b}{ml}} = \sqrt{\frac{kl - mg}{ml}}$$

the characteristic equation, in this case, has complex conjugate roots defined as

$$p_1 = i\omega, \qquad p_2 = -i\omega$$

and the solution in this case is given by

$$\theta = \Theta\sin(\omega t + \phi)$$

where Θ is the amplitude and ϕ is the phase angle. The solution in this case is a harmonic function which has a constant amplitude. A system which has this type of sustained oscillation is said to be a *critically stable* system since the amplitude of oscillation does not increase or decrease with time.

Unstable System If the constant b is identically zero, we have the following equality

$$kl = mg$$

The differential equation in this case reduces to

$$ml^2\ddot{\theta} = 0$$

or

$$\ddot{\theta} = 0$$

which upon integration yields

$$\dot{\theta} = c_1$$

$$\theta = c_1 t + c_2$$

where c_1 and c_2 are constants that can be determined from the initial conditions. Because the solution is a linear function in time, the absolute value of θ increases with time, and the system becomes *unstable*.

If, on the other hand, the constant b is negative, we have the following inequality

$$kl < mg$$

Consequently, the characteristic equation has two real roots given by

$$p_1 = \sqrt{\frac{-bl}{ml^2}} = \sqrt{\frac{mg - kl}{ml}}$$

$$p_2 = -\sqrt{\frac{-bl}{ml^2}} = -\sqrt{\frac{mg - kl}{ml}}$$

The solution can then be written as

$$\theta = A_1 e^{p_1 t} + A_2 e^{p_2 t}$$

where A_1 and A_2 are constants which can be determined from the initial conditions. The solution is a linear combination of the exponential growth $e^{p_1 t}$ and the exponential decay $e^{p_2 t}$, and the result is a case of instability.

3.4 CONTINUOUS SYSTEMS

Even though continuous systems such as rods, beams, and shafts have infinite numbers of degrees of freedom due to the fact that such systems can assume arbitrary deformation shapes, in many practical applications simple single degree of freedom models can be developed for these systems. There are two approaches which are commonly used for developing single degree of freedom models for continuous systems; in the first approach the distributed inertia of the elastic elements is neglected, while in the second approach the distributed inertia is taken into consideration using the *assumed mode method*. In this section, these two approaches are discussed, starting with the more simple approach in which the distributed inertia of the elastic member is neglected. For this purpose, we use *torsional systems* as an example.

Torsional Systems Shafts are used, to transmit torque, in many mechanical systems such as engines, turbines, and helicopter rotor systems. These systems may be subject to cyclic variations of the transmitted torque which result in torsional oscillations. In these cases, shafts as a result of their flexibility, produce torsional restoring torques which depend on the rigidity as well as on the dimensions of the shafts. The torsional system shown in Fig. 10 consists of a disk which has a mass moment of inertia I. The disk is supported by a circular shaft which has length l and diameter D. It is clear that if the disk is subjected to a rotation θ, a torque will be produced by the shaft which tends to restore the disk to its original position. The relationship between the angular displacement θ and the applied torque T which produces this displacement can be obtained from standard texts on strength of materials as

$$\theta = \frac{Tl}{GJ} \tag{3.27}$$

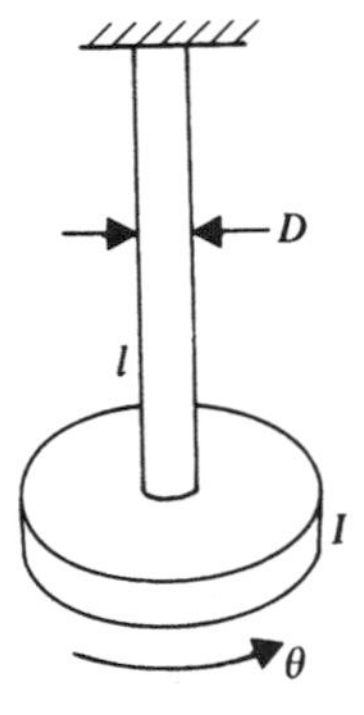

Free body diagram Inertia diagram

$k_t\theta$ = $I\ddot{\theta}$

FIG. 3.10. Torsional systems.

where l is the length of the shaft, J is the polar second moment of area of the shaft, and G is its modulus of rigidity. Similar to the case of linear springs, we define the torsional spring constant of the shaft as

$$k_t = \frac{T}{\theta} \tag{3.28}$$

Using Eqs. 27 and 28, the torsional stiffness of the shaft can be recognized as

$$k_t = \frac{GJ}{l} \tag{3.29}$$

The internal restoring torque that opposes the angular displacement θ is then given by

$$T_r = k_t\theta = \frac{GJ}{l}\theta \tag{3.30}$$

which must be equal and opposite to the applied torque T if the disk is to be in a static equilibrium position. Suppose now that the disk is given an initial angular displacement and/or angular velocity. Because of the inertia of the disk and the restoring elastic torque of the shaft, the disk oscillates. In order to obtain the differential equation of motion we apply the equation

$$M_a = M_{eff}$$

where M_a is given by

$$M_a = -k_t\theta = -\frac{GJ}{l}\theta$$

and M_{eff} is

$$M_{eff} = I\ddot{\theta}$$

Therefore, the differential equation of motion is given by

$$-k_t\theta = I\ddot{\theta}$$

or

$$I\ddot{\theta} + k_t\theta = 0 \tag{3.31}$$

from which the natural frequency is defined as

$$\omega = \sqrt{\frac{k_t}{I}} = \sqrt{\frac{GJ}{Il}}$$

The solution of Eq. 31 is

$$\theta(t) = A_1 \sin \omega t + A_2 \cos \omega t$$

where A_1 and A_2 are constants to be determined using the initial conditions. The solution $\theta(t)$ can also be written in terms of an amplitude Θ and a phase angle ϕ as

$$\theta(t) = \Theta \sin(\omega t + \phi)$$

where Θ and ϕ also can be determined from the initial conditions.

Assumed Mode Method In the analysis of the torsional systems presented in this section, we assumed that the shaft is massless, such that it does not contribute to the system inertia. In addition to the fact that such an assumption is not necessary for developing a single degree of freedom model for continuous systems, this assumption can lead to inaccurate results if the inertia of the shaft is not negligible. A more accurate model can be developed if the distributed inertia of the elastic member is taken into consideration. As was previously pointed out, continuous systems have an infinite number of degrees of freedom because they can assume an arbitrary shape when they deform. In most mechanical and structural system applications, however, the shape of deformation of the elastic components can be predicted and approximated using simple functions such that the vibration of the infinite number of degrees of freedom continuous system can be represented accurately by a single degree of freedom system that has the same mathematical model as the mass–spring system. In order to demonstrate the use of the *assumed mode method* to obtain the single degree of freedom model for continuous systems, we consider the cantilever beam shown in Fig. 11(a). The beam is assumed to have length l, cross-sectional area A, mass m, mass density ρ, modulus of elasticity E, and second moment of area I_z. We assume that the beam bending deflection is described in terms of the axial coordinate x by the shape function

$$\phi(x) = \frac{3x^2 l - x^3}{2l^3}$$

In terms of this assumed mode of deformation, the bending displacement of the beam can be written as

$$v(x, t) = \phi(x)q(t) = \left(\frac{3x^2 l - x^3}{2l^3}\right)q(t)$$

where $q(t)$ is the amplitude of the displacement at the free end, and it is the only time-dependent unknown in the displacement equation of the beam. Using the preceding expression for the transverse displacement v, the kinetic and strain energy of the beam can be written as

$$T = \frac{1}{2}\int_0^l \rho A(\dot{v})^2\, dx, \qquad U = \frac{1}{2}\int_0^l EI_z\left(\frac{\partial^2 v}{\partial x^2}\right)^2 dx$$

Substituting the expression for v into the kinetic and strain energy equations

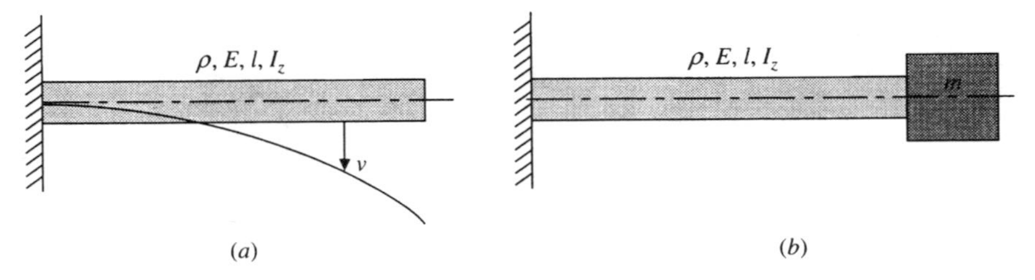

FIG. 3.11. Bending vibration.

and keeping in mind that the axial beam coordinate x does not depend on time, one obtains

$$T = \tfrac{1}{2}m_e\dot{q}^2(t), \qquad U = \tfrac{1}{2}k_e q^2(t)$$

where m_e and k_e are, respectively, the equivalent mass and stiffness coefficients defined as

$$m_e = \frac{33}{140}\rho A l, \qquad k_e = \frac{3EI_z}{l^3}$$

If we assume that $q(t)$ can be described using a simple harmonic motion, the equation of motion of the free vibration of the beam can be written as

$$m_e\ddot{q}(t) + k_e q(t) = 0$$

and the natural frequency of the system due to the distributed inertia of the beam is

$$\omega = \sqrt{\frac{k_e}{m_e}} = \frac{3.568}{l^2}\sqrt{\frac{EI_z}{\rho A}}$$

If a concentrated mass m is attached to the free end of the beam as shown in Fig. 11(b), the kinetic energy of the beam must be altered in order to take into consideration the effect of the inertia of the concentrated mass. In this case, the kinetic energy is given by

$$T = \frac{1}{2}\int_0^l \rho A(\dot{v}(x, t))^2\,dx + \frac{1}{2}m(\dot{v}(l, t))^2$$

where $v(l, t)$ is the bending displacement at the free end. Using the assumed shape function for v, the preceding expression for the kinetic energy leads to

$$T = \tfrac{1}{2}m_e\dot{q}^2(t)$$

where m_e, in this case, is given by

$$m_e = \frac{33}{140}\rho A l + m$$

A similar procedure also can be used to obtain an approximate single degree of freedom model in the case of longitudinal and torsional systems. For example, for the torsional system shown in Fig. 10, one can assume the torsional displacement of the shaft to be linear as described by the function

$$\phi(x) = \frac{x}{l}$$

where x is the axial coordinate of the shaft. The torsional displacement of the shaft then can be written using this assumed shape function as

$$\theta(x, t) = \phi(x)q(t) = \frac{x}{l}q(t)$$

where $q(t) = \theta(l, t)$ is the amplitude of the displacement at the end of the shaft to which the disk is attached. Using the assumed displacement of the preceding equation, the kinetic and strain energy expressions of the torsional system shown in Fig. 10 are given by

$$T = \frac{1}{2}\int_0^l \rho J \dot{\theta}^2(x, t)\,dx + \frac{1}{2} I \dot{\theta}^2(l, t) = \frac{1}{2}\left[\frac{\rho J l}{3} + I\right]\dot{q}^2(t)$$

$$U = \frac{1}{2}\int_0^l GJ\left(\frac{\partial \theta}{\partial x}\right)^2 dx = \frac{1}{2}\frac{GJ}{l} q^2(t)$$

from which the equivalent inertia and stiffness coefficients I_e and k_e can be recognized as

$$I_e = I + \frac{\rho A l}{3}, \qquad k_e = \frac{GJ}{l}$$

Note that if the distributed inertia of the shaft is negligible, the inertia coefficient I_e presented in the preceding equation reduces to the mass moment of inertia of the disk, and the natural frequency of the system will be the same as previously obtained in this section.

Table 1 shows the equivalent mass and stiffness coefficients for several

TABLE 3.1. Continuous Systems

System	Shape Function	m_e	k_e
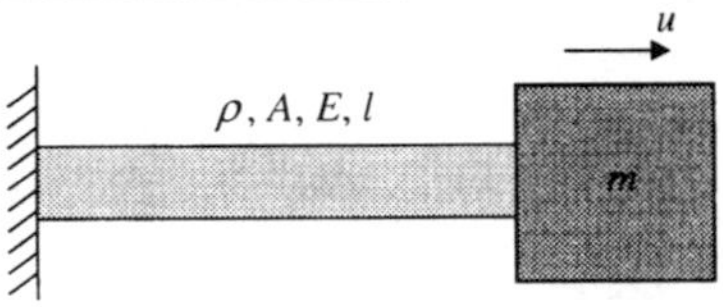 u; ρ, A, E, l; m — Longitudinal vibration	$\phi(x) = \frac{x}{l}$	$m + \frac{\rho A l}{3}$	$\frac{EA}{l}$
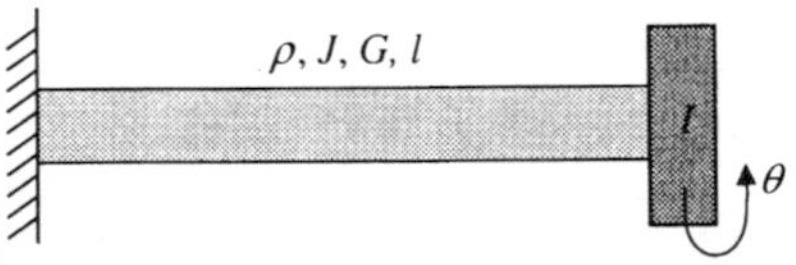 ρ, J, G, l; I; θ — Torsional vibration	$\phi(x) = \frac{x}{l}$	$I + \frac{\rho J l}{3}$	$\frac{GJ}{l}$
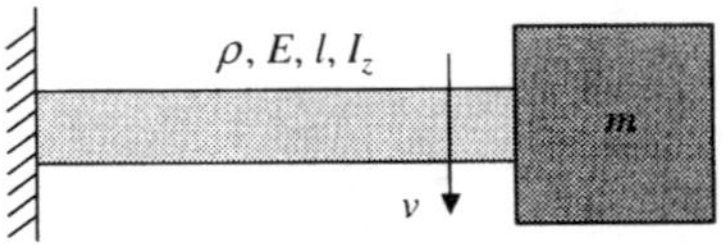 ρ, E, l, I_z; m; v — Bending of cantilever beam	$\phi(x) = \frac{3x^2 l - x^3}{2l^3}$	$m + \frac{33\rho A l}{140}$	$\frac{3EI_z}{l^3}$
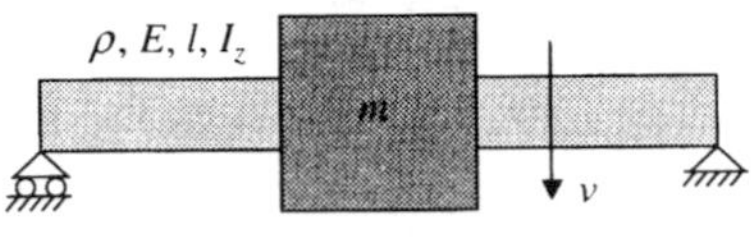 ρ, E, l, I_z; m; v — Bending of simply supported beam	$\phi(x) = \sin\left(\frac{\pi x}{l}\right)$	$m + \frac{\rho A l}{2}$	$\frac{\pi^4 EI_z}{2l^3}$

elastic elements to which concentrated masses are attached. The results presented in this table are approximate, and other approximations of the mass and stiffness coefficients can be obtained if different shape functions are used. The results obtained for the natural frequencies using different shape functions should closely match the results obtained using the mass and stiffness coefficients presented in the table so long as these shape functions provide a good approximation for the deformation shapes of the elastic members. A good choice of the shape functions for vibrating beams is the shape of deformation resulting from their own static weights.

3.5 EQUIVALENT SYSTEMS

Several elastic elements such as springs are often used in combination, and in many cases, an equivalent single elastic element can be used to simplify the mathematical model developed. In this section, we discuss two types of connections; parallel and series connections, and develop methods for obtaining the equivalent stiffness coefficients.

Parallel Connection Figure 12(a) shows a mass m supported by the two linear springs k_1 and k_2. The mass m is constrained to move only in the vertical direction; and therefore, the system has only one degree of freedom. In this case, in which the displacements of the two springs k_1 and k_2 are the same, the two springs are said to be *connected in parallel*. We wish to obtain the stiffness coefficient of a single spring k_e such that the system of Fig. 12(a) is equivalent to the system of Fig. 12(b).

From the free body diagrams shown in the figure, it is clear that the two systems are equivalent if

$$k_e \Delta = (k_1 + k_2)\Delta \tag{3.32}$$

where Δ is the static deflection due to the weight. Equation 32 leads to

$$k_e = k_1 + k_2 \tag{3.3}$$

which implies that, in the case of parallel connection, the equivalent spring

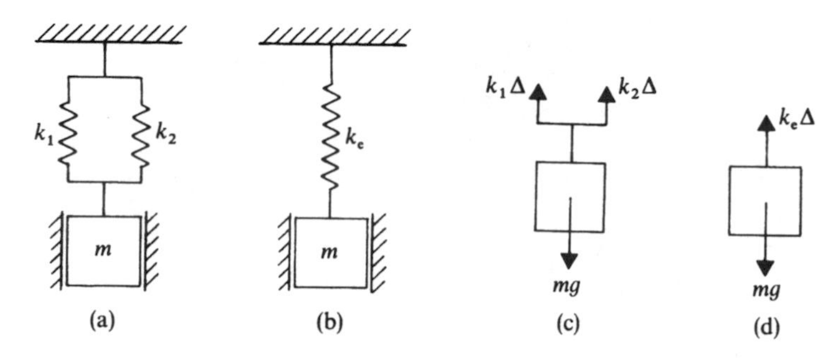

FIG. 3.12. Springs in parallel.

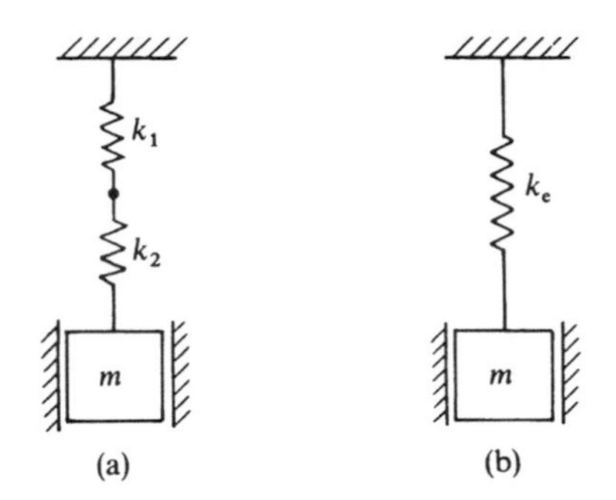

FIG. 3.13. Springs in series.

stiffness is equal to the sum of the two springs k_1 and k_2, and as such k_e is greater than k_1 and is greater than k_2.

In general, if there are n springs connected in parallel and they have stiffness coefficients $k_1, k_2, \ldots, k_n$, then the equivalent stiffness coefficient k_e is given by

$$k_e = k_1 + k_2 + \cdots + k_n = \sum_{j=1}^{n} k_j \tag{3.34}$$

Series Connection Figure 13(a) shows a mass m supported by two springs k_1 and k_2 which are connected in *series*. The mass m is allowed to move only in the vertical direction; and, therefore, the system has only one degree of freedom. In this case, in which the forces in the two springs are the same, we wish to obtain the stiffness coefficient of a single spring k_e such that the system of Fig. 13(a) is equivalent to the system of Fig. 13(b).

Since the force is the same in the two springs, we have

$$mg = k_1 \Delta_1 \tag{3.35}$$

$$mg = k_2 \Delta_2 \tag{3.36}$$

where g is the gravitational constant, and Δ_1 and Δ_2 are, respectively, the deformation of the springs k_1 and k_2. The displacement of the mass m is the sum of the deformations of the two springs, that is,

$$\Delta = \Delta_1 + \Delta_2 \tag{3.37}$$

where Δ is the displacement of the mass which is equal to

$$\Delta = \frac{mg}{k_e} \tag{3.38}$$

Substituting Eqs. 35, 36, and 38 into Eq. 37, one obtains

$$\frac{mg}{k_e} = \frac{mg}{k_1} + \frac{mg}{k_2}$$

or

$$\frac{1}{k_e} = \frac{1}{k_1} + \frac{1}{k_2} \tag{3.39}$$

This equation can be used to define the equivalent stiffness coefficient of the two springs k_1 and k_2 as

$$k_e = \frac{k_1 k_2}{k_1 + k_2}$$

Observe that, since k_1 and k_2 are assumed to be positive constants, one has

$$k_e = \frac{k_1 k_2}{k_1 + k_2} < \frac{k_1 k_2}{k_2} = k_1$$

Similarly,

$$k_e = \frac{k_1 k_2}{k_1 + k_2} < \frac{k_1 k_2}{k_1} = k_2$$

That is, the equivalent stiffness coefficient of two springs connected in series is less than the stiffness coefficient of each of the two springs.

Equation 39 can be generalized to the case of n springs connected in series as

$$\frac{1}{k_e} = \frac{1}{k_1} + \frac{1}{k_2} + \cdots + \frac{1}{k_n} = \sum_{j=1}^{n} \frac{1}{k_j} \tag{3.40}$$

where $k_1, k_2, \ldots, k_n$ are the stiffness coefficients of the springs.

Example 3.3

Obtain the differential equation of motion of the system shown in Fig. 14(a) and determine the system natural frequency.

Solution. Since the springs k_1 and k_2 are in parallel, they are equivalent to a single spring which has a stiffness coefficient k_{e1} defined by the equation

$$k_{e1} = k_1 + k_2$$

The system in Fig. 14(a) is then equivalent to the system shown in Fig. 14(b). In Fig. 14(b), the two springs k_{e1} and k_3 are connected in series, and therefore, they are

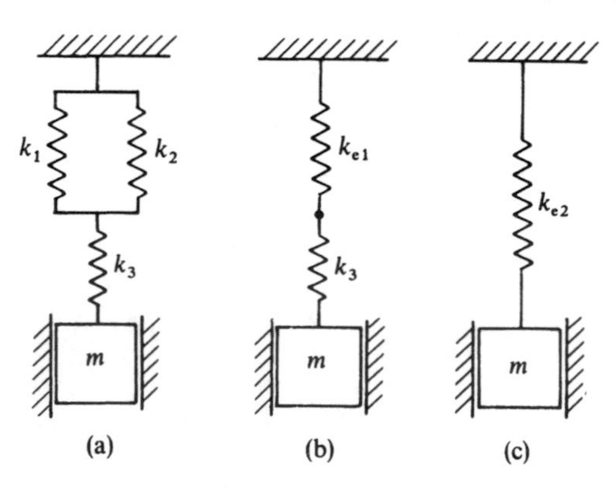

FIG. 3.14. Equivalent springs.

equivalent to one spring which has a stiffness coefficient k_{e2} defined by the equation

$$\frac{1}{k_{e2}} = \frac{1}{k_{e1}} + \frac{1}{k_3}$$

or

$$k_{e2} = \frac{k_{e1}k_3}{k_{e1} + k_3} = \frac{(k_1 + k_2)k_3}{k_1 + k_2 + k_3}$$

It follows that the system in Fig. 14(b) is equivalent to the system shown in Fig. 14(c), which has the following differential equation of motion

$$m\ddot{x} + k_{e2}x = 0$$

The natural frequency is given by

$$\omega = \sqrt{\frac{k_{e2}}{m}} = \sqrt{\frac{(k_1 + k_2)k_3}{(k_1 + k_2 + k_3)m}}$$

If the springs are not connected in parallel or series, an equivalent stiffness coefficient still can be obtained by examining the spring forces as demonstrated by the following example.

Example 3.4

Figure 15(a) shows a pendulum which is supported by two springs which have stiffness coefficient k_1 and k_2. The two springs are connected to the pendulum rod at points which are at distances a and b from the pin joint, as shown in the figure. If the two springs shown in Fig. 15(a) are to be equivalent to a single spring which is connected to the rod at a distance d from the pin joint as shown in the figure, determine the stiffness coefficient of this single spring in terms of the constants k_1 and k_2.

Solution. As shown in Fig. 15(c), the force produced by the springs k_1 and k_2, as the results of an angular displacement θ, are given by

$$F_1 = -k_1 a \sin\theta$$

$$F_2 = -k_2 b \sin\theta$$

The resultant moment due to the forces F_1 and F_2 about point O is given by

$$M = F_1 a \cos\theta + F_2 b \cos\theta = -(k_1 a^2 + k_2 b^2)\sin\theta\cos\theta$$

The force F_e produced by the equivalent spring is given by

$$F_e = -k_e d \sin\theta$$

and the moment about O produced by this force is

$$M_e = F_e d \cos\theta = -k_e d^2 \sin\theta\cos\theta$$

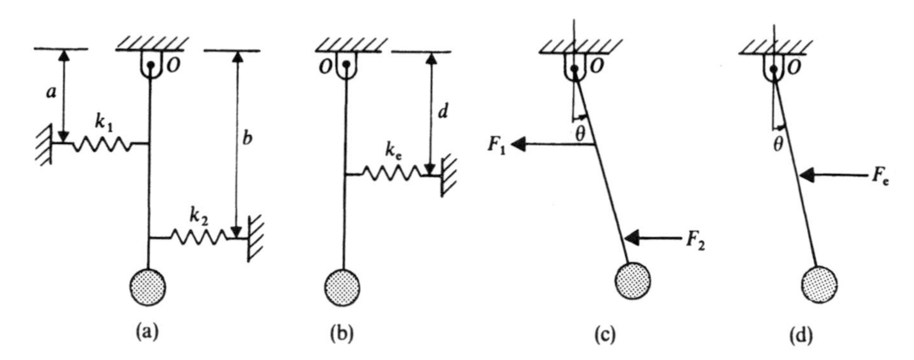

FIG. 3.15. Equivalent springs in the case of angular oscillations.

If the two systems in Fig. 15(a) and 15(b) are to be equivalent, the moment M produced by the original system must be equal to the moment M_e produced by the equivalent system, that is,

$$M = M_e$$

or

$$-(k_1 a^2 + k_2 b^2) \sin \theta \cos \theta = -k_e d^2 \sin \theta \cos \theta$$

from which the equivalent stiffness coefficient k_e is defined as

$$k_e = \frac{k_1 a^2 + k_2 b^2}{d^2}$$

A moment equation is used in this example instead of a force equation to determine k_e, because this is the case of angular oscillations in which the system degree of freedom is the angular orientation θ.

3.6 FREE DAMPED VIBRATION

Thus far, we have only considered the free vibration of the undamped single degree of freedom systems. As was shown in the preceding sections, the response of such undamped systems can be represented by a harmonic function which has a constant amplitude, which is the case of a sustained oscillation. In this section, we study the effect of viscous damping on the free vibration of single degree of freedom systems, and develop, solve, and examine their differential equation. It will be seen from the theoretical development and the examples presented in this section that the damping force has a pronounced effect on the stability of the systems.

Figure 16(a) depicts a single degree of freedom system which consists of a mass m supported by a spring and a damper. The stiffness coefficient of the spring is k and the viscous damping coefficient of the damper is c. If the system is set in motion because of an initial displacement and/or an initial

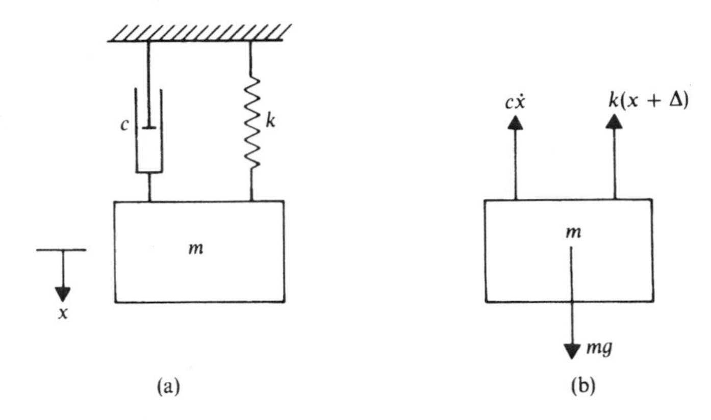

FIG. 3.16. Damped single degree of freedom system.

velocity, the mass will vibrate freely. At an arbitrary position x of the mass from the equilibrium position, the restoring spring force is equal to kx and the viscous damping force is proportional to the velocity and is equal to $c\dot{x}$, where the displacement x is taken as positive downward from the equilibrium position. Using the free body diagram shown in Fig. 16(b), the differential equation of motion can be written as

$$m\ddot{x} = mg - c\dot{x} - k(x + \Delta) \tag{3.41}$$

where Δ is the static deflection at the equilibrium position. Since the damper does not exert force at the static equilibrium position, the condition for the static equilibrium remains

$$mg = k\Delta \tag{3.42}$$

Substituting Eq. 42 into Eq. 41 yields

$$m\ddot{x} = -c\dot{x} - kx$$

or

$$m\ddot{x} + c\dot{x} + kx = 0 \tag{3.43}$$

This is the standard form of the second-order differential equation of motion that governs the linear vibration of damped single degree of freedom systems. A solution of this equation is in the form

$$x = Ae^{pt} \tag{3.44}$$

Substituting this solution into the differential equation yields

$$(mp^2 + cp + k)Ae^{pt} = 0$$

From which the characteristic equation is defined as

$$mp^2 + cp + k = 0 \tag{3.45}$$

The roots of this equation are given by

$$p_1 = -\frac{c}{2m} + \frac{1}{2m}\sqrt{c^2 - 4mk} \tag{3.46}$$

$$p_2 = -\frac{c}{2m} - \frac{1}{2m}\sqrt{c^2 - 4mk} \tag{3.47}$$

Define the following dimensionless quantity

$$\xi = \frac{c}{C_c} \tag{3.48}$$

where ξ is called the *damping factor* and C_c is called the *critical damping coefficient* defined as

$$C_c = 2m\omega = 2\sqrt{km} \tag{3.49}$$

where ω is the system natural frequency defined as

$$\omega = \sqrt{k/m} \tag{3.50}$$

The roots p_1 and p_2 of the characteristic equation can be expressed in terms of the damping factor ξ as

$$p_1 = -\xi\omega + \omega\sqrt{\xi^2 - 1} \tag{3.51}$$

$$p_2 = -\xi\omega - \omega\sqrt{\xi^2 - 1} \tag{3.52}$$

If ξ is greater than one, the roots p_1 and p_2 are real and distinct. If ξ is equal to one, the root p_1 is equal to p_2 and both roots are real. If ξ is less than one, the roots p_1 and p_2 are complex conjugates. The damping factor ξ is greater than one if the damping coefficient c is greater than the critical damping coefficient C_c. This is the case of an *overdamped* system. The damping factor ξ is equal to one when the damping coefficient c is equal to the critical damping coefficient C_c, and in this case, the system is said to be *critically damped*. The damping factor ξ is less than one if the damping coefficient c is less than the critical damping coefficient C_c, and in this case, the system is said to be *underdamped*. In the following, the three cases of overdamped, critically damped, and underdamped systems are discussed in more detail.

Overdamped System In the overdamped case the roots p_1 and p_2 of Eqs. 51 and 52 are real. The response of the single degree of freedom system can be written as

$$x(t) = A_1 e^{p_1 t} + A_2 e^{p_2 t} \tag{3.53}$$

where A_1 and A_2 are arbitrary constants. Thus the solution, in this case, is the sum of two exponential functions and the motion of the system is non-

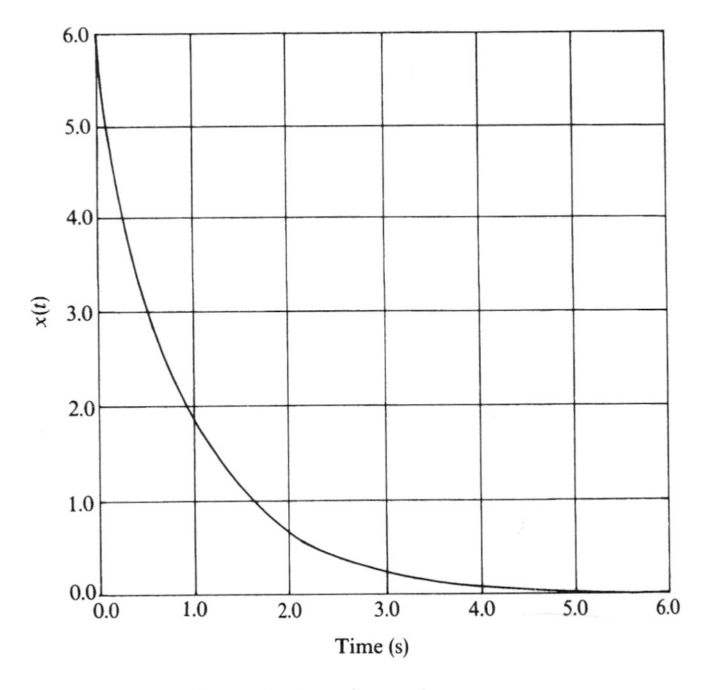

FIG. 3.17. Overdamped systems.

oscillatory, as shown in Fig. 17. The velocity can be obtained by differentiating Eq. 53 with respect to time, that is,

$$\dot{x}(t) = p_1 A_1 e^{p_1 t} + p_2 A_2 e^{p_2 t} \tag{3.54}$$

The extremum of the displacement occurs at time t_m when the velocity $\dot{x}(t)$ is equal to zero, that is, the maximum displacement occurs when

$$p_1 A_1 e^{p_1 t_m} + p_2 A_2 e^{p_2 t_m} = 0$$

or

$$e^{(p_1 - p_2)t_m} = -\frac{p_2 A_2}{p_1 A_1}$$

This equation can be used to determine the time t_m at which the displacement is maximum as

$$t_m = \frac{1}{p_1 - p_2} \ln\left(-\frac{p_2 A_2}{p_1 A_1}\right) \tag{3.55}$$

The constants A_1 and A_2 can be determined from the initial conditions. For instance, if x_0 and $\dot{x}_0$ are, respectively, the initial displacement and velocity, one has from Eqs. 53 and 54

$$x_0 = A_1 + A_2$$

$$\dot{x}_0 = p_1 A_1 + p_2 A_2$$

from which A_1 and A_2 are

$$A_1 = \frac{x_0 p_2 - \dot{x}_0}{p_2 - p_1} \tag{3.56}$$

$$A_2 = \frac{\dot{x}_0 - p_1 x_0}{p_2 - p_1} \tag{3.57}$$

provided that $(p_1 - p_2)$ is not equal to zero. The displacement $x(t)$ can then be written in terms of the initial conditions as

$$x(t) = \frac{1}{p_2 - p_1}[(x_0 p_2 - \dot{x}_0)e^{p_1 t} + (\dot{x}_0 - p_1 x_0)e^{p_2 t}] \tag{3.58}$$

The time t_m at which the maximum displacement occurs can also be written in terms of the initial conditions as

$$t_m = \frac{1}{p_2 - p_1} \ln\left[\frac{p_2(\dot{x}_0 - p_1 x_0)}{p_1(\dot{x}_0 - p_2 x_0)}\right] \tag{3.59}$$

provided that the natural logarithmic function is defined, that is, t_m of Eq. 59 is defined only when the argument of the natural logarithmic function in this equation is positive. If the system has initial displacement and zero initial velocity, t_m is given by

$$t_m = \frac{1}{p_2 - p_1} \ln 1 = 0$$

and the maximum absolute displacement is equal to the initial displacement, as shown in Fig. 18. It is important, however, to point out that there are also cases in which the response curve does not have an extremum, as shown in Fig. 17. This case corresponds to the case in which the argument of the ln function in Eq. 59 is negative.

Example 3.5

The damped mass–spring system shown in Fig. 15 has mass $m = 10$ kg, stiffness coefficient $k = 1000$ N/m, and damping coefficient $c = 300$ N·s/m. Determine the displacement of the mass as a function of time.

Solution. The natural frequency ω of the system is

$$\omega = \sqrt{\frac{k}{m}} = \sqrt{\frac{1000}{10}} = 10 \text{ rad/s}$$

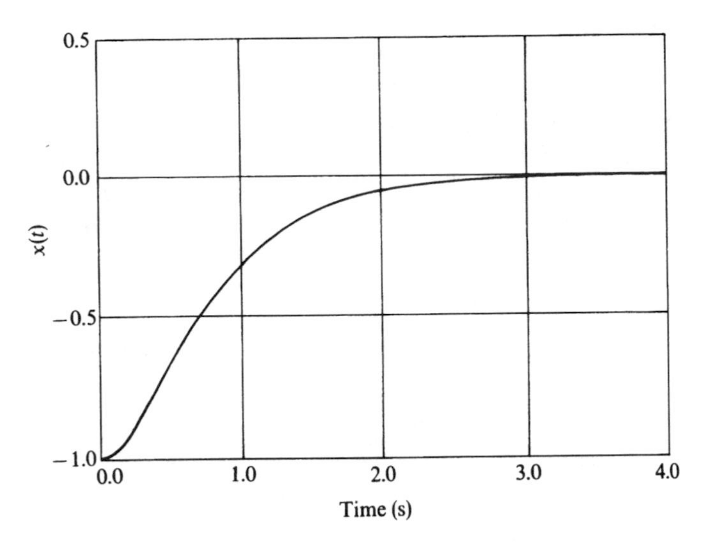

FIG. 3.18. Overdamped system with zero initial velocity.

The critical damping coefficient C_c is

$$C_c = 2m\omega = 2(10)(10) = 200 \text{ N} \cdot \text{s/m}$$

The damping factor ξ is given by

$$\xi = \frac{c}{C_c} = \frac{300}{200} = 1.5$$

Since $\xi > 1$, the system is overdamped and the solution is given by

$$x(t) = A_1 e^{p_1 t} + A_2 e^{p_2 t}$$

where p_1 and p_2 can be determined using Eqs. 51 and 52 as

$$p_1 = -\xi\omega + \omega\sqrt{\xi^2 - 1} = -(1.5)(10) + (10)\sqrt{(1.5)^2 - 1} = -3.8197$$

$$p_2 = -\xi\omega - \omega\sqrt{\xi^2 - 1} = -(1.5)(10) - (10)\sqrt{(1.5)^2 - 1} = -26.1803$$

The solution $x(t)$ is then given by

$$x(t) = A_1 e^{p_1 t} + A_2 e^{p_2 t} = A_1 e^{-3.8197t} + A_2 e^{-26.1803t}$$

The constants A_1 and A_2 can be determined from the initial conditions.

Critically Damped Systems For critically damped systems, the damping coefficient c is equal to the system's critical damping coefficient C_c, and in this case, the damping factor ξ is equal to one. The roots p_1 and p_2 of the characteristic equation are equal and are given by

$$p_1 = p_2 = p = -\omega$$

The solution, in this case, as discussed in Chapter 2, is given by

$$x(t) = (A_1 + A_2 t)e^{-\omega t} \tag{3.60}$$

where A_1 and A_2 are arbitrary constants. It is clear from the above equation that the solution $x(t)$ is nonoscillatory and it is the product of a linear function of time and an exponential decay. The form of the solution depends on the constants A_1 and A_2 or, equivalently, on the initial conditions. The velocity $\dot{x}$ can be obtained by differentiating Eq. 60 with respect to time as

$$\dot{x}(t) = [A_2 - \omega(A_1 + A_2 t)]e^{-\omega t} \tag{3.61}$$

The peak of the displacement curve occurs when $\dot{x} = 0$, that is,

$$A_2 - \omega(A_1 + A_2 t_m) = 0$$

From which the time t_m at which the peak occurs is

$$t_m = \frac{A_2 - \omega A_1}{\omega A_2} \tag{3.62}$$

The constants A_1 and A_2 can also be determined from the initial conditions. For instance, given the initial displacement x_0 and the initial velocity $\dot{x}_0$, Eqs. 60 and 61 yield

$$x_0 = A_1$$

$$\dot{x}_0 = A_2 - \omega A_1$$

which can be used to determine A_1 and A_2 as

$$A_1 = x_0 \tag{3.63}$$

$$A_2 = \dot{x}_0 + \omega x_0 \tag{3.64}$$

The displacement can then be written in terms of the initial conditions as

$$x(t) = [x_0 + (\dot{x}_0 + \omega x_0)t]e^{-\omega t} \tag{3.65}$$

This nonoscillatory solution is shown in Fig. 19.

The time t_m of Eq. 62 can also be written in terms of the initial conditions as

$$t_m = \frac{\dot{x}_0}{\omega(\dot{x}_0 + \omega x_0)} \tag{3.66}$$

For a given set of initial conditions, a critically damped system tends to approach its equilibrium position the fastest without any oscillations, that is, a critically damped system has the smallest amount of damping required for nonoscillatory motion. This property is utilized in many control and military system applications such as machine guns, which are designed as critically damped systems so that they return to their initial position, as fast as possible without vibration, after firing.

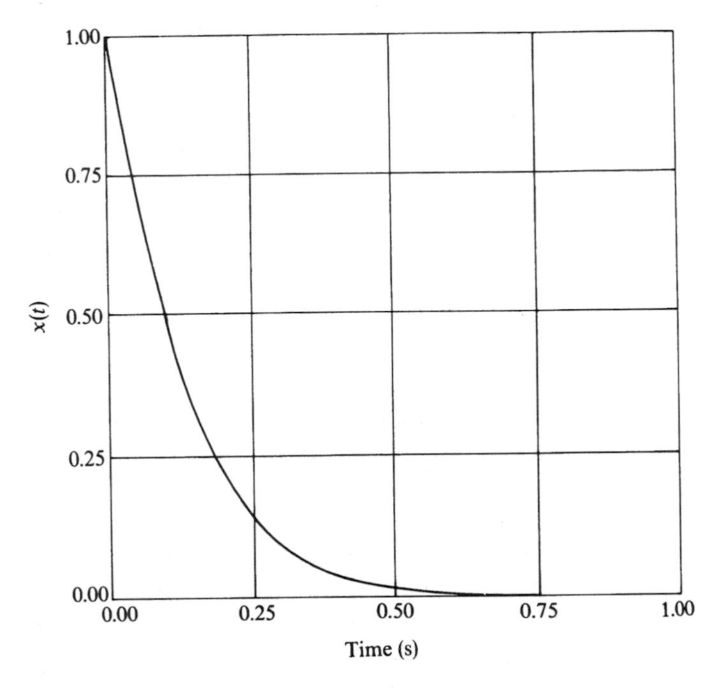

FIG. 3.19. Critically damped systems.

Example 3.6

The damped mass–spring system shown in Fig. 16 has mass $m = 10$ kg, stiffness coefficient $k = 1000$ N/m, and damping coefficient $c = 200$ N · s/m. Determine the displacement of the mass as a function of time.

Solution. The natural frequency ω of the system is

$$\omega = \sqrt{\frac{k}{m}} = \sqrt{\frac{1000}{10}} = 10 \text{ rad/s}$$

The critical damping coefficient C_c is given by

$$C_c = 2m\omega = 2(10)(10) = 200 \text{ N} \cdot \text{s/m}$$

The damping factor ξ is given by

$$\xi = \frac{c}{C_c} = \frac{200}{200} = 1$$

Since $\xi = 1$, the system is critically damped and the solution is given by Eq. 60 as

$$x(t) = (A_1 + A_2 t)e^{-10t}$$

where the constants A_1 and A_2 can be determined from the initial conditions by using Eqs. 63 and 64.

Underdamped Systems In the case of underdamped systems, the damping coefficient c is less than the critical damping coefficient C_c. In this case, the damping factor ξ is less than one and the roots of the characteristic equations p_1 and p_2, defined by Eqs. 51 and 52, respectively, are complex conjugates. Let us define the *damped natural frequency* ω_d as

$$\omega_d = \omega\sqrt{1 - \xi^2} \tag{3.67}$$

Using this equation, the roots p_1 and p_2 of the characteristic equation given by Eqs. 51 and 52 are defined as

$$p_1 = -\xi\omega + i\omega_d \tag{3.68}$$

$$p_2 = -\xi\omega - i\omega_d \tag{3.69}$$

Following the procedure described in Chapter 2, one can show that the solution $x(t)$ of the underdamped system can be written as

$$x(t) = Xe^{-\xi\omega t}\sin(\omega_d t + \phi) \tag{3.70}$$

where the amplitude X and the phase angle ϕ are constant and can be determined from the initial conditions. The solution $x(t)$ is the product of an exponential decay and a harmonic function, and unlike the preceding two cases of overdamped and critically damped systems, the motion of the underdamped system is oscillatory, as shown in Fig. 20.

The velocity $\dot{x}(t)$ is obtained by differentiating Eq. 70 with respect to time as

$$\dot{x}(t) = Xe^{-\xi\omega t}[-\xi\omega\sin(\omega_d t + \phi) + \omega_d\cos(\omega_d t + \phi)] \tag{3.71}$$

The peaks of the displacement curve shown in Fig. 20 can be obtained by setting $\dot{x}(t)$ equal to zero, that is,

$$Xe^{-\xi\omega t_i}[-\xi\omega\sin(\omega_d t_i + \phi) + \omega_d\cos(\omega_d t_i + \phi)] = 0$$

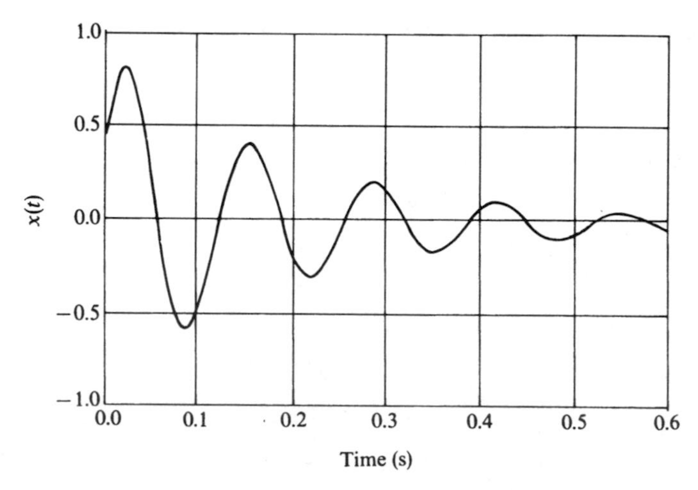

FIG. 3.20. Response of the underdamped systems.

where t_i is the time at which the peak i occurs. The above equation yields

$$\tan(\omega_d t_i + \phi) = \frac{\omega_d}{\xi\omega} = \frac{\sqrt{1-\xi^2}}{\xi} \tag{3.72}$$

Using the trigonometric identity

$$\sin\theta = \frac{\tan\theta}{\sqrt{1+\tan^2\theta}}$$

Eq. 72 yields

$$\sin(\omega_d t_i + \phi) = \sqrt{1-\xi^2} \tag{3.73}$$

Equations 70 and 73 can be used to define the displacement of the peak i as

$$x_i = \sqrt{1-\xi^2}\, X e^{-\xi\omega t_i} \tag{3.74}$$

This equation will be used in the following section to develop a technique for determining the damping coefficient experimentally.

Example 3.7

The damped mass–spring system shown in Fig. 16 has mass $m = 10$ kg, stiffness coefficient $k = 1000$ N/m, and damping coefficient $c = 10$ N·s/m. Determine the displacement of the mass as a function of time.

Solution. The circular frequency ω of the system is

$$\omega = \sqrt{\frac{k}{m}} = \sqrt{\frac{1000}{10}} = 10 \text{ rad/s}$$

and the critical damping factor ξ is given by

$$C_c = 2m\omega = 2(10)(10) = 200 \text{ N}\cdot\text{s/m}$$

Therefore, the damping factor ξ is given by

$$\xi = \frac{c}{C_c} = \frac{10}{200} = 0.05$$

The damped natural frequency ω_d is given by

$$\omega_d = \omega\sqrt{1-\xi^2} = 10\sqrt{1-(0.05)^2} = 9.9875 \text{ rad/s}$$

Substituting ω, ξ, and ω_d into Eq. 70, the solution for the damped single degree of freedom system can be expressed as

$$x = Xe^{-0.5t}\sin(9.9875t + \phi)$$

where X and ϕ are constants which can be determined from the initial conditions.

Equivalent Coefficients It is important to point out at this stage that the linear differential equation of free vibration of the damped single degree of freedom system can, in general, be written in the following form

$$m_e \ddot{x} + c_e \dot{x} + k_e x = 0 \tag{3.75}$$

where m_e, c_e, and k_e are equivalent inertia, damping, and stiffness coefficients, and the dependent variable x can be a linear or angular displacement. In this general case, m_e, c_e, and k_e must have consistent units. The natural frequency ω, the critical damping coefficient C_c, and the damping factor ξ are defined, in this general case, as

$$\omega = \sqrt{\frac{k_e}{m_e}} \tag{3.76}$$

$$C_c = 2m_e\omega = 2\sqrt{m_e k_e} \tag{3.77}$$

$$\xi = \frac{c_e}{C_c} \tag{3.78}$$

The use of Eqs. 75–78, which reduce, respectively, to Eqs. 43, 50, 49, and 48 in the simple case of damped mass–spring systems, is demonstrated by the following example.

Example 3.8

Assuming small oscillations, obtain the differential equation of the free vibration of the pendulum shown in Fig. 21. Determine the circular frequency, the critical damping coefficient, and the damping factor of this system. Assume that the rod is massless.

Solution. As shown in the figure, let R_x and R_y be the components of the reaction force at the pin joint. The moments of the externally applied forces about O are

$$M_a = -(kl \sin\theta)l\cos\theta - (cl\dot{\theta}\cos\theta)l\cos\theta - mgl\sin\theta$$

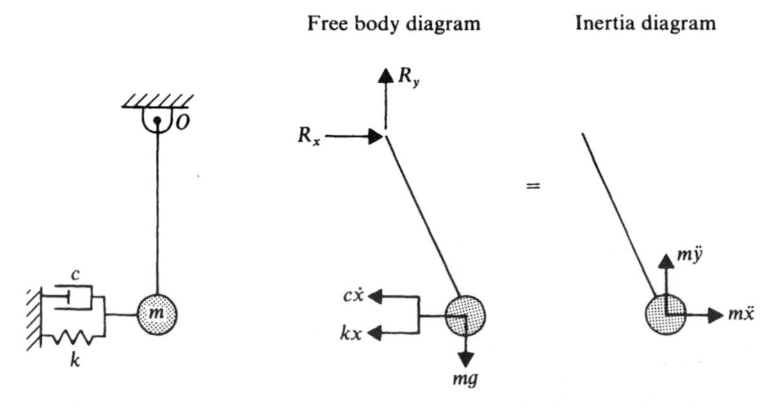

FIG. 3.21. Damped angular oscillations.

For small oscillations, $\sin\theta \approx \theta$ and $\cos\theta \approx 1$. In this case, M_a reduces to

$$M_a = -kl^2\theta - cl^2\dot{\theta} - mgl\theta$$

One can show that the moment of the inertia (effective) forces about O is given by

$$M_{eff} = ml^2\ddot{\theta}$$

Therefore, the second-order differential equation of motion of the free vibration is given by

$$-kl^2\theta - cl^2\dot{\theta} - mgl\theta = ml^2\ddot{\theta}$$

or

$$ml^2\ddot{\theta} + cl^2\dot{\theta} + (kl + mg)l\theta = 0$$

which can be written in the general form of Eq. 75 as

$$m_e\ddot{\theta} + c_e\dot{\theta} + k_e\theta = 0$$

where

$$m_e = ml^2$$

$$c_e = cl^2$$

$$k_e = (kl + mg)l$$

where the units of m_e are $kg \cdot m^2$ or, equivalently, $N \cdot m \cdot s^2$, the units of the equivalent damping coefficient c_e are $N \cdot m \cdot s$ and the units of the equivalent stiffness coefficient k_e are $N \cdot m$. The natural frequency ω is

$$\omega = \sqrt{\frac{k_e}{m_e}} = \sqrt{\frac{(kl + mg)l}{ml^2}} = \sqrt{\frac{kl + mg}{ml}} \quad \text{rad/s}$$

The critical damping coefficient C_c is

$$C_c = 2m_e\omega = 2ml^2\sqrt{\frac{kl + mg}{ml}}$$

$$= 2\sqrt{ml^3(kl + mg)}$$

The damping factor ξ of this system is

$$\xi = \frac{c_e}{C_c} = \frac{cl^2}{2\sqrt{ml^3(kl + mg)}}$$

3.7 LOGARITHMIC DECREMENT

In the preceding section it was shown that the displacement of the underdamped single degree of freedom system is oscillatory with amplitude that decreases with time. The peaks of the displacement curve shown in Fig. 20 can be determined using Eq. 74 which is reproduced here for convenience. According to this equation, the amplitude of the ith cycle of the displacement is

$$x_i = \sqrt{1 - \xi^2}\, X e^{-\xi\omega t_i} \tag{3.79}$$

This equation is used in this section to develop a method for determining experimentally the damping coefficient of the underdamped single degree of freedom systems. This can be achieved by comparing the amplitudes of successive cycles. Since Eq. 79 defines the amplitude for the ith cycle, the amplitude for the successive $(i + 1)$th cycle which occurs at time $t_i + \tau_d$ is given by

$$x_{i+1} = \sqrt{1 - \xi^2}\, X e^{-\xi\omega(t_i + \tau_d)} \tag{3.80}$$

where τ_d is the damped periodic time defined as

$$\tau_d = \frac{2\pi}{\omega_d} \tag{3.81}$$

Dividing Eq. 79 by Eq. 80 leads to

$$\frac{x_i}{x_{i+1}} = \frac{\sqrt{1 - \xi^2}\, X e^{-\xi\omega t_i}}{\sqrt{1 - \xi^2}\, X e^{-\xi\omega(t_i + \tau_d)}} = e^{\xi\omega\tau_d} \tag{3.82}$$

which indicates that the ratio of any two successive amplitudes is constant. Equation 82 can be expressed in terms of the natural logarithm as

$$\ln \frac{x_i}{x_{i+1}} = \xi\omega\tau_d = \delta \tag{3.83}$$

where δ is a constant known as the *logarithmic decrement*. By using Eq. 83 and the fact that ω_d is

$$\omega_d = \omega\sqrt{1 - \xi^2}$$

the logarithmic decrement δ can be written as

$$\delta = \ln \frac{x_i}{x_{i+1}} = \frac{2\pi\xi}{\sqrt{1 - \xi^2}} \tag{3.84}$$

If the damping factor ξ is very small, that is,

$$\xi \ll 1$$

the expression for δ reduces to

$$\delta = 2\pi\xi$$

We now consider the ratio of nonsuccessive amplitudes. If we consider the amplitudes x_i and x_{i+n}, where n is an integer, we obtain

$$\frac{x_i}{x_{i+n}} = \frac{\sqrt{1 - \xi^2}\, X e^{-\xi\omega t_i}}{\sqrt{1 - \xi^2}\, X e^{-\xi\omega(t_i + n\tau_d)}} = e^{n\xi\omega\tau_d}$$

that is,

$$\ln \frac{x_i}{x_{i+n}} = n\xi\omega\tau_d = n\delta \tag{3.85}$$

If δ can be determined by the experimental measurements of two successive or nonsuccessive amplitudes, then one can determine the damping factor ξ

using Eq. 84 as

$$\xi = \frac{\delta}{\sqrt{(2\pi)^2 + \delta^2}} \tag{3.86}$$

Energy Loss In the case of small damping, the energy loss can be expressed in terms of the logarithmic decrement. Let U_i be the energy of the system at the peak of the cycle i. Since at the peak the displacement is a local maximum, the velocity of the mass is equal to zero, and in this case, the energy is equal to the elastic energy stored in the spring, that is,

$$U_i = \tfrac{1}{2}kx_i^2$$

Similarly, the energy at the peak of the cycle $i + 1$ is

$$U_{i+1} = \tfrac{1}{2}kx_{i+1}^2$$

The loss of energy between the two cycles is, therefore, given by

$$\Delta U = U_i - U_{i+1} = \tfrac{1}{2}k(x_i^2 - x_{i+1}^2)$$

which can also be written as

$$\Delta U = \tfrac{1}{2}k(x_i - x_{i+1})(x_i + x_{i+1})$$

We define the *specific energy loss* as

$$\begin{aligned}\frac{\Delta U}{U_i} &= \frac{\frac{1}{2}k(x_i - x_{i+1})(x_i + x_{i+1})}{\frac{1}{2}kx_i^2} \\ &= \left(1 - \frac{x_{i+1}}{x_i}\right)\left(1 + \frac{x_{i+1}}{x_i}\right) \\ &= (1 - e^{-\delta})(1 + e^{-\delta}) = 1 - e^{-2\delta}\end{aligned}$$

Note that the specific energy loss increases as the logarithmic decrement increases, and if the damping is zero, δ is equal to zero and accordingly $\Delta U/U_i$ is equal to zero.

Example 3.9

A damped single degree of freedom mass–spring system has mass $m = 5$ kg and stiffness coefficient $k = 500$ N/m. From the experimental measurements, it was observed that the amplitude of vibration diminishes from 0.02 m to 0.012 m in six cycles. Determine the damping coefficient c.

Solution. In this example, n of Eq. 85 is equal to 6. Let

$$x_i = 0.02$$

$$x_{i+6} = 0.012$$

The logarithmic decrement δ is determined from

$$n\delta = 6\delta = \ln\frac{0.02}{0.012} = 0.51083$$

that is,

$$\delta = 0.085138$$

The damping factor ξ can be determined from Eq. 86 as

$$\xi = \frac{\delta}{\sqrt{(2\pi)^2 + \delta^2}} = \frac{0.085138}{\sqrt{(2\pi)^2 + (0.085138)^2}} = 0.013549$$

The damping coefficient c is defined as

$$c = \xi C_c$$

where the critical damping coefficient C_c is

$$C_c = 2\sqrt{km} = 2\sqrt{(500)(5)} = 100 \text{ N}\cdot\text{s/m}$$

Therefore,

$$c = (0.013549)(100) = 1.3549 \text{ N}\cdot\text{s/m}$$

The specific energy loss is

$$\frac{\Delta U}{U_i} = 1 - e^{-2\delta} = 1 - e^{-2(0.085138)} = 0.1566$$

3.8 STRUCTURAL DAMPING

While a viscous damping force is proportional to the velocity, in many cases, such simple expressions for the damping forces are not directly available. It is, however, possible to obtain an equivalent viscous damping coefficient by equating energy expressions that represent the dissipated energy during the motion. In this section, we consider the case of *structural damping* which is sometimes referred to as *hysteretic damping*. The influence of this type of damping can be seen in the vibration of solid materials, which in general, are not perfectly elastic. When solids vibrate, there is an energy dissipation due to internal friction, as the result of the relative motion between particles of the solids during deformation. It was observed that there is a phase lag between the applied force F and the displacement x, as shown by the hysteresis loop curve in Fig. 22, which shows that the effect of the force does not suddenly disappear when the force is removed. The energy loss during one cycle can be obtained as the enclosed area in the hysteresis loop, and can be expressed mathematically using the following integral

$$\Delta D = \int F\,dx \tag{3.87}$$

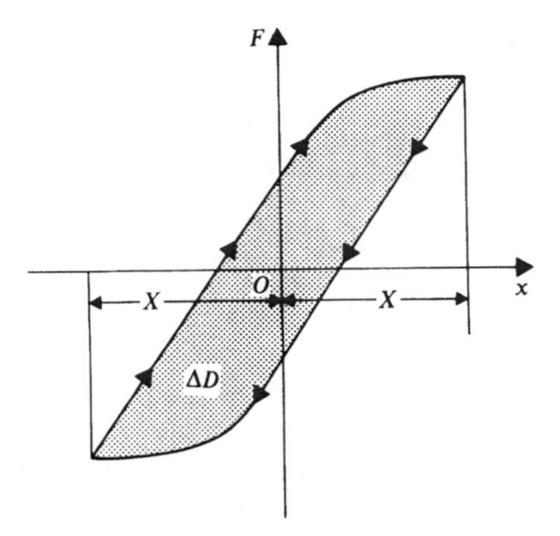

FIG. 3.22. Hysteresis loop.

It was also observed experimentally that the energy loss ΔD during one cycle is proportional to the stiffness of the material k and the square of the amplitude of the displacement X, and can be expressed in the following simple form

$$\Delta D = \pi c_s k X^2 \tag{3.88}$$

where c_s is a dimensionless *structural damping coefficient* and the factor π is included for convenience. In order to use Eq. 88 to obtain an equivalent viscous damping coefficient, we first assume simple harmonic oscillations in the form

$$x = X \sin(\omega t + \phi)$$

The force exerted by a viscous damper can be written as

$$F_d = c_e \dot{x} = c_e X \omega \cos(\omega t + \phi) \tag{3.89}$$

and the energy loss per cycle can be written as

$$\Delta D = \int F_d \, dx = \int c_e \dot{x} \, dx \tag{3.90}$$

Since $\dot{x} = dx/dt$, we have

$$dx = \dot{x} \, dt \tag{3.91}$$

Substituting Eqs. 89 and 91 into Eq. 90 yields

$$\begin{aligned} \Delta D &= \int_0^\tau c_e \dot{x}^2 \, dt = \int_0^\tau c_e \omega^2 X^2 \cos^2(\omega t + \phi) \, dt \\ &= \pi c_e \omega X^2 \end{aligned} \tag{3.92}$$

where τ is the periodic time defined as

$$\tau = \frac{2\pi}{\omega}$$

Equating Eqs. 88 and 92 yields the equivalent viscous damping coefficient as

$$c_e = \frac{kc_s}{\omega} \tag{3.93}$$

The structural damping coefficient c_s can be determined experimentally to obtian the equivalent viscous damping coefficient c_e which can be used to develop a simple mathematical model.

Experimental Methods If a structure behaves as a single degree of freedom system, the equivalent damping factor ξ_e can be defined as

$$\xi_e = \frac{c_e}{C_c} = \frac{c_e}{2m\omega}$$

Substituting Eq. 93 into the preceding equation, one obtains an expression for the equivalent damping factor ξ_e as a function of the structural damping coefficient c_s as

$$\xi_e = \frac{kc_s}{2m\omega^2} = \frac{c_s}{2}$$

Since the damping factor ξ_e can be expressed in terms of the logarithmic decrement using Eq. 86, one has

$$\xi_e = \frac{\delta}{\sqrt{(2\pi)^2 + \delta^2}} = \frac{c_s}{2}$$

which leads to

$$c_s = \frac{2\delta}{\sqrt{(2\pi)^2 + \delta^2}}$$

That is, if the ratio between two successive amplitudes can be measured experimentally, the structural damping coefficient c_s can be evaluated by using the preceding equation.

Example 3.10

A simple structure is found to vibrate as a single degree of freedom system. The spring constant is determined using static testing and is found to be 1500 N/m, and the equivalent mass of the structure is assumed to be 2 kg. By using a simple vibration test, the ratio of successive amplitudes is found to be 1.2. Determine the structural damping coefficient and the equivalent viscous damping coefficient. Determine also the energy loss per cycle for an amplitude of 0.05 m.

Solution. The logarithmic decrement δ is given by

$$\delta = \ln \frac{x_i}{x_{i+1}} = \ln 1.2 = 0.18232$$

The structural damping coefficient c_s is given by

$$c_s = \frac{2\delta}{\sqrt{(2\pi)^2 + \delta^2}} = \frac{2(0.18232)}{\sqrt{(2\pi)^2 + (0.18232)^2}} = 0.058$$

The equivalent viscous damping coefficient can then be determined using Eq. 93 as

$$c_e = \frac{kc_s}{\omega} = c_s\sqrt{km} = 3.1768 \text{ N} \cdot \text{s/m}$$

Equation 88 can be used to determine the energy loss per cycle as

$$\Delta D = \pi c_s k X^2 = \pi(0.058)(1500)(0.05)^2 = 0.6833 \text{ N} \cdot \text{m}$$

3.9 COULOMB DAMPING

In this section, we examine the effect of *Coulomb* or *dry-friction damping* on the response of the single degree of freedom mass–spring system shown in Fig. 23. In the case of Coulomb damping, the friction force always acts in a direction opposite to the direction of the motion of the mass. In this case, the friction force can be written as

$$F_f = \mu N \tag{3.94}$$

where μ is the *coefficient of sliding friction* and N is the normal reaction force. The values of the coefficient of sliding friction, which depends on the properties of the materials in contact, can be found experimentally. While Table 2 shows approximate values of this coefficient in several cases of dry surfaces, we should keep in mind that the effect of friction is significantly reduced in the case of lubricated surfaces.

If the motion of the mass shown in Fig. 23(a) is to the right, that is, $\dot{x} > 0$, the friction force F_f is negative, as shown in Fig. 23(b), while if the motion of the mass is to the left, that is, $\dot{x} < 0$, the friction force F_f is positive, as shown in Fig. 23(c). Therefore, the vibration of the system is governed by two differential equations which depend on the direction of motion. From the free body diagram shown in Fig. 23(b), it is clear that if the mass moves to the

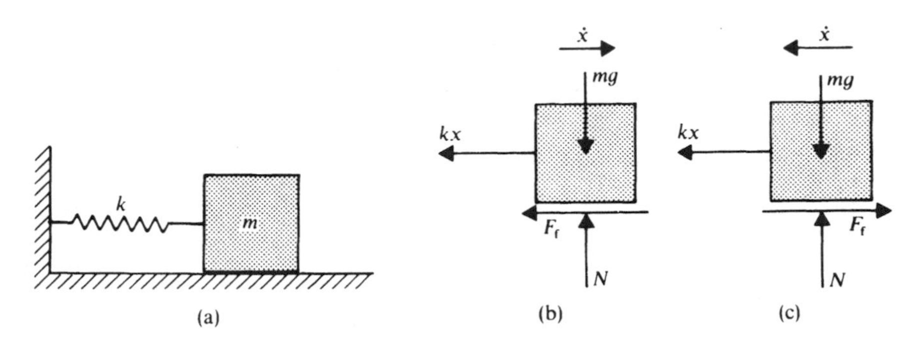

FIG. 3.23. Coulomb damping.

TABLE 3.2. Approximate Values of the Coefficient of Sliding Friction

Rubber on concrete	0.45–0.68
Metal on stone	0.25–0.55
Metal on wood	0.15–0.45
Metal on metal	0.12–0.45
Wood on wood	0.19–0.38
Stone on stone	0.30–0.53

right, the differential equation of motion is

$$m\ddot{x} = -kx - F_f, \qquad \dot{x} > 0 \tag{3.95}$$

Similarly, the free body diagram of Fig. 23(c) shows that the differential equation of motion, when the mass moves to the left, is

$$m\ddot{x} = -kx + F_f, \qquad \dot{x} < 0 \tag{3.96}$$

Equation 95 and 96 can be combined in one equation as

$$m\ddot{x} + kx = \mp F_f \tag{3.97}$$

where the negative sign is used when the mass moves to the right and the positive sign is used when the mass moves to the left. Equation 97 is a nonhomogeneous differential equation, and its solution consists of two parts, the homogeneous solution or the complementary function and the forced solution or the particular solution. Since the force F_f is constant, the particular solution x_p is assumed as

$$x_p = C$$

where C is a constant. Substituting this solution into Eq. 97 yields

$$x_p = \mp \frac{F_f}{k}$$

Therefore, the solution of Eq. 97 can be written as

$$x(t) = A_1 \sin \omega t + A_2 \cos \omega t - \frac{F_f}{k}, \qquad \dot{x} \geq 0 \tag{3.98}$$

$$x(t) = B_1 \sin \omega t + B_2 \cos \omega t + \frac{F_f}{k}, \qquad \dot{x} < 0 \tag{3.99}$$

where ω is the natural frequency defined as

$$\omega = \sqrt{\frac{k}{m}},$$

and A_1 and A_2 are constants that depend on the initial conditions of motion to the right, and B_1 and B_2 are constants that depend on the initial conditions of motion to the left.

Let us now consider the case in which the mass was given an initial displacement x_0 to the right and zero initial veloctiy. Equation 99 can then be used to yield the following algebraic equations

$$x_0 = B_2 + \frac{F_f}{k}$$

$$0 = \omega B_1$$

which imply that

$$B_1 = 0, \qquad B_2 = x_0 - \frac{F_f}{k}$$

that is,

$$x(t) = \left(x_0 - \frac{F_f}{k}\right)\cos \omega t + \frac{F_f}{k} \tag{3.100}$$

and

$$\dot{x}(t) = -\omega\left(x_0 - \frac{F_f}{k}\right)\sin \omega t \tag{3.101}$$

The direction of motion will change when $\dot{x} = 0$, and when this condition is substituted into the above equation, one obtains the time t_1 at which the velocity starts to be positive. The time t_1 then can be obtained from the equation

$$0 = -\omega\left(x_0 - \frac{F_f}{k}\right)\sin \omega t_1$$

which defines t_1 as

$$\omega t_1 = \pi$$

that is,

$$t_1 = \frac{\pi}{\omega}$$

At this time, the displacement is determined from Eq. 100 as

$$x(t_1) = x\left(\frac{\pi}{\omega}\right) = -x_0 + \frac{2F_f}{k} \tag{3.102}$$

which shows that the amplitude in the first half-cycle is reduced by the amount $2F_f/k$, as the result of dry friction.

In the second half-cycle, the mass moves to the right and the motion is governed by Eq. 98 with the initial conditions

$$x\left(\frac{\pi}{\omega}\right) = -x_0 + \frac{2F_f}{k}$$

$$\dot{x}\left(\frac{\pi}{\omega}\right) = 0$$

Substituting these initial conditions into Eq. 98 yields

$$A_1 = 0$$

and

$$A_2 = x_0 - 3\frac{F_f}{k}$$

The displacement $x(t)$ in the second half-cycle can then be written as

$$x(t) = \left(x_0 - 3\frac{F_f}{k}\right)\cos\omega t - \frac{F_f}{k} \tag{3.103}$$

and the velocity is

$$\dot{x}(t) = -\left(x_0 - 3\frac{F_f}{k}\right)\omega\sin\omega t \tag{3.104}$$

The velocity is zero at time t_2 when $t_2 = 2\pi/\omega = \tau$, where τ is the periodic time of the natural oscillations. At time t_2, the end of the first cycle, the displacement is

$$x(t_2) = x\left(\frac{2\pi}{\omega}\right) = x_0 - \left(\frac{4F_f}{k}\right) \tag{3.105}$$

which shows that the amplitude decreases in the second half-cycle by the amount $2F_f/k$, as shown in Fig. 24. By continuing in this manner, one can verify that there is a constant decrease in the amplitude of $2F_f/k$ every half-cycle of motion. Furthermore, unlike the case of viscous damping, the frequency of oscillation is not affected by the Coulomb damping. It is also important to point out, in the case of Coulomb friction, that it is not necessary that the system comes to rest at the undeformed spring position. The final position will be at an amplitude X_f, at which the spring force $F_s = kX_f$ is less than or equal to the friction force.

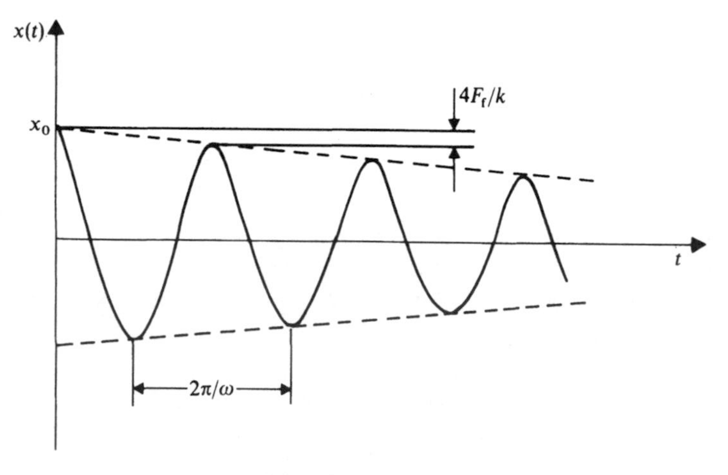

FIG. 3.24. Effect of the Coulomb damping.

Example 3.11

The single degree of freedom mass–spring system shown in Fig. 23 has mass $m = 5$ kg, stiffness coefficient of the spring $k = 5 \times 10^3$ N/m, coefficient of dry friction $\mu = 0.1$, initial displacement $x_0 = 0.03$ m, and initial veloctiy $\dot{x}_0 = 0$. Determine the number of cycles of oscillation of the mass before it comes to rest.

Solution. The friction force F_f is defined by Eq. 94 as

$$F_f = \mu N$$

where N is the normal reaction force given by

$$N = mg$$

Therefore, the force F_f is

$$F_f = \mu mg = (0.1)(5)(9.81) = 4.905 \text{ N}$$

The motion will stop if the amplitude of the cycle is such that the spring force is less than or equal to the friction force, that is,

$$kX_f \leq 4.905$$

or

$$X_f \leq \frac{4.905}{k} = \frac{4.905}{5 \times 10^3} = 0.981 \times 10^{-3} \text{ m}$$

The amplitude loss per half-cycle is

$$2F_f/k = \frac{2(4.905)}{5 \times 10^3} = 1.962 \times 10^{-3} \text{ m}$$

The number of half-cycles n completed before the mass comes to rest can be obtained from the following equation

$$x_0 - n\left(\frac{2F_f}{k}\right) \leq 0.981 \times 10^{-3}$$

where

$$x_0 = 0.03$$

It follows that

$$0.03 - n(1.962 \times 10^{-3}) \leq 0.981 \times 10^{-3}$$

The smallest n that satisfies this inequality is $n = 15$ half-cycles, and the number of cycles completed before the mass comes to rest is 7.5.

3.10 SELF-EXCITED VIBRATION

In some applications, the force that produces the motion may be velocity- or displacement-dependent, such that the force that sustains the motion is created by the motion itself, and when the motion of the system stops, the force no longer exists. The motions of such systems, which are said to be *self-excited*, are encountered in many applications such as the chatter vibration of the tool in machine-tool systems, the shimmy of automobile wheels, and airplane wing flutter. In self-excited vibration, the motion tends to increase the system energy, and as a result, the amplitude of vibration may grow drastically and the system becomes unstable. In order to better understand the effect of self-excited vibration on the system stability, we consider a simple case in which the force is proportional to the velocity. In this case, the differential equation of motion of the damped mass–spring single degree of freedom system can be written as

$$m\ddot{x} + c\dot{x} + kx = F \tag{3.106}$$

where m is the mass, c is the viscous damping coefficient, k is the spring constant, and the force F can be written as

$$F = b\dot{x} \tag{3.107}$$

in which b is a proportionality constant. Substituting Eq. 107 into Eq. 106 yields

$$m\ddot{x} + c\dot{x} + kx = b\dot{x}$$

or

$$m\ddot{x} + (c - b)\dot{x} + kx = 0 \tag{3.108}$$

The self-excited vibration can be considered as a free vibration with a negative damping. In this case, the damping force which is proportional to the velocity has the same direction as the velocity.

If $c = b$ in Eq. 108, the coefficient of $\dot{x}$ in the above equation is identically zero, and this equation reduces to the differential equation of motion of the undamped single degree of freedom system, which conserves energy.

A solution of Eq. 108 can be assumed in the form

$$x = A_1 e^{pt}$$

which yields the following characteristic equation

$$p^2 m + p(c - b) + k = 0 \tag{3.109}$$

This equation has the following two roots

$$p_1 = -\frac{c - b}{2m} + \frac{1}{2m}\sqrt{(c - b)^2 - 4mk} \tag{3.110}$$

$$p_2 = -\frac{c - b}{2m} - \frac{1}{2m}\sqrt{(c - b)^2 - 4mk} \tag{3.111}$$

If $c > b$, we have the case of positive damping discussed in the preceding sections, where it was shown that this case corresponds to stable systems in which the amplitude decreases with time. If, however, $c < b$, the velocity coefficient in Eq. 108 is negative and the system is said to exhibit negative damping. In the case of negative damping, if the roots p_1 and p_2 are real, at least one of the roots will be positive. If the roots p_1 and p_2 are complex conjugates, the solution can be written as the product of an exponential function multiplied by a harmonic function. Since the exponential function in this case will be increasing with time, the amplitude of vibration will drastically increase, and the system is said to be dynamically unstable. In this case, the damping force does positive work, which is converted into additional kinetic energy, and, as a result, the effect of the damping force is to increase the displacement instead of diminishing it.

In order to provide another explanation for the instability caused by the negative damping, we use the sum of the kinetic and strain energies of the single degree of freedom system defined as

$$E = \tfrac{1}{2}m\dot{x}^2 + \tfrac{1}{2}kx^2$$

which, upon differentiation with respect to time, yields

$$\frac{dE}{dt} = m\dot{x}\ddot{x} + kx\dot{x} = (m\ddot{x} + kx)\dot{x}$$

Since, from the differential equation, one has

$$m\ddot{x} + kx = -(c - b)\dot{x}$$

it follows that

$$\frac{dE}{dt} = -(c - b)\dot{x}^2$$

This expression for the time rate of change of energy shows that, in the case of positive damping coefficient $(c - b)$, the energy continuously decreases and, as a result, the amplitude eventually becomes small as demonstrated in the preceding sections. On the other hand, in the case of negative damping, the rate of change of energy is positive and, as a result, the energy and the amplitude continuously increase causing the instability problem discussed in this section.

Example 3.12

The following data are given for a damped single degree of freedom mass–spring system, mass $m = 5$ kg, damping coefficient $c = 20$ N·s/m, and stiffness coefficient $k = 5 \times 10^3$ N/m. The mass is subjected to a force which depends on the velocity and can be written as $F = b\dot{x}$, where b is a constant. Determine the system response in the following two cases of the forcing functions

$$(1)\ F = 50\dot{x}\ \text{N}, \qquad (2)\ F = 400\dot{x}\ \text{N}$$

Solution. The equation of motion of the system is

$$m\ddot{x} + c\dot{x} + kx = b\dot{x}$$

$$m\ddot{x} + (c - b)\dot{x} + kx = 0$$

Case 1. $F = 50\dot{x}$
In this case, the differential equation of motion is

$$5\ddot{x} + (20 - 50)\dot{x} + (5)(10^3)x = 0$$

This is the case of negative damping in which the characteristic equation is given by

$$5p^2 - 30p + 5 \times 10^3 = 0$$

The roots of the characteristic equation are

$$p_1 = \frac{30}{10} + \frac{1}{10}\sqrt{(30)^2 - 4(5)(5 \times 10^3)} = 3.0 + 31.48i$$

$$p_2 = \frac{30}{10} - \frac{1}{10}\sqrt{(30)^2 - 4(5)(5 \times 10^3)} = 3.0 - 31.48i$$

The roots are complex conjugates and the solution can then be written as

$$x(t) = Xe^{3.0t}\sin(31.48t + \phi)$$

where X and ϕ are constants that depend on the initial conditions. Because of the exponential growth in the solution, the system is unstable and the motion is oscillatory with amplitude that increases with time, as shown in Fig. 25.

Case 2. $F = 400\dot{x}$
In this case, the differential equation of motion is given by

$$5\ddot{x} + (20 - 400)\dot{x} + 5 \times 10^3 x = 0$$

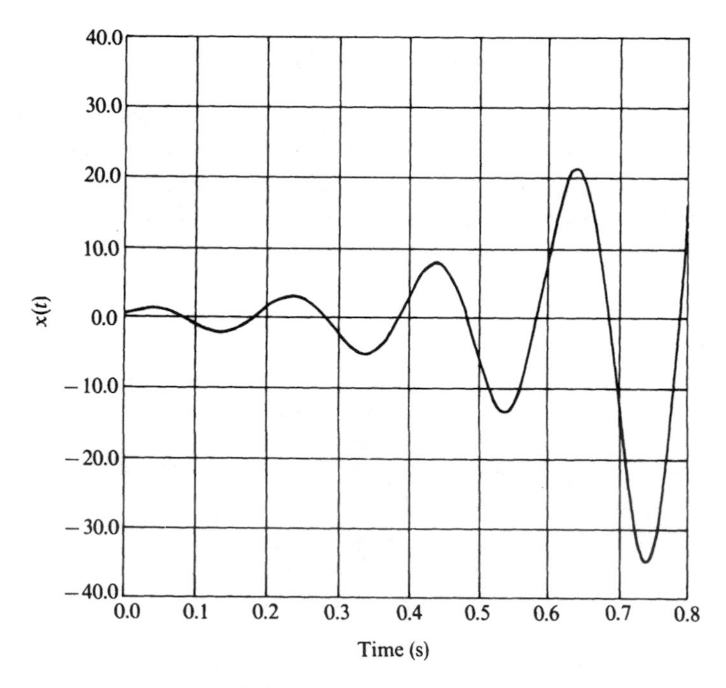

FIG. 3.25. Case 1, $F = 50\dot{x}$.

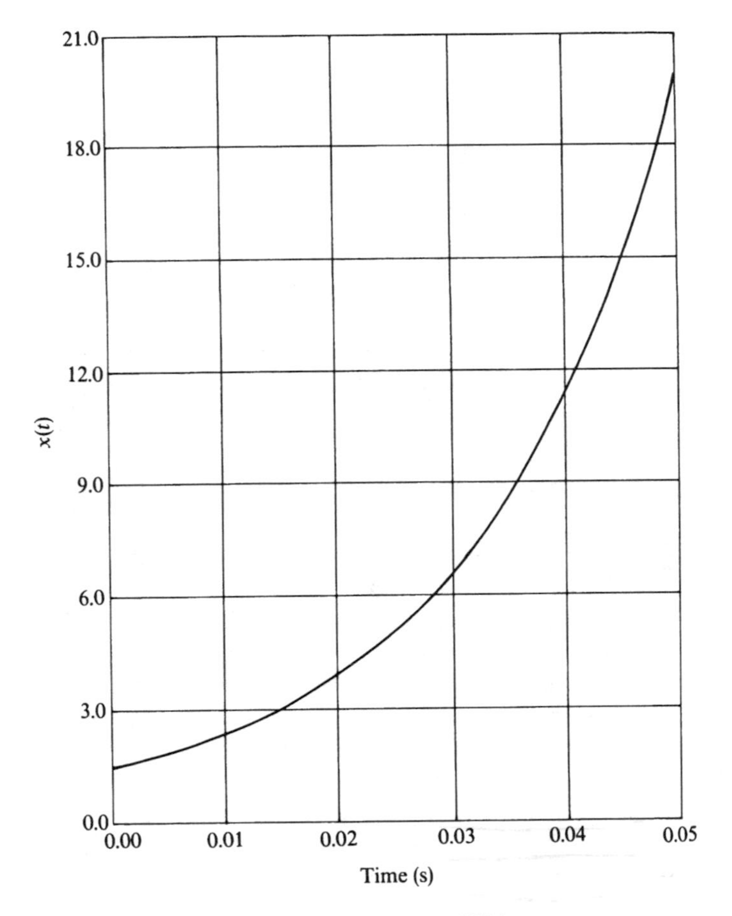

FIG. 3.26. Case 2, $F = 400\dot{x}$.

or

$$5\ddot{x} - 380\dot{x} + 5 \times 10^3 x = 0$$

The coefficient of $\dot{x}$ in this equation is also negative. The characteristic equation is

$$5p^2 - 380p + 5 \times 10^3 = 0$$

which has the roots

$$p_1 = \frac{380}{10} + \frac{1}{10}\sqrt{(380)^2 - 4(5)(5 \times 10^3)} = 59.0713$$

$$p_2 = \frac{380}{10} - \frac{1}{10}\sqrt{(380)^2 - 4(5)(5 \times 10^3)} = 16.9287$$

Since the roots are real and distinct, the solution can be written as

$$x(t) = A_1 e^{p_1 t} + A_2 e^{p_2 t} = A_1 e^{59.0713t} + A_2 e^{16.9287t}$$

where A_1 and A_2 are constants that depend on the initial conditions. The solution in this case is nonoscillatory exponential growth, as shown in Fig. 26.

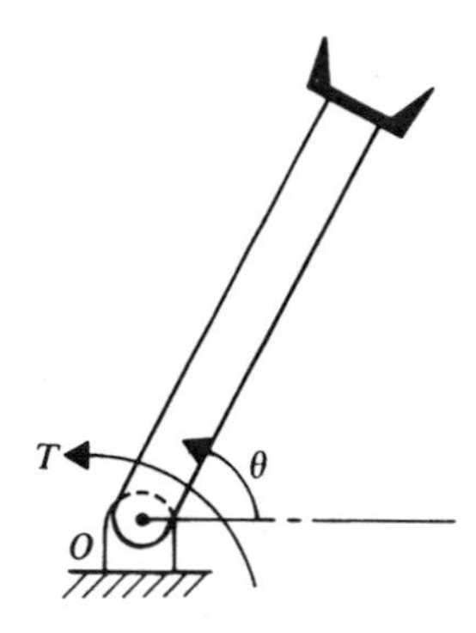

FIG. 3.27. Motion control.

3.11 MOTION CONTROL

Another important application of the theory of vibration developed in this chapter is in the area of *feedback control systems*. In many industrial applications, a system is designed to perform, with high precision, a specified task or follow a desired motion. Due, however, to disturbances or the effect of unknown parameters such as friction, wear, clearances in the joints, etc., the desired motion of the system cannot be achieved. The actual motion may deviate significantly from the desired motion, and as the result of this deviation, the performance, precision, and accuracy of the system may no longer be acceptable. It is, therefore, important to be able to deal with this problem by proper design of a control system that automatically reduces this deviation and if possible eliminates it. In order to demonstrate the use of the theory of such control systems, we consider the simple system shown in Fig. 27, which can be used to represent an industrial single robotic arm. Let T be the torque applied by a motor placed at the joint at O. The equation of motion of this system is given in terms of the angular orientation of the arm as

$$I_O\ddot{\theta} + mg\frac{l}{2}\cos\theta = T \tag{3.112}$$

where I_O is the mass moment of inertia of the robotic arm defined with respect to point O, m and l are, respectively, the mass and length of the arm, g is the gravitational constant, and θ is the angular orientation of the arm, as shown in the figure. For any applied torque, Eq. 112 represents the actual motion of the robotic arm and if θ_a denotes this actual motion, Eq. 112 can be written as

$$I_O\ddot{\theta}_a + mg\frac{l}{2}\cos\theta_a = T \tag{3.113}$$

As pointed out earlier, as the result of disturbances, and the effect of unknown parameters whose effect is not accounted for in Eq. 113, the actual motion θ_a

may deviate from the desired motion θ_d. In order to obtain the desired motion θ_d, we may choose the motor torque T in the following form

$$T = I_O\ddot{\theta}_d - k_v(\dot{\theta}_a - \dot{\theta}_d) - k_p(\theta_a - \theta_d) + mg\frac{l}{2}\cos\theta_a \tag{3.114}$$

where k_v and k_p are, respectively, velocity and position gains, which will be selected in such a manner that the deviation of the actual displacement from the desired displacement is eliminated. In Eq. 114, θ_d, $\dot{\theta}_d$, and $\ddot{\theta}_d$ are assumed to be known since the desired motion is assumed to be specified. The inertia properties and dimensions of the robotic arm are also assumed to be known. The actual displacement θ_a and its time derivatives can be obtained using proper sensors that measure the angular orientation, velocity, and acceleration of the robotic arm during the actual motion. Therefore, at any given point in time, the torque of Eq. 114 can be calculated and a proper signal is given to the motor in order to produce this torque. Now, let us substitute the expression of the torque given by Eq. 114 into the differential equation given by Eq. 113. This leads to

$$I_O(\ddot{\theta}_a - \ddot{\theta}_d) + k_v(\dot{\theta}_a - \dot{\theta}_d) + k_p(\theta_a - \theta_d) = 0 \tag{3.115}$$

We define the error as the result of the deviation of the actual motion from the desired one as

$$e = \theta_a - \theta_d \tag{3.116}$$

In terms of this error, Eq. 115 can be written compactly as

$$I_O\ddot{e} + k_v\dot{e} + k_p e = 0 \tag{3.117}$$

This equation is in the same form as the equation obtained for the damped single degree of freedom system. The velocity and position gains k_v and k_p can be selected such that the error e, that represents the deviation of the actual motion from the desired motion, goes to zero the fastest. To this end, k_v and k_p are selected such that the system of Eq. 117 is critically damped. Note that the natural frequency of oscillation ω of the error e of Eq. 117 is given by

$$\omega = \sqrt{\frac{k_p}{I_O}}$$

and the critical damping coefficient C_c is

$$C_c = 2I_O\omega = 2\sqrt{I_O k_p}$$

The damping factor ξ is given by

$$\xi = \frac{k_v}{C_c} = \frac{k_v}{2\sqrt{I_O k_p}} \tag{3.118}$$

If the system is to be critically damped, $\xi = 1$, and Eq. 118 leads to the following relationship

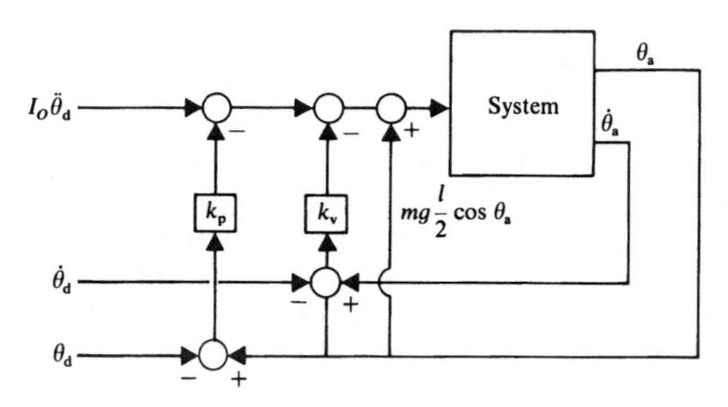

FIG. 3.28. Block diagram.

$$1 = \frac{k_v}{2\sqrt{I_O k_p}}$$

or

$$k_v^2 = 4I_O k_p \tag{3.119}$$

Therefore, in order that the error e approaches zero in a relatively short time, the gains k_v and k_p are selected such that Eq. 119 is satisfied. This leads to a critically damped system for the error equation.

The concept introduced in this section is often demonstrated using the block diagram shown in Fig. 28, and the torque T given by Eq. 114 is referred to as the *control law*.

The type of control discussed in this section is called *proportional-derivative (PD) feedback control.* The natural frequency of the error equation that governs the speed of the amplitude decay is selected based on the performance objectives and also based on the natural frequency of the system. The frequency of the error oscillations must not be close to the natural frequency of the system in order to avoid the resonance phenomenon, which will be discussed in the following chapter. Furthermore, the gain k_p must not be selected very large in order to avoid exceeding the actuator capacity, a phenomenon known in control system design as *actuator saturation* (Lewis et al., 1993).

The PD control is very effective when all of the system parameters are known and when there are no disturbances. If some of the system parameters are not known in addition to the existence of constant disturbances, the PD control will result in a non-zero state error. Another type of feedback control system is the *proportional-integral-derivative (PID) feedback control.* This control system is more effective when the disturbance is not equal to zero. In this type of control, an additional term $k_i\varepsilon$ is added to the torque T, where ε is defined as

$$\dot{\varepsilon} = e$$

The resulting error equation for the PID control system is

$$I_0 \ddot{e} + k_v \dot{e} + k_p e + k_i \varepsilon = 0$$

The preceding two equations must be solved simultaneously in order to determine the error as a function of time.

3.12 IMPACT DYNAMICS

In the preceding two sections, the self-excited vibration and motion control of mechanical systems were examined using the techniques of the free vibration analysis. There are several other important applications that can be analyzed using the techniques presented in this chapter; some of these applications are in the area of *impact dynamics*. Collision between two bodies results in a force of relatively large magnitude that acts on the two bodies over a relatively short interval of time. The common normal to the surfaces in contact during the collision between the two bodies is called the *line of impact*. If the centers of mass of the two colliding bodies are located on the line of impact, the impact is said to be *central impact*, otherwise, the impact is said to be *eccentric*.

Impact can also be classified as a *direct impact* or *oblique impact*. In direct impact the velocity of the bodies are along the line of impact. If, on the other hand, the velocity of one or both bodies are not along the line of impact, the impact is said to be oblique.

Conservation of Momentum Figure 29 shows two particles m_1 and m_2. The velocities of the two particles before collision are assumed to be $\dot{x}_1$ and $\dot{x}_2$. Let the external force that acts on the mass m_1 be $F_1(t)$ while the external force that acts on m_2 be $F_2(t)$. If $\dot{x}_1$ is assumed to be greater than $\dot{x}_2$, the particle m_1 approaches m_2 and collision will eventually occur. During the period of impact an impulsive force F_i acts on the two particles, and in the case of direct central impact, the equations of motion of the two particles can be written as

$$m_1 \ddot{x}_1 = F_1(t) - F_i(t) \tag{3.120}$$

$$m_2 \ddot{x}_2 = F_2(t) + F_i(t) \tag{3.121}$$

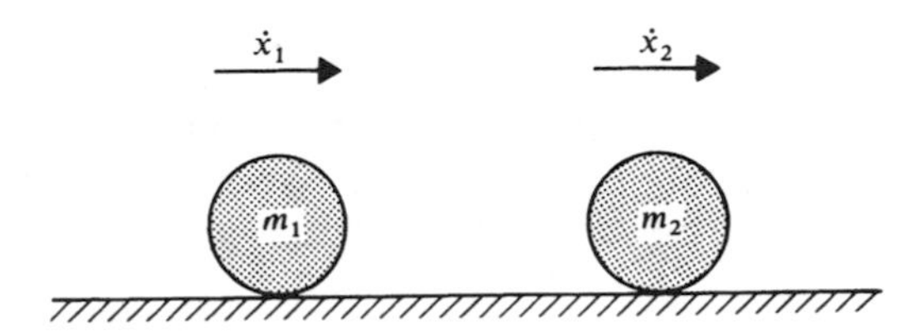

FIG. 3.29. Impact dynamics.

Since the accelerations $\ddot{x}_1 = d\dot{x}_1/dt$ and $\ddot{x}_2 = d\dot{x}_2/dt$, Eqs. 120 and 121 can be written as

$$m_1\, d\dot{x}_1 = [F_1(t) - F_i(t)]\, dt$$

$$m_2\, d\dot{x}_2 = [F_2(t) + F_i(t)]\, dt$$

Integrating these two equations over the short period of impact $\Delta t = t_2 - t_1$, one obtains

$$\int_{v_1}^{v_1'} m_1\, d\dot{x}_1 = \int_{t_1}^{t_2} [F_1(t) - F_i(t)]\, dt \tag{3.122}$$

$$\int_{v_2}^{v_2'} m_2\, d\dot{x}_2 = \int_{t_1}^{t_2} [F_2(t) + F_i(t)]\, dt \tag{3.123}$$

where v_1 and v_2 are the velocities of the two particles before collision, and v_1' and v_2' are the velocities of the two particles after collision.

During the short interval of impact, all the external forces are assumed to be small compared to the impact force $F_i(t)$. Using this assumption, Eqs. 122 and 123 reduce to

$$m_1(v_1' - v_1) = -\int_{t_1}^{t_2} F_i(t)\, dt \tag{3.124}$$

$$m_2(v_2' - v_2) = \int_{t_1}^{t_2} F_i(t)\, dt \tag{3.125}$$

which upon adding leads to the equation of *conservation of momentum*

$$m_1 v_1' + m_2 v_2' = m_1 v_1 + m_2 v_2 \tag{3.126}$$

This equation implies that the linear momentum of the system before impact is equal to the linear momentum of the system after impact.

Restitution Condition In order to be able to determine the velocities of the two masses m_1 and m_2 after impact another relationship between v_1' and v_2' must be obtained. In the impact analysis, the relationship between the relative velocities between the two bodies before and after collision is defined by the *coefficient of restitution*. The coefficient of restitution which can be determined experimentally accounts for the deformation and energy dissipation in the region of contact during the impact process. The coefficient of restitution e is defined using the kinematic relationship

$$v_2' - v_1' = e(v_1 - v_2) \tag{3.127}$$

Therefore, if the relative velocities between the two masses is measured before and after impact, the coefficient of restitution e can be determined for different materials using Eq. 127.

Special Cases In solving impact problems, the coefficient of restitution is frequently assumed to be constant for given materials and contact surface geometries. In reality, the coefficient of restitution depends on the properties of the materials as well as the velocities of the two colliding bodies. If the coefficient of restitution e is equal to one, the impact is said to be *perfectly elastic*, and in the case of perfectly elastic impact the total kinetic energy of the two masses as well as the total linear momentum is conserved. The conservation of the linear momentum is guaranteed by Eq. 126. In order to show the conservation of the total kinetic energy, Eq. 127 in the case of perfectly elastic impact ($e = 1$) can be written as

$$v_1 + v_1' = v_2 + v_2' \tag{3.128}$$

Also Eq. 126 can be written as

$$m_1(v_1 - v_1') = -m_2(v_2 - v_2') \tag{3.129}$$

Multiplying Eq. 128 and 129 one obtains

$$m_1(v_1 - v_1')(v_1 + v_1') = -m_2(v_2 - v_2')(v_2 + v_2')$$

which leads to

$$m_1 v_1^2 - m_1 v_1'^2 = -m_2 v_2^2 + m_2 v_2'^2$$

This equation can be written as

$$\tfrac{1}{2}m_1 v_1^2 + \tfrac{1}{2}m_2 v_2^2 = \tfrac{1}{2}m_1 v_1'^2 + \tfrac{1}{2}m_2 v_2'^2$$

which implies that the kinetic energy before impact is equal to the kinetic energy after the impact.

Another special case is the case of *perfectly plastic impact*, in which the coefficient of restitution is equal to zero, that is $e = 0$. In this case Eq. 127 yields

$$v_2' - v_1' = 0$$

and hence the velocities of the two masses after impact are equal. In this case the condition of conservation of momentum reduces to

$$m_1 v_1 + m_2 v_2 = (m_1 + m_2)v'$$

in which $v' = v_1' = v_2'$.

In general, given the velocities of the masses before collision and the coefficient of restitution, the conservation of momentum equation and the restitution condition can be used to determine the jumps in the velocities of the two masses after impact. These jumps can be used to update the velocities of the two masses, and the updated velocities can be used as initial conditions that define the constants of integration that appear in the complementary function of the solution of the vibration equation. The use of this procedure is demonstrated by the following example.

Example 3.13

Figure 30 shows two masses m_1 and m_2 where $m_1 = 5$ kg and $m_2 = 1$ kg. The mass m_1 which is initially at rest is supported by a spring and a damper. The stiffness

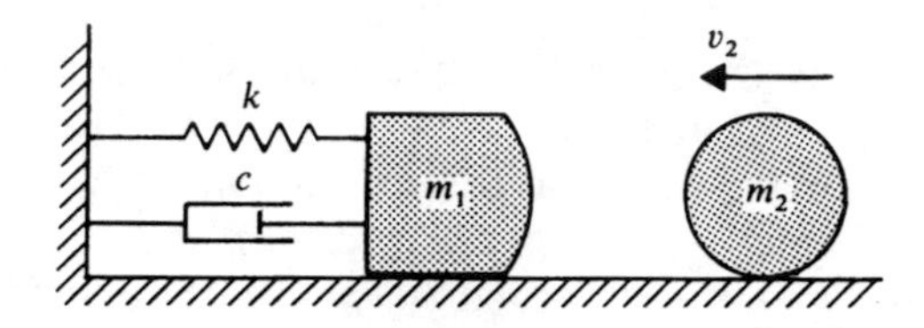

FIG. 3.30. Impact response of single degree of freedom systems.

coefficient of the spring k is assumed to be 1000 N/m, and the damping coefficient c is assumed to be 10 N·s/m. The mass m_2 is assumed to move with a constant velocity $v_2 = -5$ m/s. Assuming the coefficient of restitution $e = 0.9$, determine the displacement equations of the two masses after impact.

Solution. The conservation of momentum condition and the restitution conditions are

$$m_1 v_1' + m_2 v_2' = m_1 v_1 + m_2 v_2$$

$$v_2' - v_1' = e(v_1 - v_2)$$

where $v_1 = 0$ and $v_2 = -5$ m/s. Therefore, the conservation of momentum and restitution equations can be written as

$$5v_1' + v_2' = (5)(0) + (1)(-5)$$

$$v_2' - v_1' = 0.9\,(0 - (-5))$$

which yield

$$5v_1' + v_2' = -5$$

$$v_1' - v_2' = -4.5$$

By adding these two equations, one obtains

$$6v_1' = -9.5$$

or

$$v_1' = -1.5833 \text{ m/s}$$

and

$$v_2' = v_1' + 4.5 = -1.5833 + 4.5 = 2.9167 \text{ m/s}$$

The equations of motion of the two masses after impact can be written as

$$m_1\ddot{x}_1 + c\dot{x}_1 + kx_1 = 0$$

$$m_2\ddot{x}_2 = 0$$

The solutions of these two equations can be determined as

$$x_1 = Xe^{-\xi\omega t}\sin(\omega_d t + \phi)$$

$$x_2 = A_1 t + A_2$$

where X, ϕ, A_1, and A_2 are constants that can be determined using the initial conditions $x_{10} = 0$, $\dot{x}_{10} = v_1' = -1.5833$ m/s, $x_{20} = 0$, $\dot{x}_{20} = v_2' = 2.9167$ m/s. The damping factor ξ, natural frequency ω, and the damped natural frequency ω_d are

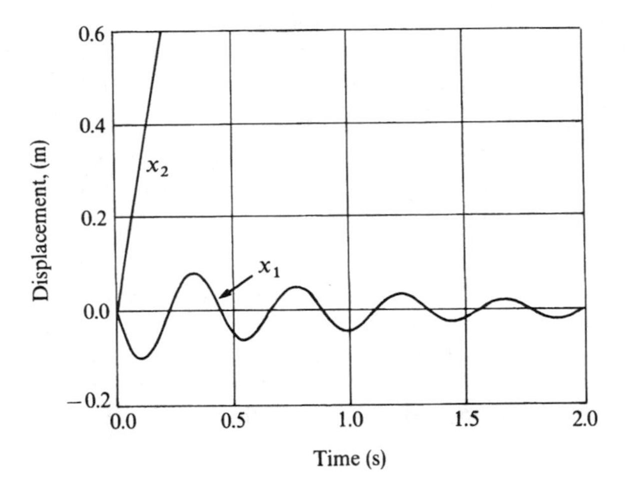

FIG. 3.31. Displacements of the two masses.

given by

$$\xi = \frac{c}{C_c} = \frac{c}{2\sqrt{km}} = \frac{10}{2\sqrt{(1000)(5)}} = 0.0707$$

$$\omega = \sqrt{\frac{k}{m}} = \sqrt{\frac{1000}{5}} = 14.142 \text{ rad/s}$$

$$\omega_d = \omega\sqrt{1 - \xi^2} = 14.1066 \text{ rad/s}$$

By using the initial conditions one can show that the displacements of the two masses as functions of time can be written as

$$x_1 = -0.1122e^{-0.9998t} \sin(14.1066t)$$

$$x_2 = 2.9167t$$

The results presented in Fig. 31, which shows x_1 and x_2 as functions of time, clearly demonstrate that the two masses do not encounter a second impact.

Problems

3.1. A single degree of freedom mass–spring system consists of a 10 kgs mass suspended by a linear spring which has a stiffness coefficient of 6×10^3 N/m. The mass is given an initial displacement of 0.04 m and it is released from rest. Determine the differential equation of motion, and the natural frequency of the system. Determine also the maximum velocity.

3.2. The oscillatory motion of an undamped single degree of freedom system is such that the mass has maximum acceleration of 50 m/s^2 and has natural frequency of 30 Hz. Determine the amplitude of vibration and the maximum velocity.

3.3. A single degree of freedom undamped mass–spring system is subjected to an impact loading which results in an initial velocity of 5 m/s. If the mass is equal to 10 kg and the spring stiffness is equal to 6×10^3 N/m, determine the system response as a function of time.

3.4. The undamped single degree of freedom system of Problem 1 is subjected to the initial conditions $x_0 = 0.02$ m and $\dot{x}_0 = 3$ m/s. Determine the system response as a function of time. Also determine the maximum velocity and the total energy of the system.

3.5. A single degree of freedom system consists of a mass m which is suspended by a linear spring of stiffness k. The static equilibrium deflection of the spring was found to be 0.02 m. Determine the system natural frequency. Also determine the response of the system as a function of time if the initial displacement is 0.03 m and the initial velocity is zero. What is the total energy of the system if the mass m is equal to 5 kg?

3.6. The system shown in Fig. P1 consists of a mass m and a massless rod of length l. The system is supported by two springs which have stiffness coefficients k_1 and k_2, as shown in the figure. Derive the system differential equation of motion assuming small oscillations. Determine the natural frequency of the system.

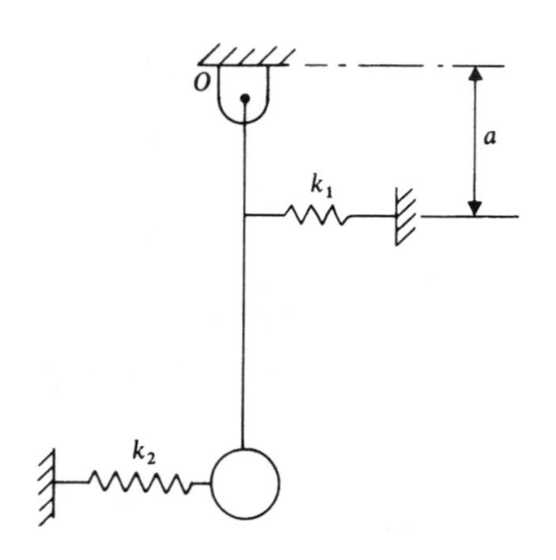

FIG. P3.1

3.7. If the two springs k_1 and k_2 in Problem 6 are to be replaced by an equivalent spring which is connected at the middle of the rod, determine the stiffness coefficient k_e of the new spring system.

3.8. If the system in Problem 6 is given an initial angular displacement θ_0 counterclockwise, determine the system response as a function of time assuming that the initial angular velocity is zero. Determine also the maximum angular velocity.

3.9. In the system shown in Fig. P2, $m = 5$ kg, $k_1 = k_5 = k_6 = 1000$ N/m, $k_3 = k_4 = 1500$ N/m, and $k_2 = 3000$ N/m. The motion of the mass is assumed to be in the

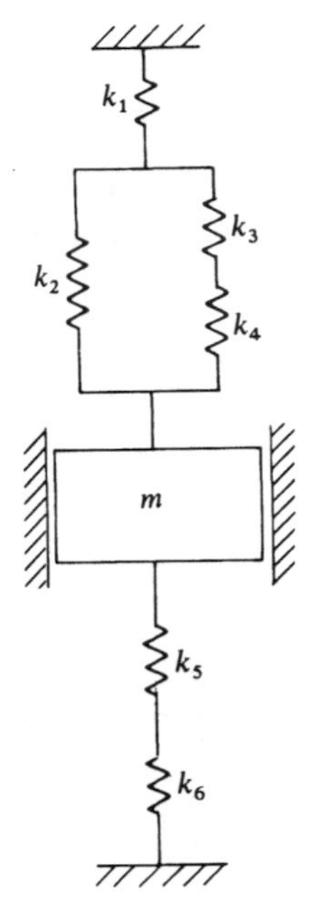

FIG. P3.2

vertical direction. If the mass is subjected to an impact such that the motion starts with an initial upward velocity of 5 m/s, determine the displacement, velocity, and acceleration of the mass after 2 s.

3.10. The system shown in Fig. P3 consists of a mass m and a massless rod of length l. The system is supported by a spring which has a stiffness coefficient k. Obtain the differential equation of motion and discuss the stability of the system. Determine the stiffness coefficient k at which the system becomes unstable.

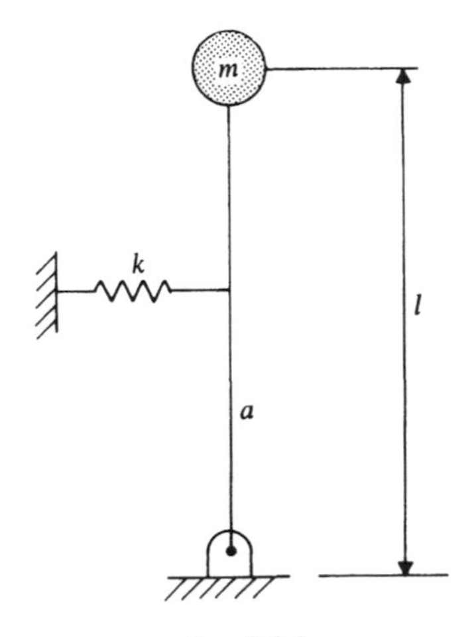

FIG. P3.3

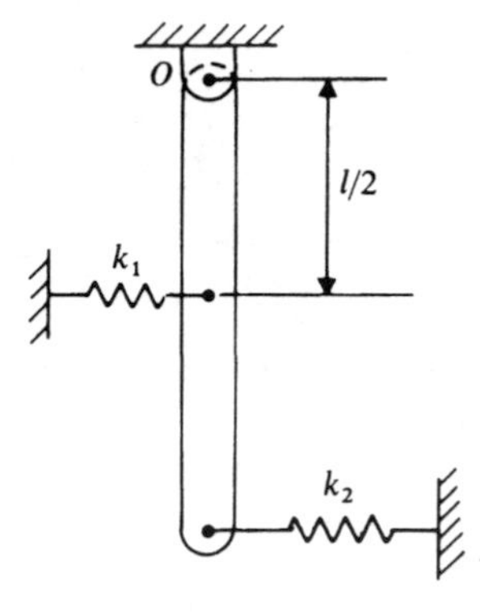

FIG. P3.4

3.11. The system shown in Fig. P4 consists of a uniform rod which has length l, mass m, and mass moment of inertia about its mass center I. The rod is supported by two springs which have stiffness coefficients k_1 and k_2, as shown in the figure. Determine the system differential equation of motion for small oscillations. Determine also the system natural frequency.

3.12. In the inverted pendulum shown in Fig. P5, the uniform rod has length l, mass m, and mass moment of inertia I about its mass center. The rod is supported by a spring which has a stiffness coefficient k. Determine the critical value of the spring coefficient at which the system becomes unstable.

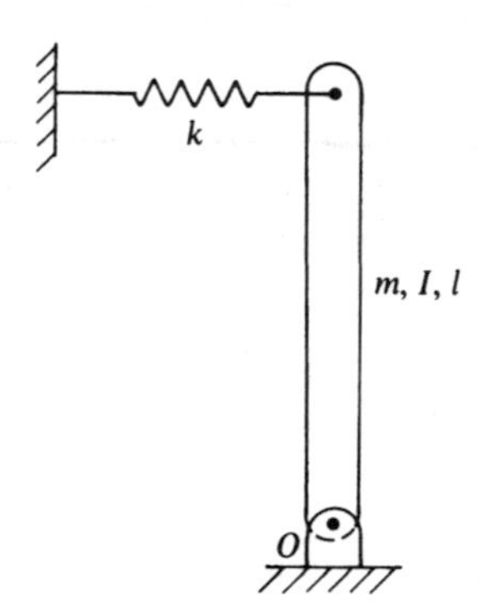

FIG. P3.5

3.13. The system shown in Fig. P6 consists of a mass m and a uniform circular rod of mass m_r, length l, and mass moment of inertia I about its mass center. The rod is connected to the ground by a spring which has a stiffness coefficient k. Assuming small oscillations, derive the system differential equation of motion and determine the natural frequency of the system. Determine the system response as a function of time.

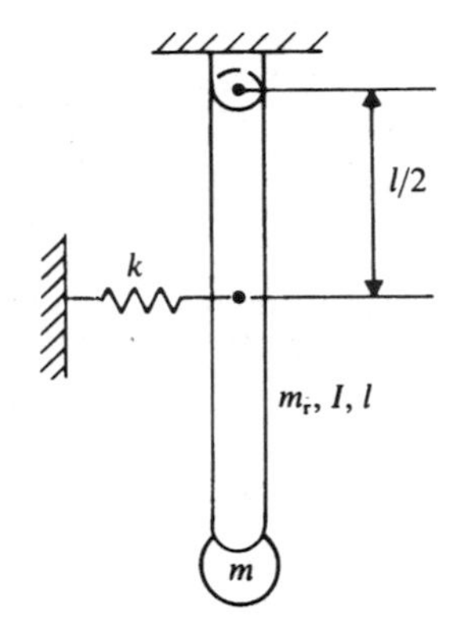

FIG. P3.6

3.14. If the shafts shown in Fig. P7 have modulus of rigidity G_1 and G_2, derive the differential equation of the system and determine the system natural frequency.

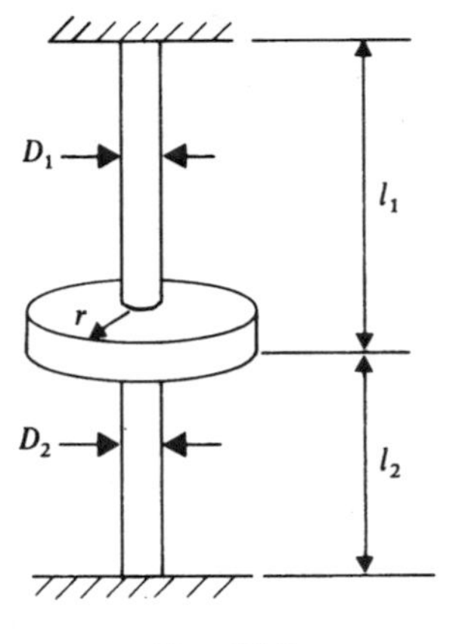

FIG. P3.7

3.15. Derive the differential equation of motion of the system shown in Fig. P8. Determine the natural frequency of the system.

3.16. Determine the natural frequency of the system shown in Fig. P9.

3.17. The uniform bar shown in Fig. P10 has mass m, length l, and mass moment of inertia I about its mass center. The bar is supported by two springs k_1 and k_2, as shown in the figure. Obtain the differential equation of motion and determine the natural frequency of the system for small oscillation.

3.18. The system shown in Fig. P11 consists of a uniform bar of length l, mass m, and mass moment of inertia I. The bar is supported by two linear springs which have stiffness coefficients k_1 and k_2. Derive the differential equation of this system assuming small oscillations. Determine also the natural frequency.

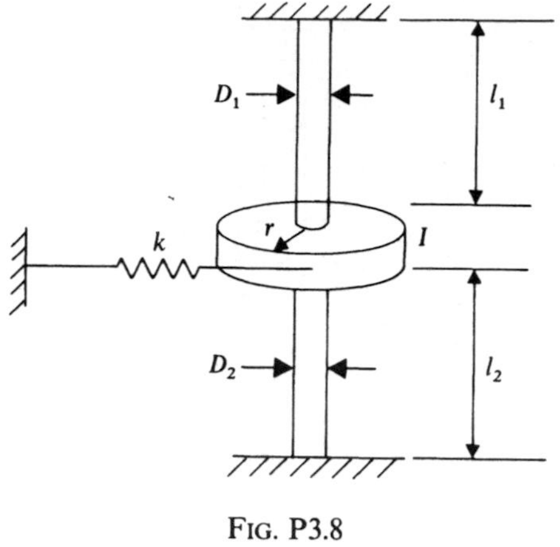

FIG. P3.8

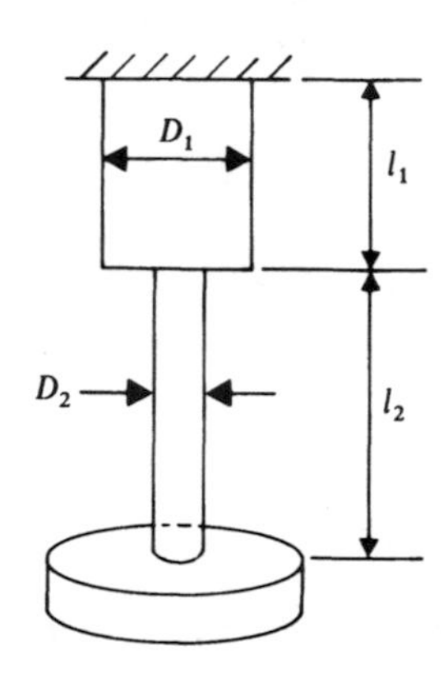

FIG. P3.9

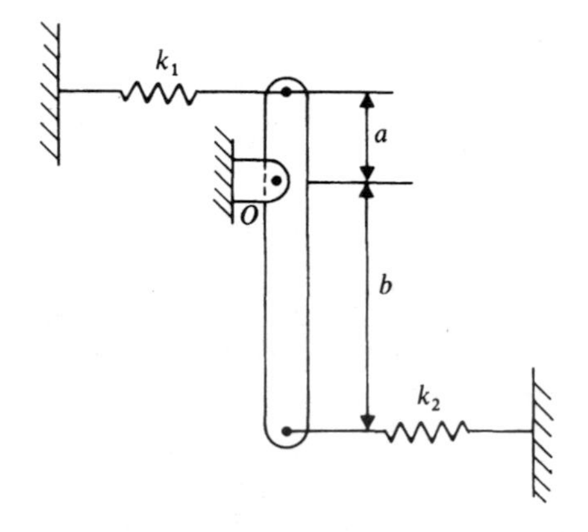

FIG. P3.10

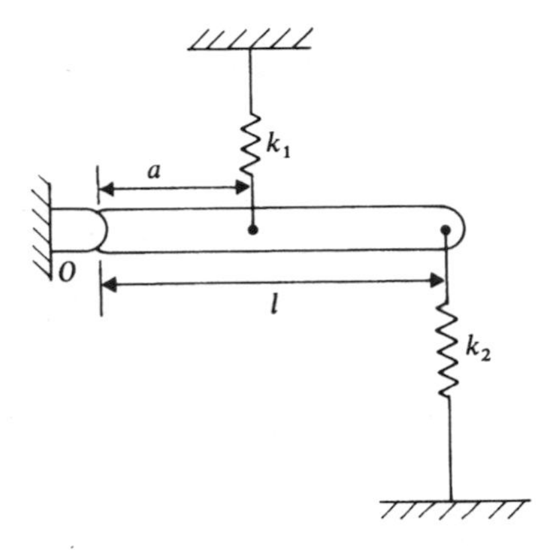

FIG. P3.11

3.19. In Problem 18, let $m = 1$ kg, $l = 0.5$ m, $a = 0.1$ m, $k_1 = 1000$ N/m, and $k_2 = 3000$ N/m. Determine the system response as a function of time if the bar is given an initial angular displacement of 3° counterclockwise.

3.20. The system shown in Fig. P12 consists of a uniform bar and a mass m_1 rigidly attached to one end of the bar. The bar is connected to the ground by a pin joint at O. The system is supported by two springs which have stiffness coefficients k_1 and k_2. The bar has length l, mass m, and mass moment of inertia I. Derive the system equation of motion and determine the natural frequency.

3.21. In Problem 20, let $m = 1$ kg, $l = 0.5$ m, $a = 0.2$ m, $b = 0.3$ m, $m_1 = 0.5$ kg, $k_1 = 2000$ N/m, and $k_2 = 5000$ N/m. Determine the system response as a function of time if the bar is given an initial angular displacement of 2° in the clockwise direction.

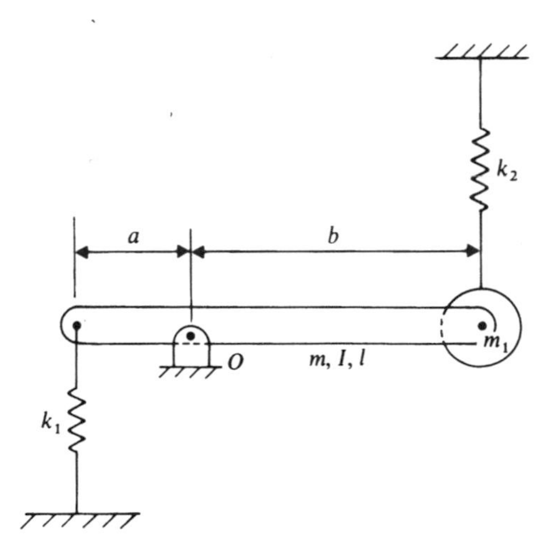

FIG. P3.12

3.22. A damped single degree of freedom mass–spring system has $m = 4$ kg, $k = 3000$ N/m, and $c = 300$ N·s/m. Determine the equation of motion of the system.

3.23. A damped single degree of freedom mass–spring system has $m = 0.5$ kg, $k = 1000$ N/m, and $c = 10$ N·s/m. The mass is set in motion with initial conditions $x_0 = 0.05$ m and $\dot{x}_0 = 0.5$ m/s. Determine the displacement, velocity, and acceleration of the mass after 0.3 s.

3.24. A viscously damped single degree of freedom mass–spring system has a mass m of 2 kg, a spring coefficient k of 2000 N/m, and a damping constant c of 5 N·s/m. Determine (a) the damping factor ξ, (b) the natural frequency ω, (c) the damped natural frequency ω_d, and (d) the spring coefficient needed to obtain critically damped system.

3.25. An overdamped single degree of freedom mass–spring system has a damping factor $\xi = 1.5$ and a natural frequency $\omega = 20$ rad/s. Determine the equation of motion of the displacement and plot the displacement and velocity versus time for the following initial conditions: (a) $x_0 = 0$, $\dot{x}_0 = 1$ m/s; (b) $x_0 = 0.05$ m, $\dot{x}_0 = 0$; and (c) $x_0 = 0.05$ m, $\dot{x}_0 = 1$ m/s.

3.26. Repeat Problem 25 if the system is critically damped.

3.27. A 0.5-kg mass attached to a linear spring elongates it 0.008 m. Determine the system natural frequency.

3.28. An unknown mass m is attached to the end of a linear spring with unknown stiffness coefficient k. The system has natural frequency of 30 rad/s. When a 0.5-kg mass is added to the unknown mass m, the natural frequency is lowered to 20 rad/s. Determine the mass m and the stiffness coefficient k.

3.29. Derive the differential equation of motion of the system shown in Fig. P13. Determine the natural frquency of the system.

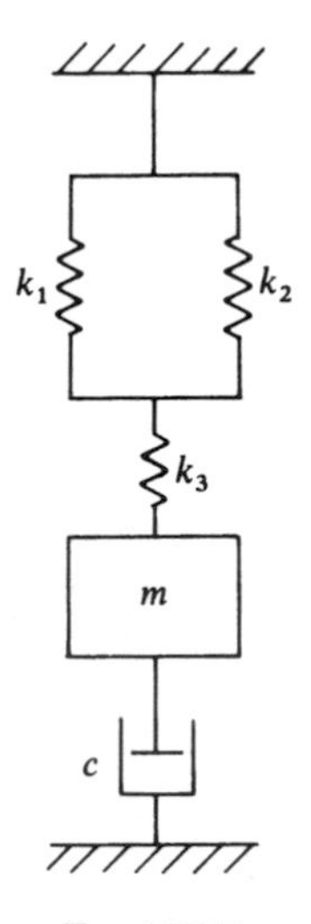

FIG. P3.13

3.30. In Problem 29, let $m = 0.5$ kg, $k_1 = k_2 = k_3 = 1000$ N/m, and $c = 10$ N·s/m. Determine the displacement and velocity of the mass after 0.2 s if the initial conditions are $x_0 = 0.05$ m and $\dot{x}_0 = 2$ m/s.

3.31. Derive the system differential equation of motion of the pendulum shown in Fig. P14 for small oscillations. Assume that the rod is massless.

3.32. For the system shown in Fig. P14, let $m = 0.5$ kg, $l = 0.5$ m, $a = 0.2$ m, $k = 1000$ N/m, and $c = 10$ N·s/m. If the system is given an initial angular displacement $\theta_0 = 4°$ in the clockwise direction, determine the angular displacement and angular velocity after 0.5 s.

3.33. For the system shown in Fig. P14, let $m = 0.5$ kg, $l = 0.5$ m, $a = 0.2$ m, $k = 1000$ N/m. Determine the damping coefficient c if the system is to be critically damped. If the system has an initial angular velocity of 5 rad/s countclockwise, determine the angular displacement and angular velocity after 0.3 s. Assume small oscillations.

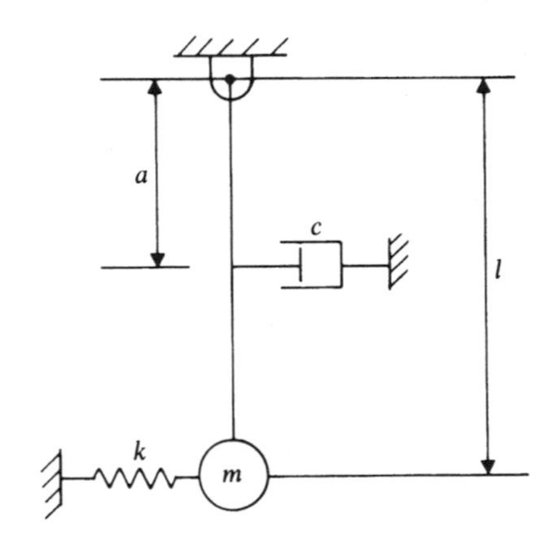

FIG. P3.14

3.34. Derive the differential equation of motion of the inverted pendulum shown in Fig. P15.

3.35. For the system shown in Fig. P15, let $m = 0.5$ kg, $l = 0.5$ m, $a = 0.2$ m, and $k = 3000$ N/m. Determine the damping coefficient c if: (a) the system is underdamped with $\xi = 0.09$, (b) the system is critically damped; and (c) the system is overdamped with $\xi = 1.2$.

3.36. In Problem 35, determine the angular displacement and velocity after 0.4 s if the system has zero initial velocity and initial displacement of 4° counterclockwise.

3.37. The system shown in Fig. P16 consists of a uniform bar of length l, mass m, and mass moment of inertia I. The bar is supported by a spring and damper which have stiffness and damping coefficients k and c, respectively. Derive the differential equation of motion and determine the system natural frequency and the critical damping coefficient.

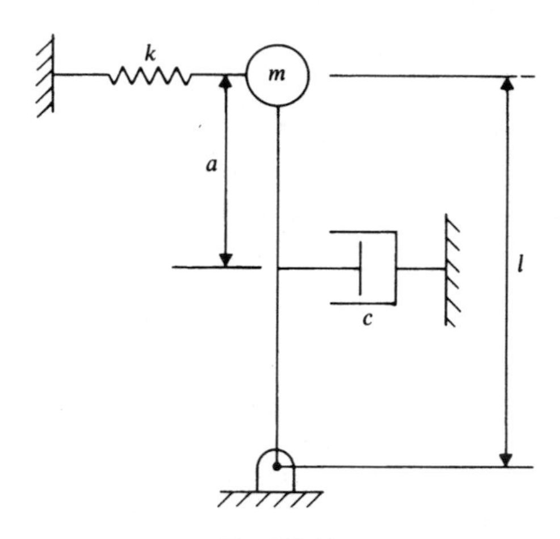

FIG. P3.15

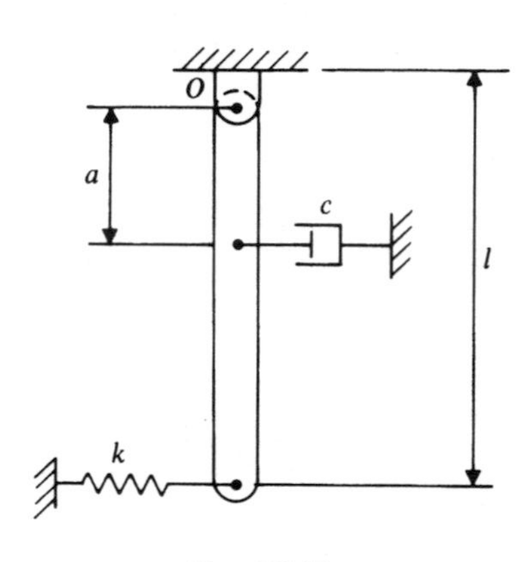

FIG. P3.16

3.38. Let the beam in Problem 37 be a slender beam with mass 1 kg and length 0.5 m. Let $a = 0.25$ m, $k = 3000$ N/m, and $c = 20$ N·s/m. Determine the angular displacement and the angular velocity of the beam after 2 s if the initial angular displacement is zero and the initial angular velocity is 3 rad/s.

3.39. Derive the differential equation of motion of the system shown in Fig. P17. Determine the natural frequency and the critical damping coefficient.

3.40. A uniform slender rod of mass m and length l is hinged at point O. The rod is attached to two springs at the top end and to a viscous damper at the other end, as shown in Fig. P18. If $k_1 = 1000$ N/m, $k_2 = 3000$ N/m, $m = 10$ kg, $l = 5$ m, and $c = 80$ N·s/m, find the complete solution provided that $\theta_0 = 0$ and $\dot{\theta}_0 = 3$ rad/s.

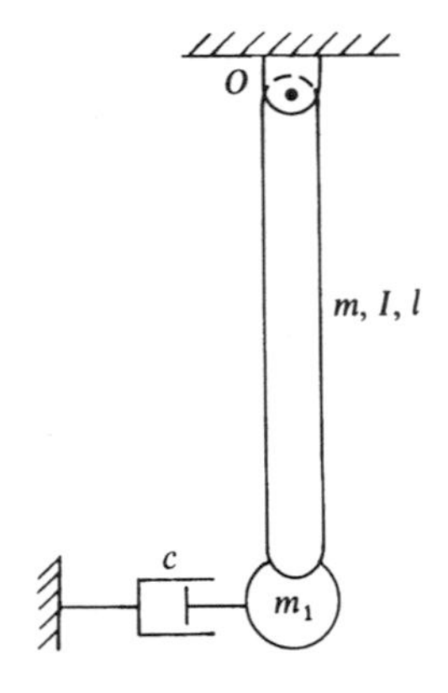

FIG. P3.17

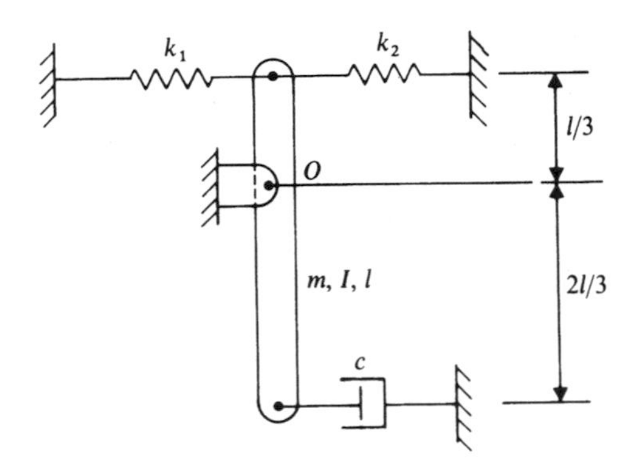

FIG. P3.18

3.41. A uniform slender rod of mass m and length l is hinged at point O, as shown in Fig. P19. The rod is attached to a spring and damper which have, respectively, a stiffness coefficient k and a damping coefficient c. If $m = 3$ kg, $l = 2$ m, $k = 4000$ N/m, and $c = 40$ N·s/m, determine the angular displacement and the velocity of the rod after 0.5 s, provided that the initial angular displacement is 5° and the initial angular velocity is zero.

3.42. A uniform slender rod of mass m and length l is hinged at point O, as shown in Fig. P20. The rod is attached to a spring and a damper at one end and to a damper at the other end, as shown in the figure. If $k = 4000$ N/m, $m = 10$ kg, $c_1 = c_2 = 20$ N·s/m and $l = 5$ m, find the complete solution provided that the rod has zero initial anglar displacement and an initial angular velocity of 5 rad/s.

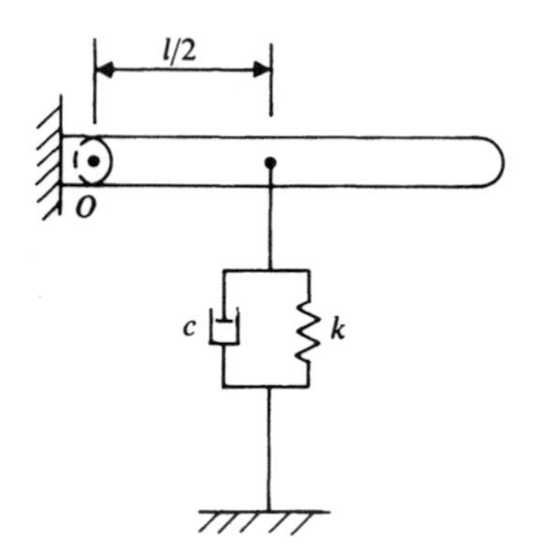

FIG. P3.19

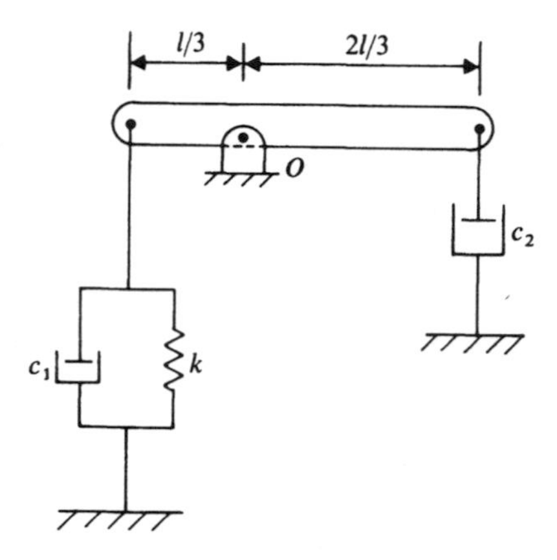

FIG. P3.20

3.43. For the system shown in Fig. P21, derive the differential equation of motion for small oscillation. If $m_1 = m_2 = 1$ kg, $k_1 = k_2 = 1000$ N/m, $c_1 = c_2 = 10$ N·s/m, $a = b = 0.5$ m, and $l = 1$ m, find the solution after 1 s provided that the initial angular displacement is zero and the initial angular velocity is 5 rad/s. Assume that the rod is massless.

3.44. In Problem 43, determine the relationship between the damping coefficients c_1 and c_2 such that the system is critically damped.

3.45. Determine the equivalent viscous damping coefficient for two viscous dampers which have damping coefficients c_1 and c_2 in the following two cases: (a) parallel connection; (b) series connection.

3.46. A damped single degree of freedom mass–spring system has a spring constant $k = 2000$ N/m. It was observed that the periodic time of free oscillations is 1.95 s and the ratio between successive amplitudes is 5.1 to 1.5. Determine the mass and the damping coefficient of the system.

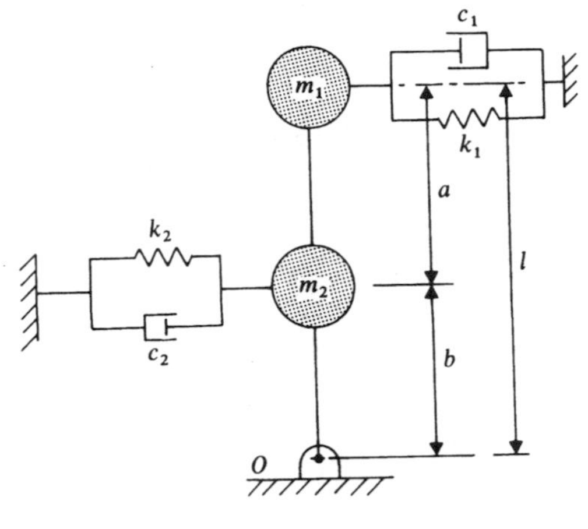

FIG. P3.21

3.47. It was observed that the damped free oscillations of a single degree of freedom system is such that the amplitude of the twelfth cycle is 48% that of the sixth cycle. Determine the damping factor ξ.

3.48. In Problem 47, if the mass of the system is 5 kg and the spring constant is 1000 N/m, determine the damping coefficient and the damped natural frequency.

3.49. Show that, if the structural damping coefficient c_s is very small, the logarithmic decrement can be defined as

$$\delta \simeq \pi c_s$$

3.50. A simple structure is found to vibrate as a single degree of freedom system. The spring constant determined using static testing is found to be 2500 N/m. The mass of the structure is 3 kg. By using a simple vibration test, the ratio of successive amplitudes is found to be 1.1. Determine the structural damping coefficient and the equivalent viscous damping coefficient. Determine also the energy loss per cycle for an amplitude of 0.08 m.

3.51. A single degree of freedom mass–spring system is such that an amplitude loss of 2% occurs in every three full cycles of oscillation. Obtain the damping factor of this system.

3.52. Develop the differential equation of motion of a single degree of freedom mass–spring system which contains both viscous damping and Coulomb damping.

3.53. A single degree of freedom mass–spring system has a mass $m = 9$ kg, a spring stiffness coefficient $k = 8 \times 10^3$ N/m, a coefficient of dry friction $\mu = 0.15$, an initial displacement $x_0 = 0.02$ m, and an initial velocity $\dot{x}_0 = 0$. Determine the number of cycles of oscillations of the mass before it comes to rest.

3.54. The following data are given for a damped single degree of freedom mass–spring system: mass $m = 6$ kg, damping coefficient $c = 25$ N·s/m, and stiffness coefficient $k = 5.5 \times 10^3$ N/m. The mass is subjected to a velocity-dependent force. The mass has zero initial displacement and initial velocity of 1 m/s. Determine the displacement and velocity of the mass after 0.1 s in the following two cases of the forcing function F

$$\text{(a)}\ F = 50\dot{x}\ \text{N}; \qquad \text{(b)}\ F = 400\dot{x}\ \text{N}.$$

3.55. A single degree of freedom mass–spring system has the following parameters, $m = 0.5$ kg, $k = 2 \times 10^3$ N/m, coefficient of dry friction $\mu = 0.15$, initial displacement $x_0 = 0.1$ m, and initial velocity $\dot{x}_0 = 0$. Determine:
(1) the decrease in amplitude per cycle;
(2) the number of half-cycles before the mass comes to rest;
(3) the displacement of the mass at time $t = 0.1$ s;
(4) the location of the mass when oscillation stops.

4
Forced Vibration

In the preceding chapter, the free undamped and damped vibration of single degree of freedom systems was discussed, and it was shown that the motion of such systems is governed by homogeneous second-order ordinary differential equations. The roots of the characteristic equations, as well as the solutions of the differential equations, strongly depend on the magnitude of the damping, and oscillatory motions are observed only in underdamped systems. In this chapter, we study the undamped and damped motion of single degree of freedom systems subjected to forcing functions which are time-dependent. Our discussion in this chapter will be limited only to the case of harmonic forcing functions. The response of the single degree of freedom system to periodic forcing functions, as well as to general forcing functions, will be discussed in the following chapter.

4.1 DIFFERENTIAL EQUATION OF MOTION

Figure 1 depicts a viscously damped single degree of freedom mass–spring system subjected to a forcing function $F(t)$. By applying Newton's second law, the differential equation of motion can be written as

$$m\ddot{x} + c\dot{x} + kx = F(t) \tag{4.1}$$

where m is the mass, c is the damping coefficient, k is the stiffness coefficient, and x is the displacement of the mass. Equation 1 is a nonhomogeneous second-order ordinary differential equation with constant coefficients. As discussed in Chapter 2, the solution of this equation consists of two parts; the complementary function x_h and the particular integral x_p, that is,

$$x = x_h + x_p \tag{4.2}$$

where the complementary function x_h is the solution of the homogeneous equation

$$m\ddot{x}_h + c\dot{x}_h + kx_h = 0 \tag{4.3}$$

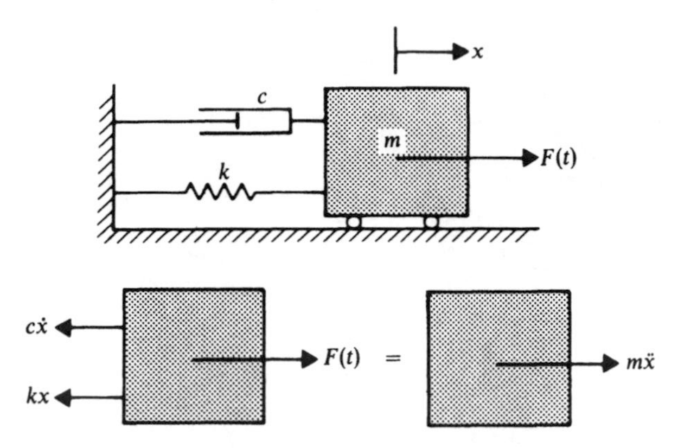

FIG. 4.1. Forced vibration of single degree of freedom systems.

The particular solution x_p represents the response of the system to the forcing function. The complementary function x_h is sometimes called the *transient solution* since in the presence of damping this solution dies out. Methods for obtaining the transient response were discussed in the preceding chapter. The particular integral x_p is sometimes called the *steady state solution*, since this solution exists long after the transient vibration disappears. Methods for obtaining the steady state response for both undamped and damped systems are discussed in the following sections. The transient solution contains two arbitrary constants, while the steady state solution does not contain any arbitrary constants, and therefore, the solution of Eq. 2 contains two arbitrary constants which can be determined by using the initial conditions.

4.2 FORCED UNDAMPED VIBRATION

In this section, we consider the forced undamped vibration of single degree of freedom systems. For undamped systems, the damping coefficient is identically zero and Eq. 1 reduces to

$$m\ddot{x} + kx = F(t) \tag{4.4}$$

In the analysis presented in this section, we consider harmonic excitation in which the forcing function $F(t)$ is given by

$$F(t) = F_0 \sin \omega_f t \tag{4.5}$$

where F_0 is the amplitude of the forcing function and ω_f is the force frequency. The harmonic forcing function $F(t)$, shown in Fig. 2, has a periodic time denoted as τ_f and is defined as

$$\tau_f = \frac{2\pi}{\omega_f} \tag{4.6}$$

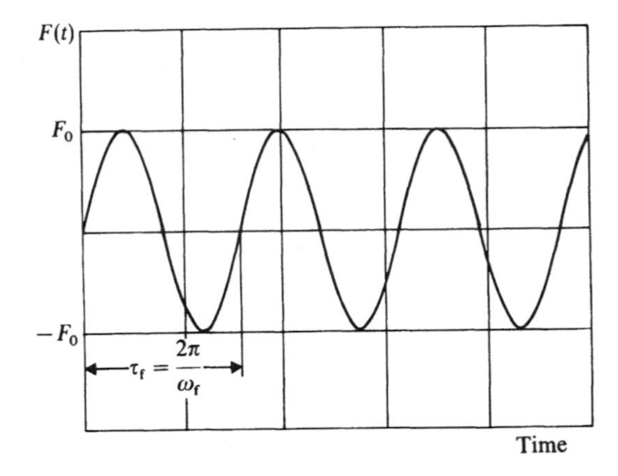

FIG. 4.2. Harmonic forcing function.

Combining Eqs. 4 and 5 yields the differential equation of the undamped single degree of freedom system under harmonic excitation as

$$m\ddot{x} + kx = F_0 \sin \omega_f t \tag{4.7}$$

It was shown in the preceding chapter that the complementary function x_h of this undamped system is given by

$$x_h = X \sin(\omega t + \phi) \tag{4.8}$$

where $\omega = \sqrt{k/m}$ is the natural frequency and X and ϕ are constants which can be determined using the complete solution and the initial conditions.

Steady State Response As discussed in Chapter 2, the steady state solution of Eq. 7 can be obtained by assuming a particular integral x_p in the form

$$x_p = A_1 \sin \omega_f t + A_2 \cos \omega_f t \tag{4.9}$$

The acceleration $\ddot{x}_p$ can be obtained by differentiating this equation twice with respect to time to yield

$$\ddot{x}_p = -\omega_f^2 x_p = -\omega_f^2(A_1 \sin \omega_f t + A_2 \cos \omega_f t) \tag{4.10}$$

Substituting Eqs. 9 and 10 into Eq. 7 yields

$$-m\omega_f^2(A_1 \sin \omega_f t + A_2 \cos \omega_f t) + k(A_1 \sin \omega_f t + A_2 \cos \omega_f t) = F_0 \sin \omega_f t$$

which can be written as

$$(k - m\omega_f^2)\, A_1 \sin \omega_f t + (k - m\omega_f^2)\, A_2 \cos \omega_f t = F_0 \sin \omega_f t$$

Equating the coefficients of the sine and cosine functions in both sides of this equation yields

$$A_1 = \frac{F_0}{k - m\omega_f^2}$$

$$A_2 = 0$$

provided that $k \neq m\omega_f^2$.

Therefore, the solution x_p of Eq. 9 can be written as

$$x_p = \frac{F_0}{k - m\omega_f^2} \sin \omega_f t \tag{4.11}$$

or

$$x_p = \frac{X_0}{1 - r^2} \sin \omega_f t \tag{4.12}$$

where X_0 and r are defined as

$$X_0 = \frac{F_0}{k} \tag{4.13}$$

$$r = \frac{\omega_f}{\omega} \tag{4.14}$$

The steady state response x_p of Eq. 11 can be written as

$$x_p = X_0 \beta \sin \omega_f t \tag{4.15}$$

where β, which is called the *magnification factor*, depends on the frequency

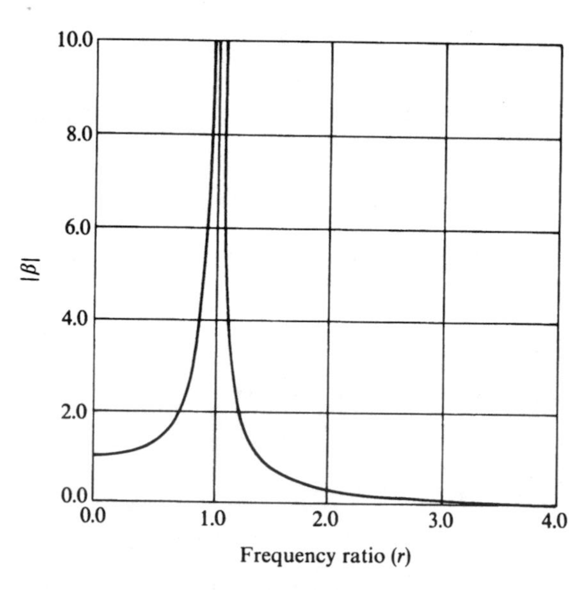

FIG. 4.3. Resonance curve.

ratio r and is defined as

$$\beta = \frac{1}{1 - r^2} \tag{4.16}$$

The magnification factor β approaches infinity when r approaches one, that is, when $\omega_f = \omega$. This is the case of *resonance* in which the frequency of the forcing function coincides with the system natural frequency. Note also that, if r is equal to zero, β is equal to one and if r approaches infinity, β approaches zero. The absolute value of the magnification factor β as a function of the frequency ratio r is shown in Fig. 3.

The steady state response x_p of Eq. 15 is a harmonic function which has the same frequency ω_f as the forcing function. Furthermore, there is no phase angle between the steady state response x_p and the harmonic forcing function $F(t)$.

Complete Solution Using Eqs. 2, 8, and 12 or 15, the complete response of the undamped single degree of freedom system under harmonic excitation can be written as

$$\begin{aligned} x(t) &= x_h + x_p \\ &= X \sin(\omega t + \phi) + X_0 \beta \sin \omega_f t \end{aligned} \tag{4.17}$$

This equation has two arbitrary constants X and ϕ which can be determined using the initial conditions.

Differentiating Eq. 17 with respect to time yields

$$\dot{x} = \omega X \cos(\omega t + \phi) + \omega_f X_0 \beta \cos \omega_f t \tag{4.18}$$

Let x_0 and $\dot{x}_0$ denote, respectively, the initial displacement and velocity. Using Eqs. 17 and 18, one obtains

$$x_0 = X \sin \phi \tag{4.19}$$

$$\dot{x}_0 = \omega X \cos \phi + \omega_f X_0 \beta \tag{4.20}$$

These are two algebraic equations which can be solved for the constants X and ϕ.

Equation 17 can also be written in an alternative form as

$$x(t) = A_1 \cos \omega t + A_2 \sin \omega t + X_0 \beta \sin \omega_f t \tag{4.21}$$

where A_1 and A_2 are constants which can be determined from the initial conditions. In terms of the initial displacement x_0 and the initial velocity $\dot{x}_0$, one can show that the constants A_1 and A_2 are given by

$$A_1 = x_0 \tag{4.22}$$

$$A_2 = \frac{\dot{x}_0}{\omega} - r X_0 \beta \tag{4.23}$$

which, upon substitution into Eq. 21, yields

$$x(t) = x_0 \cos \omega t + \left(\frac{\dot{x}_0}{\omega} - r X_0 \beta\right) \sin \omega t + X_0 \beta \sin \omega_f t \tag{4.24}$$

Example 4.1

For the undamped single degree of freedom mass–spring system shown in Fig. 1, let $m = 5$ kg, $k = 2500$ N/m, $F_0 = 20$ N, and the force frequency $\omega_f = 18$ rad/s. The mass has the initial conditions $x_0 = 0.01$ m and $\dot{x}_0 = 1$ m/s. Determine the displacement of the mass at $t = 1$ s.

Solution. The circular frequency of the system is

$$\omega = \sqrt{\frac{k}{m}} = \sqrt{\frac{2500}{5}} = 22.361 \text{ rad/s}$$

The frequency ratio r is

$$r = \frac{\omega_f}{\omega} = \frac{18}{22.361} = 0.80499$$

The steady state solution is

$$x_p = X_0 \beta \sin \omega_f t$$

where

$$X_0 = \frac{F_0}{k} = \frac{20}{2500} = 0.008 \text{ m}$$

$$\beta = \frac{1}{1 - r^2} = \frac{1}{1 - (0.80499)^2} = 2.841$$

The complete solution, which is the sum of the complementary function and the particular integral, is

$$x = X \sin(\omega t + \phi) + X_0 \beta \sin \omega_f t$$

Using the initial conditions

$$x_0 = 0.01 = X \sin \phi$$

$$\dot{x}_0 = 1 = \omega X \cos \phi + \omega_f X_0 \beta = 22.361 X \cos \phi + (18)(0.008)(2.841)$$

or

$$X \sin \phi = 0.01$$

$$X \cos \phi = 0.0264$$

from which

$$\tan \phi = 0.3784 \qquad \text{or} \qquad \phi = 20.728°$$

and the amplitude X is

$$X = 0.02825$$

Therefore, the displacement x is

$$x = 0.02825 \sin(22.361t + 20.728°) + 0.022728 \sin 18t$$

The displacement at $t = 1$ s is

$$x(t = 1) = 0.02825 \sin(22.361 + 0.3617) + 0.022728 \sin 18$$

$$= -0.0359 \text{ m}$$

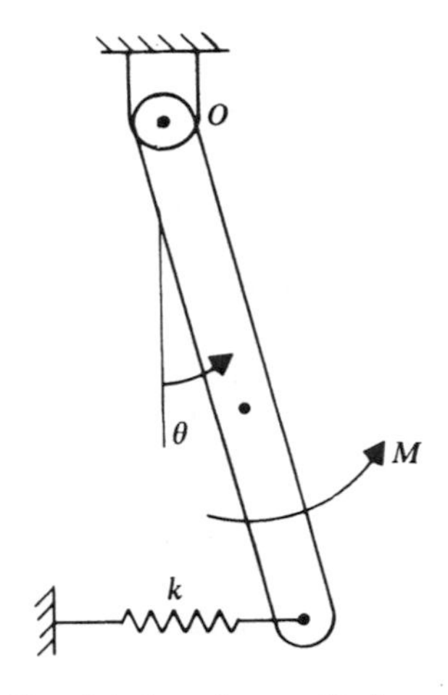

FIG. 4.4. Angular oscillations.

Example 4.2

The uniform slender bar shown in Fig. 4, which has length l and mass m, is pinned at point O and connected to a linear spring which has a stiffness coefficient k. If the bar is subjected to the moment $M = M_0 \sin \omega_f t$, obtain the steady state response of this system.

Solution. Assuming small oscillations, one can show that the linear differential equation of motion is

$$I_O \ddot{\theta} + (kl + \tfrac{1}{2}mg)\, l\theta = M_0 \sin \omega_f t \tag{4.25}$$

where I_O is the mass moment of inertia about point O and g is the gravitational constant. For a uniform slender bar, the mass moment of inertia I_O is defined as

$$I_O = \frac{ml^2}{3}$$

Equation 25 can be written in a form similar to Eq. 4 as

$$m_e \ddot{\theta} + k_e \theta = F_e \sin \omega_f t$$

where

$$m_e = I_O = \frac{ml^2}{3}$$

$$k_e = (kl + \tfrac{1}{2}mg)\, l$$

$$F_e = M_0$$

The circular frequency ω is

$$\omega = \sqrt{\frac{k_e}{m_e}} = \sqrt{\frac{(kl + \frac{1}{2}mg)\,l}{I_O}}$$

The steady state response is

$$\theta = \Theta_0 \beta \sin \omega_f t$$

where

$$\Theta_0 = \frac{F_e}{k_e} = \frac{M_0}{(kl + \frac{1}{2}mg)\,l}$$

$$\beta = \frac{1}{1 - r^2}$$

where the frequency ratio r is defined as

$$r = \frac{\omega_f}{\omega}$$

4.3 RESONANCE AND BEATING

In this section, the behavior of the undamped single degree of freedom system in the resonance region is examined. The resonance case in which the frequency ω_f of the forcing function is equal to the system natural frequency ($r = 1$) is discussed first. If the frequency ratio r is near, but not equal to one, another interesting phenomenon, known as *beating*, occurs. The beating phenomenon will be also discussed in this section.

Resonance In the case of resonance, the forcing frequency ω_f is equal to the circular frequency ω, and in this case $r = 1$. Even though Eq. 12 shows that in the resonance case the displacement goes to infinitity for any value of time t, such a behavior is not physically acceptable since the amplitude takes times to grow. Mathematically, the resonance case occurs when the forcing frequency is equal to one of the imaginary parts of the roots of the characteristic equation, and in the theory of differential equations this case is treated as a special case. Equation 12, therefore, does not represent the steady state solution at resonance since an infinite displacement cannot be attained instantaneously. Since

$$\omega_f = r\omega,$$

Eq. 12 can be written as

$$x_p(t) = \frac{X_0 \sin r\omega t}{1 - r^2} \tag{4.26}$$

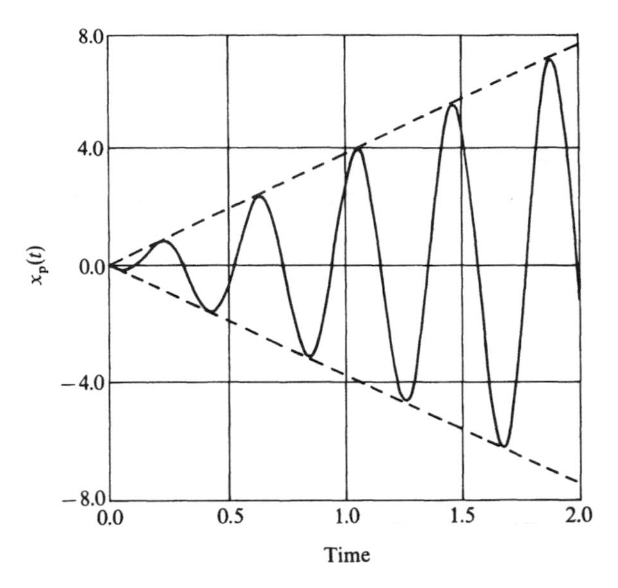

FIG. 4.5. Resonance case.

In order to obtain the correct form of the solution, we multiply the assumed particular solution by t and follow the procedure described in Chapter 2. This leads to

$$x_p(t) = -\frac{\omega t X_0}{2} \cos \omega t \qquad (4.27)$$

which indicates that the steady state solution at resonance is the product of a harmonic function with a function that depends linearly on time, and as such the displacement is oscillatory with an amplitude that increases with time. Eventually, the system attains infinite displacement, but not instantaneously, as shown in Fig. 5.

Whenever the forcing frequency coincides with the natural frequency of the system, excessive vibrations occur. An example of the devastating effect of the phenomenon of resonance is the Tacoma bridge which opened in July 1940 and collapsed in November 1940 as the result of wind induced vibration. Because of such damages, which can be caused as the result of the phenomenon of resonance, the determination of the natural frequency of the system through vibration testing has become an integral part in the design of most mechanical and structural systems.

At resonance, one can write the complete solution as

$$x(t) = X \sin(\omega t + \phi) - \frac{\omega t X_0}{2} \cos \omega t \qquad (4.28)$$

or, alternatively, as

$$x(t) = A_1 \cos \omega t + A_2 \sin \omega t - \frac{\omega t X_0}{2} \cos \omega t \tag{4.29}$$

where the constants X and ϕ or A_1 and A_2 can be determined from the initial conditions. For an initial displacement x_0 and an initial velocity $\dot{x}_0$, one can verify that

$$X = \sqrt{x_0^2 + \left(\frac{\dot{x}_0}{\omega} + \frac{X_0}{2}\right)^2} \tag{4.30}$$

$$\phi = \tan^{-1} \frac{x_0}{\left(\frac{\dot{x}_0}{\omega} + \frac{X_0}{2}\right)} \tag{4.31}$$

and

$$A_1 = x_0 \tag{4.32}$$

$$A_2 = \frac{\dot{x}_0}{\omega} + \frac{X_0}{2} \tag{4.33}$$

Beating The phenomenon of beating occurs when the forcing frequency ω_f is close, but not equal, to the system circular frequency ω. In order to understand the beating phenomenon, we consider the complete solution of Eq. 24

$$x(t) = x_0 \cos \omega t + \left(\frac{\dot{x}_0}{\omega} - rX_0\beta\right) \sin \omega t + X_0\beta \sin \omega_f t$$

If the initial conditions x_0 and $\dot{x}_0$ are zeros, the above equation yields

$$x(t) = X_0\beta(\sin \omega_f t - r \sin \omega t) \tag{4.34}$$

Using the definition of the magnification factor β given by Eq. 16, one has

$$\beta = \frac{\omega^2}{\omega^2 - \omega_f^2} = \frac{\omega^2}{(\omega + \omega_f)(\omega - \omega_f)} \tag{4.35}$$

Let

$$\alpha = \frac{\omega - \omega_f}{2} \tag{4.36}$$

Clearly, α is a very small number since ω_f is near to ω. It also follows that

$$\omega + \omega_f \approx 2\omega \tag{4.37}$$

Using Eqs. 36 and 37, the magnification factor β can be written as

$$\beta = \frac{\omega}{4\alpha} \tag{4.38}$$

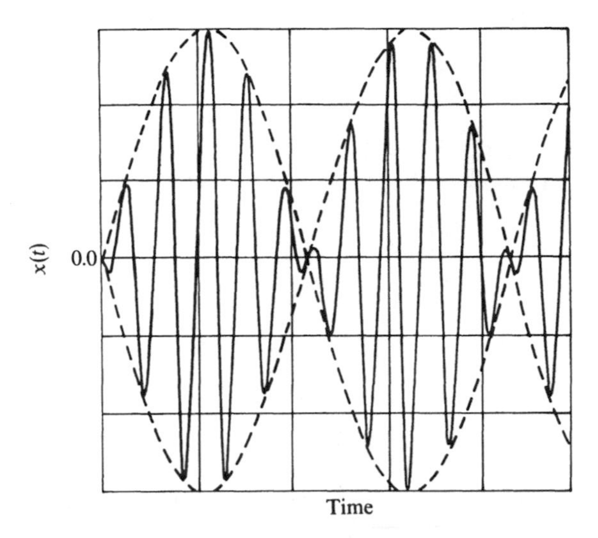

FIG. 4.6. Beating phenomenon.

Therefore, Eq. 34 can be written as

$$x(t) = \frac{X_0\omega}{4\alpha}(\sin \omega_f t - \sin \omega t) \tag{4.39}$$

If ω and ω_f are nearly equal, Eq. 39 leads to

$$x(t) = -\left(\frac{X_0\omega}{2\alpha} \cos \omega_f t\right) \sin \alpha t \tag{4.40}$$

The harmonic function $\cos \omega_f t$ has a period $2\pi/\omega_f$, while the harmonic function $\sin \alpha t$ has a period $2\pi/\alpha$. Since α is a very small number, the harmonic function in the parentheses varies more rapidly than the harmonic function $\sin \alpha t$. The result, in this case, is the function $x(t)$ shown in Fig. 6, which has an amplitude that builds up and then diminishes in a certain regular pattern. The phenomenon of beating can be observed in many cases, such as in the cases of audio or sound vibration, and beats can also be heard in the case of electric power houses when a generator is started (Den Hartog, 1956).

Work per Cycle It is clear from Eq. 12 that, for a given frequency ratio r, the amplitude of the forced vibration is constant, except in the resonance region. In order to understnad the physical reason for that, the energy input to the system, as the result of the application of the harmonic force, may be evaluated. To this end, we define the work of the force as

$$dW_e = F(t)\, dx$$

Since $dx = \dot{x}\, dt$, one has

$$dW_e = F(t)\, \dot{x}\, dt$$

Therefore, the work of the harmonic factor $F(t)$ per cycle can be written as

$$\begin{aligned} W_e &= \int_0^{2\pi/\omega_f} F(t)\, \dot{x}\, dt \\ &= \int_0^{2\pi/\omega_f} F_0 \sin \omega_f t(\omega_f X_0 \beta \cos \omega_f t)\, dt \\ &= F_0 X_0 \beta \int_0^{2\pi} \sin \omega_f t \cos \omega_f t\, d(\omega_f t) \\ &= \tfrac{1}{2} F_0 X_0 \beta \int_0^{2\pi} \sin 2\omega_f t\, d(\omega_f t) = \tfrac{1}{4} F_0 X_0 \beta \cos 2\omega_f t \Big|_0^{2\pi} \\ &= 0 \end{aligned}$$

That is, the work done by the external harmonic force per cycle is equal to zero and, accordingly, the amplitude of the steady state vibration remains constant for all values of the frequency ratio r, except at resonance where the steady state solution is defined by Eq. 27. In the case of resonance, one can show, by using Eq. 27, that the work done by the harmonic force $F(t)$ is not equal to zero, and as a consequence, the amplitude of the steady state vibration does not remain constant.

Example 4.3

For the single degree of freedom mass–spring system shown in Fig. 1, $m = 10$ kg, $k = 4000$ N/m, $F_0 = 40$ N, and $\omega_f = 20$ rad/s. The initial conditions are $x_0 = 0.02$ m and $\dot{x}_0 = 0$. Determine the displacement of the mass after $t = 0.5$ s and $t = 1$ s.

Solution. The circular frequency of the system is

$$\omega = \sqrt{\frac{k}{m}} = \sqrt{\frac{4000}{10}} = 20 \text{ rad/s}$$

In this case, $\omega = \omega_f$ and the system is at resonance. The complete solution is given by Eq. 28 as

$$x(t) = X \sin(\omega t + \phi) - \frac{\omega t X_0}{2} \cos \omega t$$

where

$$X_0 = \frac{F_0}{k} = \frac{40}{4000} = 0.01 \text{ m}$$

$$X = \sqrt{x_0^2 + \left(\frac{\dot{x}_0}{\omega} + \frac{X_0}{2}\right)^2} = \sqrt{(0.02)^2 + \left(\frac{0.01}{2}\right)^2} = 0.0206$$

$$\phi = \tan^{-1}\left(\frac{x_0}{\left(\frac{\dot{x}_0}{\omega} + \frac{X_0}{2}\right)}\right) = \tan^{-1}\left(\frac{0.02}{0 + \frac{0.01}{2}}\right) = \tan^{-1} 4 = 1.3258 \text{ rad}$$

The displacement $x(t)$ can then be written as

$$x(t) = 0.0206 \sin(20t + 1.3258) - 0.1t \cos 20t$$

at $t = 0.5$ s

$$\begin{aligned} x(t = 0.5) &= 0.0206 \sin(11.3258) - 0.05 \cos 10 \\ &= -0.01949 - (-0.04195) = 0.02246 \text{ m} \end{aligned}$$

at $t = 1$ s

$$\begin{aligned} x(t = 1) &= 0.0206 \sin(21.3258) - (0.1)(1) \cos 20 \\ &= 0.01272 - 0.04081 = -0.02809 \text{ m} \end{aligned}$$

4.4 FORCED VIBRATION OF DAMPED SYSTEMS

In this section, we study the effect of damping on the oscillatory motion of single degree of freedom systems. The differential equation of motion of such a system was shown to be (see Eq. 1)

$$m\ddot{x} + c\dot{x} + kx = F(t) \tag{4.41}$$

We consider the case of harmonic excitation in which the forcing function $F(t)$ can be expressed in the form

$$F(t) = F_0 \sin \omega_f t \tag{4.42}$$

Substituting this equation into Eq. 41 yields

$$m\ddot{x} + c\dot{x} + kx = F_0 \sin \omega_f t \tag{4.43}$$

The steady state solution x_p can be assumed in the form

$$x_p = A_1 \sin \omega_f t + A_2 \cos \omega_f t \tag{4.44}$$

which yields the following expressions for the velocity and acceleration

$$\dot{x}_p = \omega_f A_1 \cos \omega_f t - \omega_f A_2 \sin \omega_f t \tag{4.45}$$

$$\ddot{x}_p = -\omega_f^2 A_1 \sin \omega_f t - \omega_f^2 A_2 \cos \omega_f t = -\omega_f^2 x_p \tag{4.46}$$

Substituting Eqs. 44–46 into Eq. 43 and rearranging terms yield

$$\begin{aligned} &(k - \omega_f^2 m)(A_1 \sin \omega_f t + A_2 \cos \omega_f t) + c\omega_f(A_1 \cos \omega_f t - A_2 \sin \omega_f t) \\ &\quad = F_0 \sin \omega_f t \end{aligned}$$

or

$$\begin{aligned} &[(k - \omega_f^2 m) A_1 - c\omega_f A_2] \sin \omega_f t + [c\omega_f A_1 + (k - \omega_f^2 m) A_2] \cos \omega_f t \\ &\quad = F_0 \sin \omega_f t \end{aligned}$$

This equation yields the following two algebraic equations in A_1 and A_2

$$(k - \omega_f^2 m) A_1 - c\omega_f A_2 = F_0 \tag{4.47}$$

$$c\omega_f A_1 + (k - \omega_f^2 m) A_2 = 0 \tag{4.48}$$

Dividing these two equations by the stiffness coefficient k yields

$$(1 - r^2) A_1 - 2r\xi A_2 = X_0 \tag{4.49}$$

$$2r\xi A_1 + (1 - r^2) A_2 = 0 \tag{4.50}$$

where

$$r = \frac{\omega_f}{\omega} \tag{4.51}$$

$$\xi = \frac{c}{C_c} = \frac{c}{2m\omega} \tag{4.52}$$

$$X_0 = \frac{F_0}{k} \tag{4.53}$$

in which $C_c = 2m\omega$ is the *critical damping coefficient*. The two algebraic equations of Eqs. 49 and 50 can be solved using Cramer's rule in order to obtain the constants A_1 and A_2 as

$$A_1 = \frac{\begin{vmatrix} X_0 & -2r\xi \\ 0 & 1 - r^2 \end{vmatrix}}{(1 - r^2)^2 + (2r\xi)^2} = \frac{(1 - r^2) X_0}{(1 - r^2)^2 + (2r\xi)^2} \tag{4.54}$$

$$A_2 = \frac{\begin{vmatrix} 1 - r^2 & X_0 \\ 2r\xi & 0 \end{vmatrix}}{(1 - r^2)^2 + (2r\xi)^2} = \frac{-(2r\xi) X_0}{(1 - r^2)^2 + (2r\xi)^2} \tag{4.55}$$

The steady state solution x_p of Eq. 44 can then be written as

$$x_p = \frac{X_0}{(1 - r^2)^2 + (2r\xi)^2} [(1 - r^2) \sin \omega_f t - (2r\xi) \cos \omega_f t] \tag{4.56}$$

which can be written as

$$x_p = \frac{X_0}{\sqrt{(1 - r^2)^2 + (2r\xi)^2}} \sin(\omega_f t - \psi) \tag{4.57}$$

where ψ is the phase angle defined by

$$\psi = \tan^{-1}\left(\frac{2r\xi}{1 - r^2}\right) \tag{4.58}$$

Equation 57 can be written in a more compact form as

$$x_p = X_0 \beta \sin(\omega_f t - \psi) \tag{4.59}$$

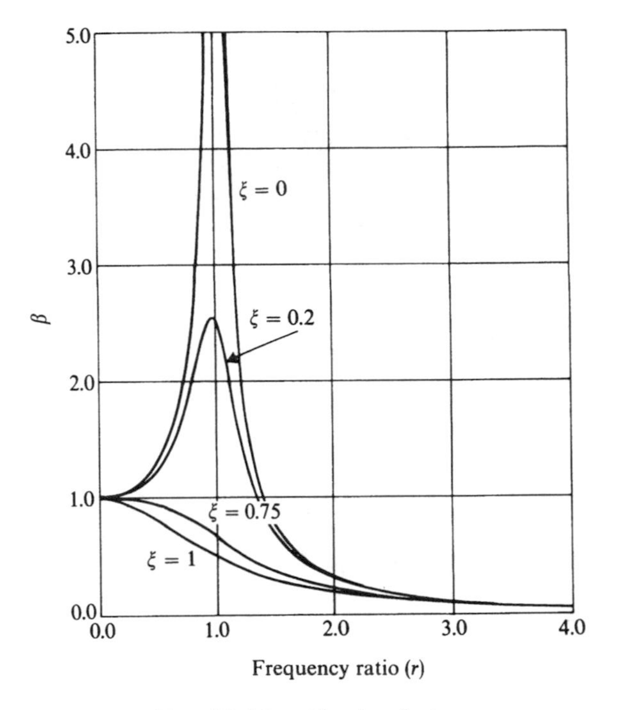

FIG. 4.7. Magnification factor.

where β is the *magnification factor* defined in the case of damped systems as

$$\beta = \frac{1}{\sqrt{(1 - r^2)^2 + (2r\xi)^2}} \tag{4.60}$$

This magnification factor reduces to the magnification factor obtained for the undamped systems when the damping factor $\xi = 0$. The magnification factor β for the damped systems and the phase angle ψ are shown, respectively, in Figs. 7 and 8 as functions of the frequency ratio r for different damping factors ξ. It is clear from these figures, that for damped systems the system does not attain infinite displacement at resonance, since for $\omega_f = \omega$, which corresponds to the case in which the frequency ratio $r = 1$, the magnification factor reduces to

$$\beta = \frac{1}{2\xi} \tag{4.61}$$

Furthermore, at resonance the magnification factor β does not have the maximum value, which can be obtained by differentiating β of Eq. 60 with respect to r and setting the result equal to zero. This leads to an algebraic equation which can be solved for the frequency ratio r at which the magnifica-

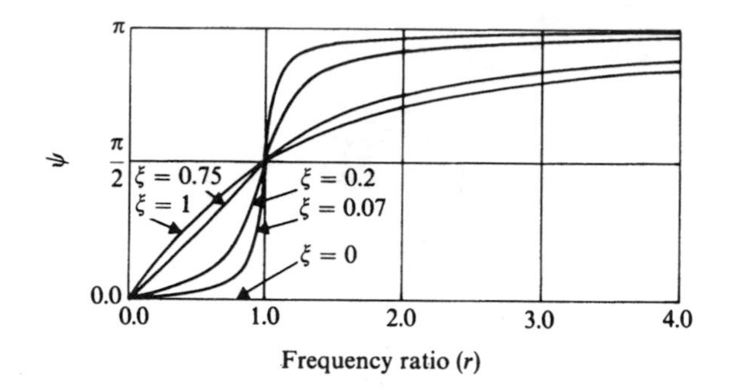

FIG. 4.8. Phase angle.

tion factor β is maximum. By so doing, one can show that the magnification factor β is maximum when

$$r = \sqrt{1 - 2\xi^2} \tag{4.62}$$

At this value of the frequency ratio, the maximum magnification factor is given by

$$\beta_{\max} = \frac{1}{2\xi\sqrt{1 - \xi^2}} \tag{4.63}$$

Force Transmission From Eq. 57 and Fig. 7 it is clear that by increasing the spring stiffness k and the damping coefficient c the amplitude of vibration decreases. The increase in the stiffness and damping coefficients, however, may have an adverse effect on the force transmitted to the support. In order to reduce the force transmitted to the support, the stiffness and damping coefficients must be properly selected. Figure 9 shows a free body diagram for the mass and the support system. The force transmitted to the support in the steady state can be written as

$$F_t = kx_p + c\dot{x}_p \tag{4.64}$$

From Eq. 59, $\dot{x}_p$ is

$$\dot{x}_p = \omega_f X_0 \beta \cos(\omega_f t - \psi)$$

Equation 64 can then be written as

$$\begin{aligned} F_t &= kX_0\beta \sin(\omega_f t - \psi) + c\omega_f X_0 \beta \cos(\omega_f t - \psi) \\ &= X_0 \beta \sqrt{k^2 + (c\omega_f)^2} \sin(\omega_f t - \bar{\psi}) \end{aligned} \tag{4.65}$$

where

$$\bar{\psi} = \psi - \psi_t \tag{4.66}$$

and ψ_t is a phase angle defined as

$$\psi_t = \tan^{-1}\left(\frac{c\omega_f}{k}\right) = \tan^{-1}(2r\xi) \tag{4.67}$$

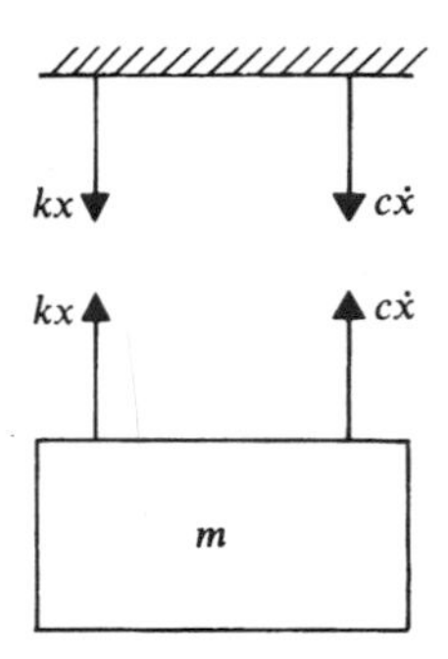

FIG. 4.9. Transmitted force.

Equation 65 can also be written as

$$F_t = X_0 k\beta\sqrt{1 + (2r\xi)^2}\,\sin(\omega_f t - \bar{\psi}) \tag{4.68}$$

Since $X_0 = F_0/k$, the above equation can be written as

$$\begin{aligned} F_t &= F_0\beta\sqrt{1 + (2r\xi)^2}\,\sin(\omega_f t - \bar{\psi}) \\ &= F_0\beta_t \sin(\omega_f t - \bar{\psi}) \end{aligned} \tag{4.69}$$

where

$$\begin{aligned} \beta_t &= \beta\sqrt{1 + (2r\xi)^2} \\ &= \frac{\sqrt{1 + (2r\xi)^2}}{\sqrt{(1 - r^2)^2 + (2r\xi)^2}} \end{aligned} \tag{4.70}$$

The coefficient β_t, which represents the ratio between the amplitude of the transmitted force and the amplitude of the applied force, is called the *transmissibility* and is plotted in Fig. 10 versus the frequency ratio r for different values of the damping factor ξ. It is clear form Fig. 10 that $\beta_t > 1$ for $r < \sqrt{2}$, and in this region the amplitude of the transmitted force is greater than the amplitude of the applied force. Furthermore, for $r < \sqrt{2}$, the transmitted force to the support can be reduced by increasing the damping factor ξ. For $r > \sqrt{2}$, $\beta_t < 1$, and as a consequence the amplitude of the transmitted force is less than the amplitude of the applied force, and the amplitude of the transmitted force increases by increasing the damping factor ξ.

Work per Cycle Equation 59 defines the steady state response, to a harmonic excitation, of the single degree of freedom system in the presence of damping. This equation implies that, for a given frequency ratio r and a given damping factor ξ, the amplitude of vibration remains constant. This can be achieved only if the energy input to the system, as the result of the work done by the external harmonic force, is equal to the energy dissipated as the

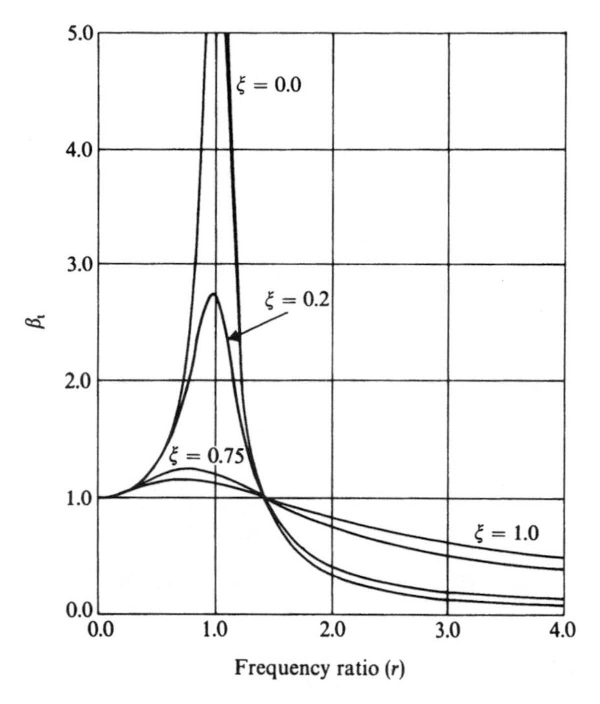

FIG. 4.10. Transmissibility.

result of the presence of damping. In order to see this, we first evaluate the work of the harmonic force. To this end, we write

$$dW_e = F(t)\,dx = F(t)\dot{x}\,dt$$

where W_e is the work done by the external force per cycle. It follows that

$$\begin{aligned} W_e &= \int_0^{2\pi/\omega_f} F(t)\dot{x}\,dt \\ &= \int_0^{2\pi/\omega_f} (F_0 \sin \omega_f t) X_0 \beta \omega_f \cos(\omega_f t - \psi)\,dt \\ &= F_0 X_0 \beta \int_0^{2\pi} \sin \omega_f t \cos(\omega_f t - \psi)\,d(\omega_f t) \end{aligned}$$

which upon integration yields

$$W_e = \pi F_0 X_0 \beta \sin \psi$$

Similarly, one can evaluate the energy dissipated per cycle, as the result of the damping force, as

$$W_d = \int_0^{2\pi/\omega_f} c\dot{x}\dot{x}\, dt$$

$$= cX_0^2\beta^2\omega_f \int_0^{2\pi} \cos^2(\omega_f t - \psi)\, d(\omega_f t)$$

which upon integration yields

$$W_d = \pi c X_0^2 \beta^2 \omega_f$$

Note that the input energy to the system is a linear function of the amplitude of the steady state vibration $X_0\beta$, while the energy dissipated as the result of the damping force is a quadratic function of the amplitude. Since at the steady state they must be equal, one has

$$W_e = W_d$$

or

$$\pi F_0 X_0 \beta \sin\psi = \pi c X_0^2 \beta^2 \omega_f$$

which defines the magnification factor β as

$$\beta = \frac{F_0/X_0}{c\omega_f} \sin\psi$$

Using the definition of X_0 and the phase angle ψ given, respectively, by Eqs. 53 and 58, the magnification factor β can be written as

$$\beta = \frac{F_0/X_0}{c\omega_f} \sin\psi = \frac{k}{c\omega_f} \frac{2r\xi}{\sqrt{(1-r^2)^2 + (2r\xi)^2}}$$

Since $c\omega_f/k = 2r\xi$, the above equation reduces to

$$\beta = \frac{1}{\sqrt{(1-r^2)^2 + (2r\xi)^2}}$$

which is the same definition of the magnification factor obtained by solving the differential equation. It is obtained here from equating the input energy, resulting from the work done by the harmonic force, to the energy dissipated as the result of the damping force. In fact, this must be the case, since the change in the strain energy in a complete cycle must be equal to zero, owing to the fact that the spring takes the same elongation after a complete cycle. This can also be demonstrated mathematically by using the definition of the work done by the spring force as

$$W_s = \int_0^{2\pi/\omega_f} kx\dot{x}\, dt$$

$$= kX_0^2\beta^2 \int_0^{2\pi} \sin(\omega_f t - \psi)\cos(\omega_f t - \psi)\, d(\omega_f t)$$

which upon integration yields

$$W_s = 0$$

Example 4.4

A damped single degree of freedom mass–spring system has mass $m = 10$ kg, spring coefficient $k = 4000$ N/m, and damping coefficient $c = 40$ N·s/m. The amplitude of the forcing function $F_0 = 60$ N, and the forcing frequency $\omega_f = 40$ rad/s. Determine the displacement of the mass as a function of time, and determine also the transmissibility and the amplitude of the force transmitted to the support.

Solution. The circular frequency of the system is

$$\omega = \sqrt{\frac{k}{m}} = \sqrt{\frac{4000}{10}} = 20 \text{ rad/s}$$

The frequency ratio r is given by

$$r = \frac{\omega_f}{\omega} = \frac{40}{20} = 2$$

The critical damping coefficient C_c is defined as

$$C_c = 2m\omega = 2(10)(20) = 400 \text{ N}\cdot\text{s/m}$$

The damping factor ξ is then given by

$$\xi = \frac{c}{C_c} = \frac{40}{400} = 0.1$$

which is the case of an underdamped system, and the complete solution can be written in the following form

$$\begin{aligned} x(t) &= x_h + x_p \\ &= Xe^{-\xi\omega t}\sin(\omega_d t + \phi) + X_0\beta\sin(\omega_f t - \psi) \end{aligned}$$

where ω_d is the damped circular frequency

$$\omega_d = \omega\sqrt{1 - \xi^2} = 20\sqrt{1 - (0.1)^2} = 19.8997 \text{ rad/s}$$

The constants X_0, β, and ψ are

$$X_0 = \frac{F_0}{k} = \frac{60}{4000} = 0.015 \text{ m}$$

$$\beta = \frac{1}{\sqrt{(1 - r^2)^2 + (2r\xi)^2}} = \frac{1}{\sqrt{(1 - (2)^2)^2 + (2 \times 2 \times 0.1)^2}} = \frac{1}{\sqrt{9 + 0.16}}$$
$$= 0.3304$$

$$\psi = \tan^{-1}\left(\frac{2r\xi}{1 - r^2}\right) = \tan^{-1}\frac{2(2)(0.1)}{1 - (2)^2} = \tan^{-1}(-0.13333) = -0.13255 \text{ rad}$$

The displacement can then be written as a function of time as

$$x(t) = Xe^{-2t}\sin(19.8997t + \phi) + 0.004956\sin(40t + 0.13255)$$

The constants X and ϕ can be determined using the initial conditions. The transmissibility β_t is defined by Eq. 70 as

$$\beta_t = \frac{\sqrt{1 + (2r\xi)^2}}{\sqrt{(1 - r^2)^2 + (2r\xi)^2}} = \frac{\sqrt{1 + (2 \times 2 \times 0.1)^2}}{\sqrt{[1 - (2)^2]^2 + (2 \times 2 \times 0.1)^2}} = 0.35585$$

The amplitude of the force transmitted is given by

$$|F_t| = F_0\beta_t = (60)(0.35585) = 21.351 \text{ N}$$

4.5 ROTATING UNBALANCE

In many rotating mechanical systems, gears, wheels, shafts, and disks, which are not perfectly uniform, produce unbalance forces which cause excessive vibrations which may lead to failure of the system components. Rotating unbalance may occur in systems such as rotors, flywheels, blowers, and fans. Figure 11 depicts a damped single degree of freedom machine which is supported elastically by a spring and damper which have, respectively, stiffness coefficient k and damping constant c. The machine, which has total mass M, has a rotor which is mounted on a bearing whose center is defined by point O. The rotor, which rotates counterclockwise with an angular velocity ω_f rad/s,

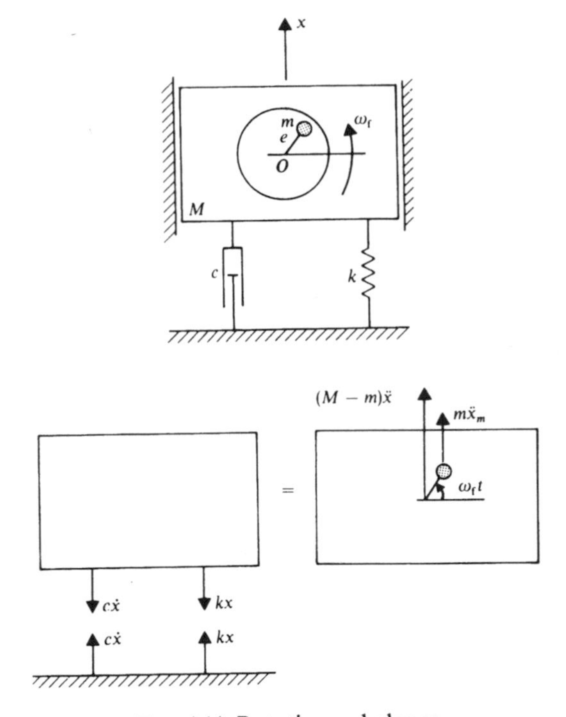

FIG. 4.11. Rotating unbalance.

does not have a uniform mass distribution, resulting in a rotor unbalance which is equivalent to an eccentric mass m located at a distance e from the center of the rotor as shown in Fig. 11. One can show from the static equilibrium analysis that the weight of the machine cancels with the static deflection of the spring. If the displacement of the machine in the vertical direction is denoted as x, the component of the displacement of the eccentric mass m in the vertical direction, denoted as x_m, is given by

$$x_m = x + e \sin \omega_f t \tag{4.71}$$

and the acceleration $\ddot{x}_m$ is

$$\ddot{x}_m = \ddot{x} - \omega_f^2 e \sin \omega_f t \tag{4.72}$$

Since M is assumed to be the total mass of the machine including the eccentric mass, the mass of the machine which has acceleration $\ddot{x}$ is $(M - m)$; therefore, the total inertia forces of the machine can be written as

$$\text{inertia force} = (M - m)\,\ddot{x} + m\ddot{x}_m$$

According to Newton's second law, this inertia force must be equal to the applied forces. Hence,

$$(M - m)\ddot{x} + m\ddot{x}_m = -c\dot{x} - kx$$

Using Eq. 72 and rearranging the terms in the above equation yields

$$M\ddot{x} + c\dot{x} + kx = me\omega_f^2 \sin \omega_f t \tag{4.73}$$

which is similar to Eq. 43 with F_0 defined as

$$F_0 = me\omega_f^2 \tag{4.74}$$

This case, however, is different from the case discussed in the preceding section since F_0 depends on the frequency of the rotor. The circular frequency of the system is

$$\omega = \sqrt{\frac{k}{M}} \tag{4.75}$$

and the steady state solution is defined as

$$x_p(t) = \frac{X_0}{\sqrt{(1 - r^2)^2 + (2r\xi)^2}} \sin(\omega_f t - \psi) \tag{4.76}$$

where ψ is the phase angle defined by Eq. 58 and the constant X_0, in this case, is defined as

$$X_0 = \frac{F_0}{k} = \frac{me\omega_f^2}{k} = \frac{me\omega_f^2}{\omega^2 M} = \left(\frac{m}{M}e\right)r^2 \tag{4.77}$$

where r is the frequency ratio

$$r = \frac{\omega_f}{\omega}$$

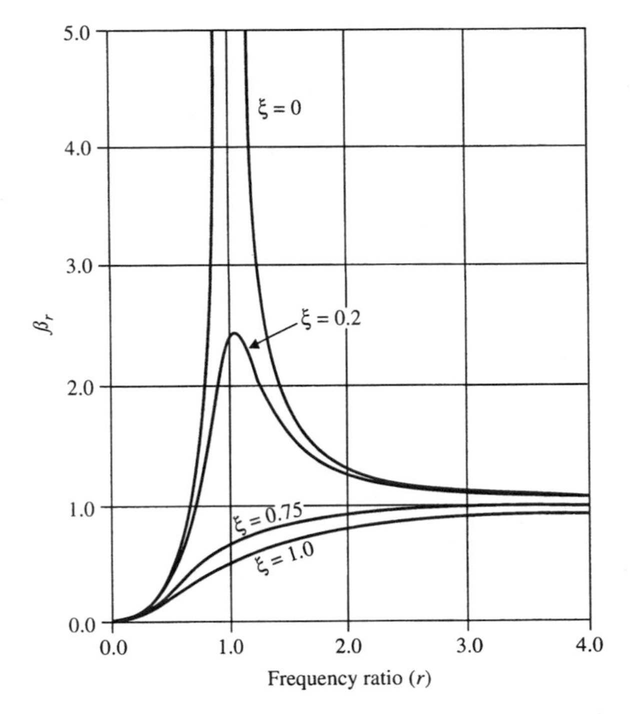

FIG. 4.12. Magnification factor in the case of rotating unbalance.

Substituting Eq. 77 into Eq. 76 yields

$$x_p(t) = \frac{(me/M)r^2}{\sqrt{(1 - r^2)^2 + (2r\xi)^2}} \sin(\omega_f t - \psi) \tag{4.78}$$

which can be written in a more compact form as

$$x_p(t) = \left(\frac{me}{M}\right) \beta_r \sin(\omega_f t - \psi) \tag{4.79}$$

where β_r is a magnification factor defined in this case as

$$\beta_r = \frac{r^2}{\sqrt{(1 - r^2)^2 + (2r\xi)^2}} \tag{4.80}$$

The magnification factor β_r is plotted in Fig. 12 versus the frequency ratio r for different values of the damping factor ξ.

Force Transmission It is clear from the free body diagram shown in Fig. 11 that the force transmitted to the support in the steady state, in the case of rotating unbalance, is

$$F_t = kx_p + c\dot{x}_p \tag{4.81}$$

where $\dot{x}_p$ can be obtained by differentiating Eq. 79 with respect to time as

$$\dot{x}_p = \omega_f \left(\frac{me}{M}\right) \beta_r \cos(\omega_f t - \psi) \tag{4.82}$$

Substituting Eqs. 79 and 82 into Eq. 81 leads to

$$F_t = \left(\frac{me}{M}\right) \beta_r [k \sin(\omega_f t - \psi) + c\omega_f \cos(\omega_f t - \psi)]$$

which can be written as

$$F_t = \left(\frac{me}{M}\right) \beta_r \sqrt{k^2 + (c\omega_f)^2} \sin(\omega_f t - \bar{\psi}) \tag{4.83}$$

where

$$\bar{\psi} = \psi - \psi_t \tag{4.84}$$

and ψ_t is a phase angle defined as

$$\psi_t = \tan^{-1}\left(\frac{c\omega_f}{k}\right) = \tan^{-1}(2r\xi) \tag{4.85}$$

Equation 83 can also be expressed as

$$F_t = \left(\frac{me}{M}\right) k\beta_r \sqrt{1 + (2r\xi)^2} \sin(\omega_f t - \bar{\psi}) \tag{4.86}$$

Since $k/M = \omega^2$, the above equation can be written as

$$F_t = (me\omega^2)\beta_t \sin(\omega_f t - \bar{\psi}) \tag{4.87}$$

where β_t is defined as

$$\beta_t = \beta_r \sqrt{1 + (2r\xi)^2} = \frac{r^2 \sqrt{1 + (2r\xi)^2}}{\sqrt{(1 - r^2) + (2r\xi)^2}} \tag{4.88}$$

Example 4.5

It was observed that a machine with a mass $M = 600$ kg has an amplitude of vibration of 0.01 m when the operating speed is 10 Hz. If the machine is critically damped with an equivalent stiffness coefficient k of 10×10^3 N/m, determine the amount of unbalance me.

Solution. The forcing frequency ω_f is

$$\omega_f = 2\pi f = 2\pi(10) = 20\pi = 62.832 \text{ rad/s}$$

The circular frequency ω of the machine is

$$\omega = \sqrt{\frac{k}{M}} = \sqrt{\frac{10 \times 10^3}{600}} = 4.082 \text{ rad/s}$$

The frequency ratio r is then given by

$$r = \frac{\omega_f}{\omega} = \frac{62.832}{4.082} = 15.392$$

Since the machine is critically damped, $\xi = 1$. The magnification factor β_r can be determined by using Eq. 80 as

$$\beta_r = \frac{r^2}{\sqrt{(1 - r^2)^2 + (2r\xi)^2}} = \frac{(15.392)^2}{\sqrt{[1 - (15.392)^2]^2 + [2(15.392)(1)]^2}}$$
$$= 0.9958$$

Thus, the amplitude is

$$0.01 = \left(\frac{me}{M}\right)\beta_r = \left(\frac{me}{M}\right)(0.9958)$$

which implies that

$$\frac{me}{M} = 0.01004$$

That is, the amount of unbalance me is

$$me = 0.01004(M) = 0.01004 \times 600 = 6.0253 \text{ kg}\cdot\text{m}$$

Balancing The case of rotating unbalance is an example where the inertia forces can vary significantly during one cycle of operation. In order to reduce or eliminate these unwanted inertia forces, *balancing techniques* are often used in order to correct or eliminate inertia forces and moments which are the result of manufacturing errors or production tolerances.

In order to demonstrate the use of balancing techniques, consider the rigid body shown in Fig. 13. The motion of this rigid body is assumed to be a noncentroidal rotation, that is, the rigid body is rotating about an axis passing through point O which is not the center of mass of the body. The location of

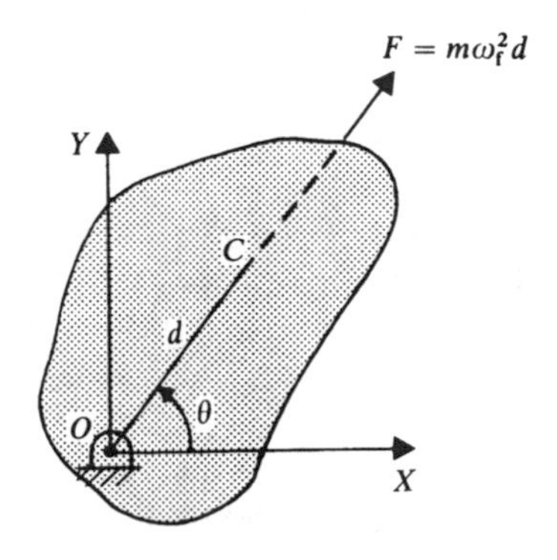

FIG. 4.13. Noncentroidal rotation.

the center of mass C with respect to the fixed point O can be defined as

$$x = d \cos \theta$$

$$y = d \sin \theta$$

The velocity of the center of mass can be obtained by differentiating the preceding two equations with respect to time. This yields

$$\dot{x} = -d\dot{\theta} \sin \theta$$

$$\dot{y} = d\dot{\theta} \cos \theta$$

which upon differentiation yield the accelerations as

$$\ddot{x} = -d\ddot{\theta} \sin \theta - d\dot{\theta}^2 \cos \theta$$

$$\ddot{y} = d\ddot{\theta} \cos \theta - d\dot{\theta}^2 \sin \theta$$

If the angular velocity of the body is assumed to be constant, such that $\dot{\theta} = \omega_f$, the expressions for the accelerations of the center of mass reduce to

$$\ddot{x} = -d\dot{\theta}^2 \cos \theta = -d\omega_f^2 \cos \theta$$

$$\ddot{y} = -d\dot{\theta}^2 \sin \theta = -d\omega_f^2 \sin \theta$$

which can also be written as

$$\ddot{x} = -\omega_f^2 x$$

$$\ddot{y} = -\omega_f^2 y$$

Since the angular velocity is assumed to be constant, the inertia or effective moment $I\ddot{\theta}$ is equal to zero, while the inertia forces reduce to the *centrifugal force* components defined as

$$F_x = -m\ddot{x} = m\omega_f^2 x = m\omega_f^2 d \cos \theta$$

$$F_y = -m\ddot{y} = m\omega_f^2 y = m\omega_f^2 d \sin \theta$$

The magnitude and direction of the centrifugal force can then be given by

$$F = \sqrt{F_x^2 + F_y^2} = m\omega_f^2 d$$

$$\phi = \tan^{-1} \frac{F_y}{F_x} = \tan^{-1} \frac{m\omega_f^2 d \sin \theta}{m\omega_f^2 d \cos \theta} = \tan^{-1}\theta = \theta$$

Note that in this case, the centrifugal force resulting from the rotation of the body is equivalent to the centrifugal force produced by a single mass rotor as the one shown in Fig. 14(a). The mass in this rotor system is equal to the total mass of the rigid body and it is located at a distance d from the center of the rotation.

The centrifugal forces resulting from the noncentroidal rotation of bodies in mechanical systems can be main sources of excessive vibrations. If the

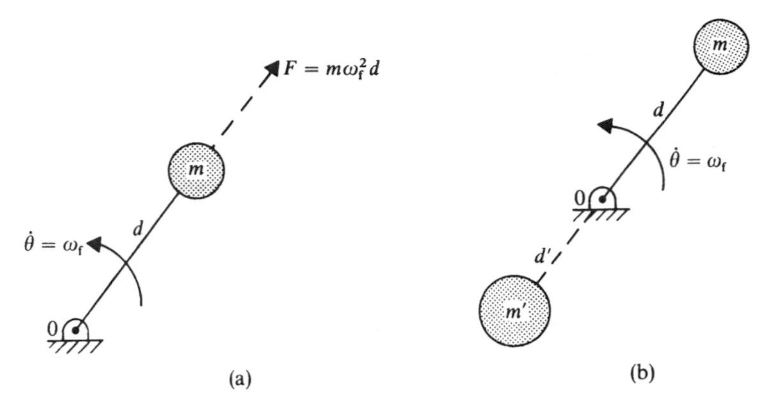

FIG. 4.14. Balancing.

rotation is centroidal, the body rotates about an axis passing through its center of mass; $d = 0$, and consequently the centrifugal forces are equal to zero. Obviously, one method for achieving this is to use the balancing techniques by adding a counterweight of mass m' located at radius d' such that $m'd' = md$ as shown in Fig. 14(b). The centrifugal force resulting from the rotation of the counterweight is equal in magnitude and opposite in direction to the centrifugal force produced as the result of the body rotation, and as a consequence, the resulting net inertia force of the balanced system is equal to zero, and the vibration produced by the rotating unbalance is eliminated.

In the analysis presented in this section, the unbalance force has two components, and this force can produce vibrations in two different directions. In this chapter, however, only the special case in which the motion is in one direction is considered, since we are mainly concerned with the vibration of one degree of freedom systems, where an assumption is made that vibrations occur only in one direction. This assumption, which is made in order to simplify the mathematical model, may not be a valid assumption in some applications. If the system has more than one degree of freedom, the inertia forces resulting from the rotating unbalance can be a source of excessive vibration in more than one direction.

4.6 BASE MOTION

The forced vibration of mechanical systems can be caused by the support motion, as in the case of vehicles, aircraft, and ships. Figure 15 shows a damped single degree of freedom mass–spring system with a moving support. The support motion is assumed to be harmonic and expressed in the form

$$y = Y_0 \sin \omega_f t \tag{4.89}$$

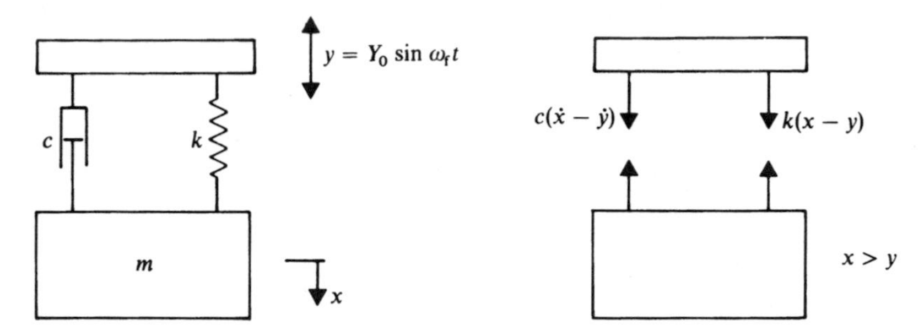

FIG. 4.15. Support motion.

From the free body diagram shown in Fig. 15, it is clear that if we assume that x is greater than y, the equation of motion of the mass m is

$$m\ddot{x} = -k(x - y) - c(\dot{x} - \dot{y}) \tag{4.90}$$

which can be written as

$$m\ddot{x} + c\dot{x} + kx = ky + c\dot{y}$$

By using Eq. 89, the above equation can be written as

$$m\ddot{x} + c\dot{x} + kx = kY_0 \sin \omega_f t + c\omega_f Y_0 \cos \omega_f t \tag{4.91}$$

The right-hand side of this equation can be written as

$$kY_0 \sin \omega_f t + c\omega_f Y_0 \cos \omega_f t = Y_0 \sqrt{k^2 + (c\omega_f)^2} \sin(\omega_f t + \psi_b) \tag{4.92}$$

where the phase angle ψ_b is

$$\psi_b = \tan^{-1}\left(\frac{c\omega_f}{k}\right) = \tan^{-1}(2r\xi) \tag{4.93}$$

Substituting Eq. 92 into Eq. 91 yields

$$m\ddot{x} + c\dot{x} + kx = Y_0 \sqrt{k^2 + (c\omega_f)^2} \sin(\omega_f t + \psi_b)$$

or

$$m\ddot{x} + c\dot{x} + kx = Y_0 k \sqrt{1 + (2r\xi)^2} \sin(\omega_f t + \psi_b) \tag{4.94}$$

This equation is in a similar form to Eq. 43, therefore the steady state solution x_p is

$$x_p(t) = \frac{Y_0 \sqrt{1 + (2r\xi)^2}}{\sqrt{(1 - r^2)^2 + (2r\xi)^2}} \sin(\omega_f t - \psi + \psi_b) \tag{4.95}$$

where the phase angle ψ is

$$\psi = \tan^{-1}\left(\frac{2r\xi}{1 - r^2}\right) \tag{4.96}$$

Equation 95 can be written in a compact form as

$$x_p(t) = Y_0 \beta_b \sin(\omega_f t - \psi + \psi_b) \tag{4.97}$$

where β_b is called the magnification factor defined, in this case, as

$$\beta_b = \frac{\sqrt{1 + (2r\xi)^2}}{\sqrt{(1 - r^2)^2 + (2r\xi)^2}} \tag{4.98}$$

Transmitted Force In the case of the support motion, the force carried by the support can be obtained, by using the free body diagram shown in Fig. 15, as

$$F_t = k(x - y) + c(\dot{x} - \dot{y}) \tag{4.99}$$

which by using Eq. 90 can be written as

$$F_t = -m\ddot{x} \tag{4.100}$$

In the case of steady state oscillations, the acceleration $\ddot{x}$ can be obtained by differentiating Eq. 97. This leads to the following expression for the transmitted force

$$\begin{aligned} F_t &= mY_0\beta_b\omega_f^2 \sin(\omega_f t - \psi + \psi_b) \\ &= \left(\frac{k}{\omega^2}\right) Y_0\beta_b\omega_f^2 \sin(\omega_f t - \psi + \psi_b) \\ &= Y_0 k\beta_b r^2 \sin(\omega_f t - \psi + \psi_b) \end{aligned} \tag{4.101}$$

where $r = \omega_f/\omega$ is the frequency ratio. Using Eq. 98, the force transmitted to the support can be written in an explicit form as

$$F_t = Y_0 k \frac{r^2\sqrt{1 + (2r\xi)^2}}{\sqrt{(1 - r^2)^2 + (2r\xi)^2}} \sin(\omega_f t - \psi + \psi_b) \tag{4.102}$$

Relative Motion Sometimes, the interest is focused on studying the motion of the mass relative to the support. This relative displacement denoted as z can be expressed as

$$z = x - y \tag{4.103}$$

which, upon differentiation, leads to

$$\dot{z} = \dot{x} - \dot{y} \qquad \text{and} \qquad \ddot{z} = \ddot{x} - \ddot{y} \tag{4.104}$$

Equation 90 can be written in terms of z as

$$m(\ddot{z} + \ddot{y}) = -kz - c\dot{z} \tag{4.105}$$

which can be written as

$$\begin{aligned} m\ddot{z} + c\dot{z} + kz &= -m\ddot{y} \\ &= mY_0\omega_f^2 \sin \omega_f t \end{aligned} \tag{4.106}$$

The particular solution of this differential equation is

$$z = \frac{mY_0\omega_f^2/k}{\sqrt{(1-r^2)^2 + (2r\xi)^2}} \sin(\omega_f t - \psi)$$

$$= \frac{Y_0 r^2}{\sqrt{(1-r^2)^2 + (2r\xi)^2}} \sin(\omega_f t - \psi) \tag{4.107}$$

where the phase angle ψ is defined by Eq. 96.

Example 4.6

The damped single degree of freedom mass–spring system shown in Fig. 15 has a mass $m = 25$ kg, and a spring stiffness coefficient $k = 2500$ N/m. Determine the damping coefficient of the system knowing that the mass exhibits an amplitude of 0.01 m when the support oscillates at the natural frequency of the system with amplitude $Y_0 = 0.005$ m. Determine also the amplitude of dynamic force carried by the support and the displacement of the mass relative to the support.

Solution. From Eq. 97, the amplitude of the oscillation of the mass is given by

$$X_p = Y_0\beta_b$$

$$= Y_0 \frac{\sqrt{(1 + (2r\xi)^2}}{\sqrt{(1-r^2)^2 + (2r\xi)^2}}$$

Since in this example, $\omega_f = \omega$, that is, $r = 1$, the above equation reduces to

$$X_p = Y_0 \frac{\sqrt{(1 + (2\xi)^2}}{2\xi}$$

From which the damping factor ξ is

$$\xi = \frac{1}{2}\frac{1}{\sqrt{\left(\frac{X_p}{Y_0}\right)^2 - 1}} = \frac{1}{2}\frac{1}{\sqrt{\left(\frac{0.01}{0.005}\right)^2 - 1}} = 0.28868$$

The critical damping coefficient C_c of the system is

$$C_c = 2m\omega = 2\sqrt{km} = 2\sqrt{(2500)(25)} = 500 \text{ N}\cdot\text{s/m}$$

The damping coefficient c is

$$c = \xi C_c = 0.28868(500) = 144.34 \text{ N}\cdot\text{s/m}$$

From Eq. 101, the amplitude of the force transmitted is

$$(F_t)_{max} = Y_0 k r^2 \beta_b = Y_0 k \frac{r^2\sqrt{1 + (2r\xi)^2}}{\sqrt{(1-r^2)^2 + (2r\xi)^2}}$$

At resonance $r = 1$, therefore,

$$(F_t)_{max} = Y_0 k \frac{\sqrt{1 + (2\xi)^2}}{2\xi} = (0.005)(2500)\frac{\sqrt{1 + (2 \times 0.28868)^2}}{2(0.28868)}$$

$$= 24.9997 \text{ N}$$

The amplitude of the displacement of the mass relative to the support can be obtained using Eq. 107 as

$$Z = \frac{Y_0 r^2}{\sqrt{(1 - r^2)^2 + (2r\xi)^2}}$$

At resonance

$$Z = \frac{Y_0}{2\xi} = \frac{0.005}{2(0.28868)} = 8.66 \times 10^{-3} \text{ m}$$

Observe that $Z \neq X_p - Y_0$ because of the phase shift angles ψ and ψ_b.

Example 4.7

Figure 16 depicts a simple pendulum with a support that has a specified motion $y = Y_0 \sin \omega_f t$. Assuming small oscillations, determine the differential equation of motion and the steady state solution.

Solution. Assuming small angular oscillations, the displacement of the mass m in the horizontal direction is given by

$$x = y + l\theta = Y_0 \sin \omega_f t + l\theta$$

The velocity and acceleration of the mass are then

$$\dot{x} = \dot{y} + l\dot{\theta} = \omega_f Y_0 \cos \omega_f t + l\dot{\theta}$$

$$\ddot{x} = \ddot{y} + l\ddot{\theta} = -\omega_f^2 Y_0 \sin \omega_f t + l\ddot{\theta}$$

Let R_x and R_y denote the reaction forces, as shown in the figure. By taking the moments of the applied and inertia forces about O, one can obtain the dynamic equation

$$-mgl \sin \theta - c\dot{x}l \cos \theta = m\ddot{x}l \cos \theta$$

For small angular oscillations $\sin \theta \approx \theta$ and $\cos \theta \approx 1$, thus

$$m\ddot{x}l + c\dot{x}l + mgl\theta = 0$$

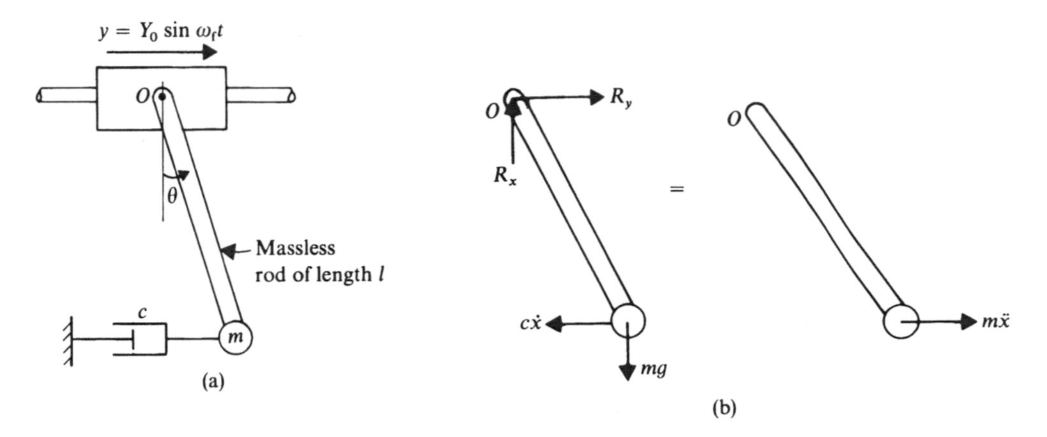

FIG. 4.16. Support oscillations.

Using the expressions for x, $\dot{x}$, and $\ddot{x}$, one obtains

$$m(-\omega_f^2 Y_0 \sin \omega_f t + l\ddot{\theta})\, l + c(\omega_f Y_0 \cos \omega_f t + l\dot{\theta})\, l + mgl\theta = 0$$

which can be written as

$$\begin{aligned} ml^2\ddot{\theta} + cl^2\dot{\theta} + mgl\theta &= m\omega_f^2 Y_0 l \sin \omega_f t - c\omega_f Y_0 l \cos \omega_f t \\ &= \omega_f Y_0 l[m\omega_f \sin \omega_f t - c \cos \omega_f t] \\ &= \omega_f Y_0 l \sqrt{(m\omega_f)^2 + c^2} \sin(\omega_f t - \psi_b) \end{aligned}$$

This equation can be written in a simple form as

$$m_e\ddot{\theta} + c_e\dot{\theta} + k_e\theta = F_e \sin(\omega_f t - \psi_b) \tag{4.108}$$

where

$$m_e = ml^2$$

$$c_e = cl^2$$

$$k_e = mgl$$

$$F_e = \omega_f Y_0 l \sqrt{(m\omega_f)^2 + c^2}$$

$$\psi_b = \tan^{-1}\left(\frac{c}{m\omega_f}\right)$$

Equation 108 is in a form similar to Eq. 43, therefore its steady state solution can be expressed as

$$\theta_p = \frac{F_e/k_e}{\sqrt{(1 - r^2)^2 + (2r\xi)^2}} \sin(\omega_f t - \psi - \psi_b)$$

where

$$r = \frac{\omega_f}{\omega}$$

$$\omega = \sqrt{\frac{k_e}{m_e}} = \sqrt{\frac{mgl}{ml^2}} = \sqrt{\frac{g}{l}}$$

$$\xi = \frac{c_e}{C_c} = \frac{cl^2}{C_c}$$

$$C_c = 2m_e\omega = 2ml^2\sqrt{\frac{g}{l}}$$

$$\psi = \tan^{-1}\left(\frac{2r\xi}{1 - r^2}\right)$$

4.7 MEASURING INSTRUMENTS

The analysis of the motion of damped single degree of freedom systems subject to base excitation shows that the differential equation of such systems can be expressed in terms of the relative displacement between the mass and the oscillating base. The solution of the vibration equation showed that the relative displacement z can be expressed in terms of the frequency ratio r and the damping factor ξ as (Eq. 107)

$$z = \frac{Y_0 r^2}{\sqrt{(1-r^2)^2 + (2r\xi)^2}} \sin(\omega_f t - \psi) \tag{4.109}$$

where Y_0 is the amplitude of vibration of the base, ω_f is the frequency of oscillation of the base, and ψ is the phase angle defined as

$$\psi = \tan^{-1}\left(\frac{2r\xi}{1-r^2}\right) \tag{4.110}$$

Equation 109 can be written as

$$z = Y_0 \beta_r \sin(\omega_f t - \psi) \tag{4.111}$$

or

$$z = Y_0 r^2 \beta \sin(\omega_f t - \psi) \tag{4.112}$$

where the magnification factors β_r and β are defined as

$$\beta_r = \frac{r^2}{\sqrt{(1-r^2)^2 + (2r\xi)^2}} \tag{4.113}$$

$$\beta = \frac{1}{\sqrt{(1-r^2)^2 + (2r\xi)^2}} \tag{4.114}$$

The magnification factor β_r and β are plotted in Figs. 12 and 7 versus the frequency ratio r for different values for the damping factor ξ.

Equation 109, or its equivalent forms of Eqs. 111 and 112, represent the basic equation in designing vibration instruments for measuring the displacements, velocities, and accelerations.

Vibrometer The vibrometer is used to measure the displacement of systems which exhibit oscillatory motion. Figure 17 shows the basic elements used in many vibration measuring instruments. The relative displacement of the mass m with respect to the moving base is described by Eq. 109. In this equation, if

$$\beta_r = \frac{r^2}{\sqrt{(1-r^2)^2 + (2r\xi)^2}} \approx 1 \tag{4.115}$$

the relative displacement z can be expressed as

$$z \approx Y_0 \sin(\omega_f t - \psi) \tag{4.116}$$

That is, the amplitude of the relative displacement is approximately the same as the amplitude of the base Y_0. It can be seen from Fig. 12 that the condition of Eq. 115 is satisfied for a frequency ratio $r > 3$, that is, when $\omega_f > 3\omega$. This

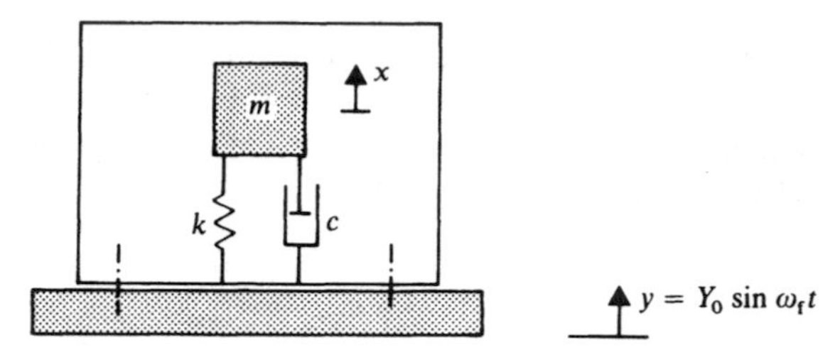

FIG. 4.17. Vibration measurements.

is the case in which the natural frequency of the measuring instrument ω is lower than the excitation frequency ω_f. The natural frequency of the vibrometer can be made lower by increasing its mass or decreasing the stiffness of its spring, and therefore, the large size of the vibrometer is one of its main disadvantages. Furthermore, Eq. 116 indicates that the amplitude of the relative displacement is of the same order as the amplitude Y_0 of the base, and as a consequence, if the oscillation of the base has large amplitude, the amplitude of the gauge is also large. It is also clear from Fig. 12 that damping does improve the range of application of the vibrometer. A viscous damping coefficient $\xi = 0.7$ is recommended.

Accelerometer The same basic elements shown in Fig. 17 can be used to measure the acceleration of the vibrating systems. Equation 112, which can be used in the design of such measuring instruments, can be rewritten as

$$\omega^2 z = \omega_f^2 Y_0 \beta \sin(\omega_f t - \psi) \tag{4.117}$$

where $\omega_f^2 Y_0$ is the amplitude of the acceleration of the vibrating system. Therefore, if the measuring instrument is designed such that

$$\beta = \frac{1}{\sqrt{(1-r^2)^2 + (2r\xi)^2}} \approx 1, \tag{4.118}$$

Eq. 117 can be written as

$$-\omega^2 z \approx -\omega_f^2 Y_0 \sin(\omega_f t - \psi) \tag{4.119}$$

That is, except for the phase lag ψ, the left-hand side of Eq. 119, $\omega^2 z$ is a measure of the acceleration $\ddot{y}$ of the vibrating system. Since the natural frequency $\omega = \sqrt{k/m}$ is a constant, the gauge can be calibrated to read directly the acceleration $\ddot{y}$. It is clear from Eq. 119 that the time lag between the gauge and the motion of the vibrating system can be determined from the equation

$$\omega_f t_1 = \psi \tag{4.120}$$

or

$$t_1 = \frac{\psi}{\omega_f} \tag{4.121}$$

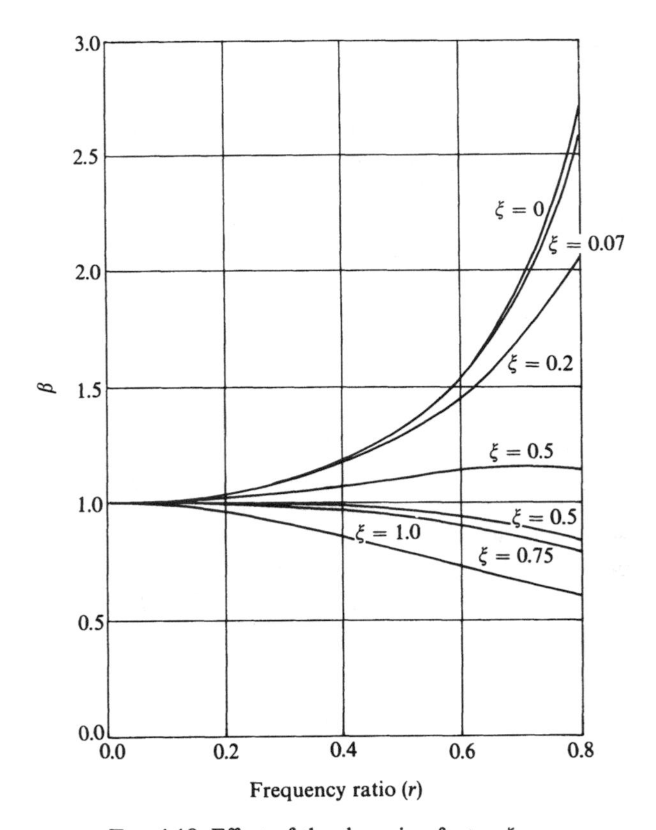

FIG. 4.18. Effect of the damping factor ξ.

Note that Eq. 118 is satisfied if r is small (Fig. 7), that is, if the natural frequency ω of the measuring instrument is much higher than the excitation frequency ω_f. Furthermore, as shown in Fig. 18, the damping factor ξ affects the frequency range in which the instrument must be used.

Vibrometers and accelerometers are designed such that the output of the measuring instrument is the relative displacement. This displacement in most commerical pickups is converted into an electrical signal using a conversion device known as *transducer*. There are different arrangements for the transducers; in one of them, a permanent magnet is attached to the base of the instrument and a coil is wound around the mass such that when the mass moves with respect to the base an electric voltage proportional to the velocity of the mass with respect to the base is generated in the coil as the result of the cut of the flux lines. Another type of transducer uses strain gauges that can bc calibrated so as to produce a strain proportional to the relative displacement of the mass. A third type of transducer uses piezoelectric material, such as quartz, which produces electric voltage when it deforms.

The accelerometer, which is one of the most important pickups for shock and vibration measurements, is commerically available in different types to meet diverse application requirements. While accelerometers can be designed to measure accelerations in wide frequency ranges, the displacement and velocity can be obtained easily by electrical integration of the acceleration signal using electrical integrators such as *operational amplifiers.* At present, piezoelectric accelerometers, because of their small size, reliability, and stable characteristics, are the most widely used transducers in vibration and shock measurements. The mass of the accelerometer can be selected small such that the natural frequency of the instrument is more than 100,000 HZ, thus allowing its use in lightweight applications and over a wide frequency range.

For a given vibration pickup, the error in the measurement can be obtained by calculating the magnification factor for the given value of the frequency ratio r, as demonstrated by the following example.

Example 4.8

A vibrometer is used to measure the amplitude of a vibrating machine. It was observed that the machine is vibrating at a frequency of 16 Hz. The natural frequency and damping ratio of the vibrometer are, respectively, 8 Hz and 0.7. The reading of the vibrometer indicates that the amplitude of vibration is 1.5 cm. Determine the correct value for the amplitude of vibration of the machine.

Solution. The magnification factor β_r is

$$\beta_r = \frac{r^2}{\sqrt{(1 - r^2)^2 + (2r\xi)^2}}$$

where $\xi = 0.7$ and

$$r = \frac{\omega_f}{\omega} = \frac{16}{8} = 2$$

It follows that

$$\beta_r = \frac{(2)^2}{\sqrt{[1 - (2)^2]^2 + (2 \times 2 \times 0.7)^2}} = 0.97474$$

Since

$$z = Y_0 \beta_r \sin(\omega_f t - \psi)$$

the amplitude of oscillation of the machine is

$$Y_0 = \frac{1.5}{\beta_r} = \frac{1.5}{0.97474} = 1.538872$$

The percentage error is 2.5915%. The error in measurement can be reduced by using another vibrometer with lower natural frequency.

4.8 EXPERIMENTAL METHODS FOR DAMPING EVALUATION

The analysis presented in this chapter and the preceding chapter shows that damping has a significant effect on the dynamic behavior and the force transmission in mechanical systems, and it also has an effect on the design of the measuring instruments, as demonstrated in the preceding section. In the analysis presented thus far, we assumed that all the physical properties of the system such as mass, stiffness, and damping are known. While in most applications, the mass and stiffness of the system can easily be determined, relatively more elaborate experiments usually are required for evaluating the damping coefficients. Because of the significant effect of the damping, we devote this section to discussing some of the techniques which can be used to determine experimentally the damping coefficients.

Logarithmic Decrement The use of the logarithmic decrement in the experimental determination of the viscous damping coefficient of underdamped single degree of freedom systems was discussed in the preceding chapter. This is probably the simplest and most frequently used technique, since equipment and instrumentation requirements are minimal. In this technique, the free vibration of the system can be initiated and the ratio between successive or nonsuccessive displacement amplitudes can be measured, and used to define the logarithmic decrement δ as

$$\delta = \frac{1}{n} \ln \frac{x_i}{x_{i+n}} \tag{4.122}$$

where x_i and x_{i+n} are two displacement amplitudes n cycles apart. Once δ has been determined, the damping factor ξ can be determined according to

$$\xi = \frac{\delta}{\sqrt{(2\pi)^2 + \delta^2}} \tag{4.123}$$

The equivalent viscous damping coefficient c can then be determined as

$$c = \xi C_c = \xi(2m\omega) = 2\xi m\omega \tag{4.124}$$

where C_c is the critical damping coefficient, m is the mass, and ω is the natural frequency of the system.

Frequency Response The method of the logarithmic decrement, discussed in this section for the experimental determination of the damping coefficient, requires only the free vibration of the system, which can be initiated using an *impact hammer*, and there is no need in this case to use force generator equipment. In the remainder of this section, we discuss some techniques which require a means of applying a harmonic forcing function to the system in order to obtain the frequency response curve, from which some information can be obtained and used in evaluating the damping coefficient.

Figure 19 shows a frequency response curve for a moderately damped single degree of freedom system. The magnification factor β, as the result of application of a harmonic forcing function, is given by Eq. 60. At resonance, when $\omega_f = \omega$, the case in which the frequency ratio $r = 1$, the magnification factor β is given by

$$\beta_{r=1} = \frac{1}{2\xi} \tag{4.125}$$

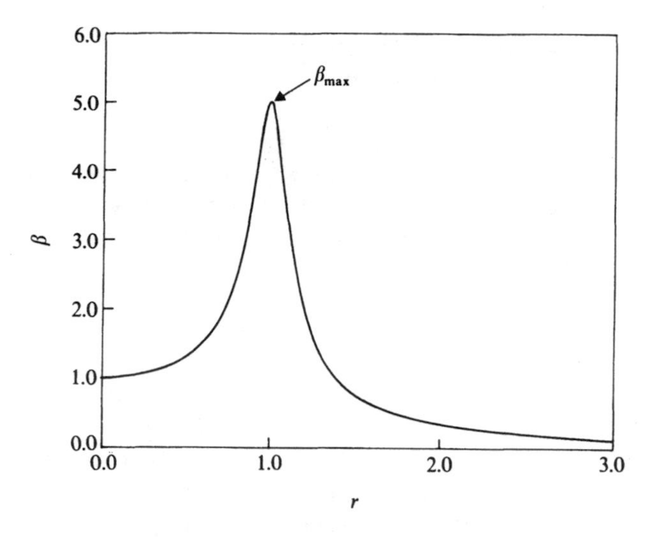

FIG. 4.19. Frequency response curve.

At resonance, one can use the frequency response curve to measure $\beta_{r=1}$, and use Eq. 125 to determine the damping factor ξ as

$$\xi = \frac{1}{2\beta_{r=1}} \tag{4.126}$$

The equivalent viscous damping coefficient c can then be determined by using Eq. 124.

In practice, it may be difficult to determine the exact resonance frequency, and in many cases, it is much easier to measure the maximum magnification factor β_{max}. It was shown in Section 4 that the maximum magnification factor β_{max} is given by (Eq. 63)

$$\beta_{max} = \frac{1}{2\xi\sqrt{1-\xi^2}} \tag{4.127}$$

If we assume that the damping is small, such that $\sqrt{1-\xi^2} \approx 1$, Eq. 127 yields an approximation for the damping factor ξ as

$$\xi \approx \frac{1}{2\beta_{max}} \tag{4.128}$$

By comparing Eqs. 126 and 128, it is clear that the use of Eq. 128 implies the assumption that the maximum magnification factor is equal to the resonance magnification factor, that is,

$$\beta_{max} \approx \beta_{r=1}$$

This approximation is acceptable when the system is lightly damped. The advantage of using Eq. 126 or Eq. 128 to evaluate the damping factor ξ is that only simple instrumentations are required for measuring the steady state amplitudes. The disadvantage, however, is that one must be able to evaluate the static deflection $X_0 = F_0/k$, in order to be able to plot the dimensionless magnification factor β versus the frequency ratio r. This may be a source of problems since many exciters cannot be operated at zero frequency. In the following, a method which alleviates this problem is discussed.

Bandwidth Method This method also utilizes the frequency response curve, but it does not require the use of the dimensionless magnification factor β. Note that the curve $X = \beta X_0$ has the same shape as the frequency response curve in which β and r are the coordinates, and this shape very much depends on the amount of damping in the system. In order to understand the bandwidth method, Eq. 60 is reproduced here for convenience

$$\beta = \frac{1}{\sqrt{(1 - r^2)^2 + (2r\xi)^2}} \tag{4.129}$$

The *bandwidth method*, which is sometimes referred to as the *half-power method*, is one of the most convenient techniques for determining the amount of damping in the system. In this technique, the damping factor is determined from the frequencies at which the displacement amplitudes are equal to $(1/\sqrt{2})\beta_{r=1}$. In order to determine these frequencies, we use Eqs. 125 and 129, which lead to

$$\frac{1}{\sqrt{2}}\frac{1}{2\xi} = \frac{1}{\sqrt{(1 - r^2)^2 + (2r\xi)^2}} \tag{4.130}$$

Squaring both sides of this equation yields

$$\frac{1}{8\xi^2} = \frac{1}{(1 - r^2)^2 + (2r\xi)^2}$$

which yields the following quadratic equation in r^2

$$r^4 + 2(2\xi^2 - 1)r^2 + (1 - 8\xi^2) = 0$$

This equation has two roots r_1^2 and r_2^2 which are given by

$$r_1^2 = 1 - 2\xi^2 - 2\xi(1 + \xi^2)^{1/2} \tag{4.131a}$$

$$r_2^2 = 1 - 2\xi^2 + 2\xi(1 + \xi^2)^{1/2} \tag{4.131b}$$

By using the binomial theorem, Eqs. 131a and 131b can be written as

$$r_1^2 = 1 - 2\xi^2 - 2\xi(1 + \tfrac{1}{2}\xi^2 + \cdots)$$

$$r_2^2 = 1 - 2\xi^2 + 2\xi(1 + \tfrac{1}{2}\xi^2 + \cdots)$$

Assuming that the damping is small, higher-order terms of ξ can be neglected, and

$$r_1 = (1 - 2\xi - 2\xi^2)^{1/2} \tag{4.132a}$$

$$r_2 = (1 + 2\xi - 2\xi^2)^{1/2} \tag{4.132b}$$

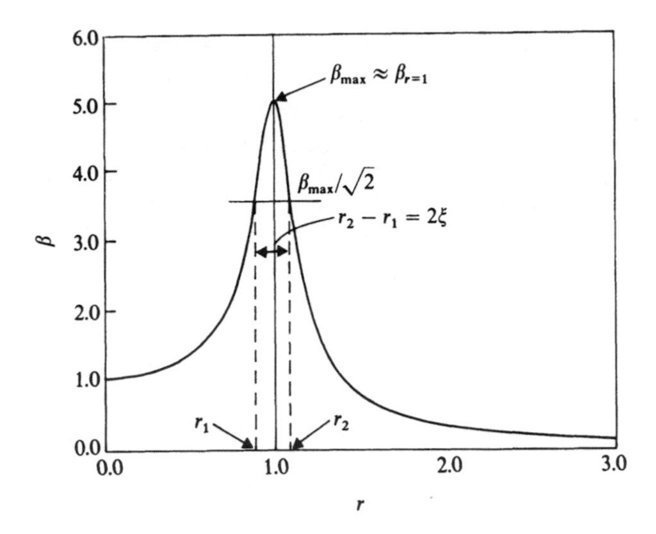

FIG. 4.20. Bandwidth method.

which upon using the binomial theorem and neglecting higher-order terms lead to

$$r_1 = 1 - \xi - \xi^2 \tag{4.133a}$$

$$r_2 = 1 + \xi - \xi^2 \tag{4.133b}$$

Subtracting the first equation from the second, one obtains the following simple expression for the damping factor ξ

$$\xi = \tfrac{1}{2}(r_2 - r_1) \tag{4.134}$$

Figure 20 shows the use of this method in evaluating the damping factor ξ. Using the frequency response curve, one can draw a horizontal line at a distance $\beta = (1/\sqrt{2})\beta_{r=1}$ from the r-axis. This horizontal line intersects the frequency response curve at two points which define the frequencies r_1 and r_2, which can be used in Eq. 134 to determine the damping factor ξ.

Energy Dissipated In Section 4, it was shown that the steady state response of the system, as the result of harmonic excitation, is given by Eq. 59 as

$$x_p = X_0\beta \sin(\omega_f t - \psi) \tag{4.135a}$$

where $X_0 = F_0/k$, β is the magnification factor, and ψ is the phase angle defined as

$$\psi = \tan^{-1}\frac{2r\xi}{1 - r^2}$$

At resonance, $\psi = \pi/2$ and $\omega_f = \omega$, and Eq. 135a can be written as

$$x_p = -X_0(\beta_{r=1}) \cos \omega t \tag{4.135b}$$

It follows that

$$\dot{x}_p = X_0(\beta_{r=1})\omega \sin \omega t$$

$$\ddot{x}_p = X_0(\beta_{r=1})\omega^2 \cos \omega t$$

Substituting these equations into the differential equation of motion of Eq. 43, one obtains

$$mX_0(\beta_{r=1})\omega^2 \cos \omega t + cX_0(\beta_{r=1})\omega \sin \omega t - kX_0(\beta_{r=1}) \cos \omega t = F_0 \sin \omega t$$

Equating the coefficients of the cosine function on both sides and equating the coefficients of the sine function on both sides of this equation, one obtains

$$mX_0(\beta_{r=1})\omega^2 = kX_0(\beta_{r=1}) \tag{4.136a}$$

and

$$cX_0(\beta_{r=1})\omega = F_0 \tag{4.136b}$$

Equation 136a implies that at resonance, the inertia force is equal to the elastic force, while Eq. 136b implies that at resonance the external force is balanced by the damping force. Observe that the coefficient of c in Eq. 136b is the maximum velocity $\dot{x}_{pm}$ defined as

$$\dot{x}_{pm} = X_0(\beta_{r=1})\omega, \tag{4.137}$$

and therefore, Eq. 136b can be used to determine the damping coefficient c as

$$c = \frac{F_0}{\dot{x}_{pm}} = \frac{F_0}{X_0(\beta_{r=1})\omega} \tag{4.138}$$

This equation shows that the damping coefficient is the ratio of the maximum force to the maximum velocity, and as a result, the damping coefficient c can be evaluated from a test run only at the resonance frequency, thus eliminating the need to construct the entire frequency response curve. One may experimentally obtain the case of resonance by adjusting the input frequency until the response is $\pi/2$ out of phase with the applied force.

In Section 4, we have also shown that the energy dissipated per cycle is given by

$$W_d = \pi c X_0^2 \beta^2 \omega_f^2 \tag{4.139}$$

If one constructs the force-displacement relationship curve at resonance for one cycle, the energy loss W_d can be determined as the area under this curve, since at resonance the damping force is equal to the applied force. In this case, an equivalent viscous damping coefficient can be determined as

$$c = \frac{W_d}{\pi X_0^2 \beta^2 \omega^2} \tag{4.140}$$

Problems

4.1. A spring–mass undamped single degree of freedom system is subjected to a harmonic forcing function which has an amplitude of 40 N and frequency 30 rad/s. The mass of the system is 2 kg and the spring coefficient is 1000 N/m. Find the amplitude of the forced vibration.

4.2. In Problem 1 find the complete solution in the following cases:

(a) $x_0 = 0$, $\dot{x}_0 = 0$.
(b) $x_0 = 0.01$ m, $\dot{x}_0 = 0$.
(c) $x_0 = 0$, $\dot{x}_0 = 2$ m/s.
(d) $x_0 = 0.01$ m, $\dot{x}_0 = 2$ m/s.

4.3. In a single degree of freedom undamped spring–mass system, the mass $m = 3$ kg, and the stiffness coefficient $k = 2 \times 10^3$ N/m. The system is subjected to a harmonic forcing function which has an amplitude 60 N and frequency 30 rad/s. The initial conditions are such that $x_0 = 0.0$ m and $\dot{x}_0 = 1$ m/s. Determine the displacement, velocity, and acceleration of the mass after $t = 1$ s.

4.4. A spring–mass system is subjected to a harmonic force which has an amplitude 30 N and frequency 20 rad/s. The system has mass $m = 5$ kg, and stiffness coefficient $k = 2 \times 10^3$ N/m. The initial conditions are such that $x_0 = 0$, $\dot{x}_0 = 2$ m/s. Determine the displacement, velocity, and acceleration of the mass after 0.5, 1, 1.5 s.

4.5. Determine the steady state response of a single degree of freedom system subjected to a harmonic forcing function $F(t) = F_0 \cos \omega_f t$.

4.6. Determine the steady state response of a single degree of freedom system subjected to the forcing function $F_1 \sin \omega_1 t + F_2 \sin \omega_2 t$.

4.7. The system shown in Fig. P1 consists of a uniform slender rod of mass m and length l. The rod is connected at its center of mass to a linear spring which has a stiffness coefficient k. The system is subjected to a harmonic forcing function $F(t) = F_0 \sin \omega_f t$, as shown in the figure. Determine the differential equation of motion and the complete solution in terms of the initial conditions θ_0 and $\dot{\theta}_0$. Assume small angular oscillations.

4.8. Find the differential equation of motion of the torsional system shown in Fig. P2. If the torque is given by $T = T_0 \sin \omega_f t$, determine the complete solution for zero initial conditions.

4.9. Assuming small oscillations, derive the differential equation of motion of the system shown in Fig. P3. Determine the complete solution in terms of the initial conditions θ_0 and $\dot{\theta}_0$.

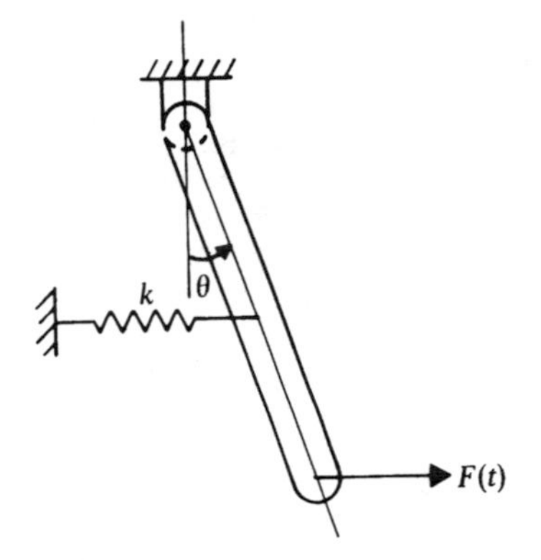

FIG. P4.1

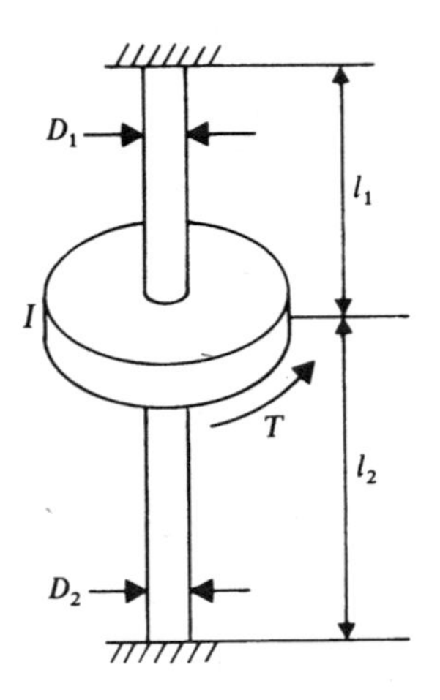

FIG. P4.2

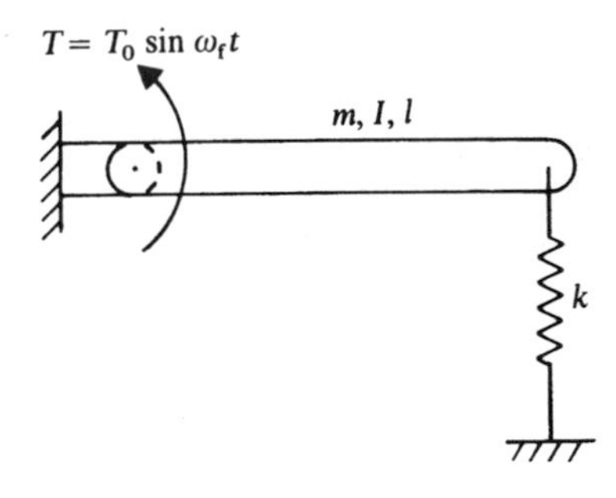

FIG. P4.3

4.10. The mass of an undamped single degree of freedom mass–spring system is subjected to a harmonic forcing function having an amplitude of 40 N and frequency 30 rad/s. The mass m is 4 kg and exhibits a forced displacement amplitude of 5 mm. Determine the stiffness coefficient of the spring.

4.11. A damped single degree of freedom mass–spring system has a mass $m = 3$ kg, a spring stiffness $k = 2700$ N/m, and a damping coefficient $c = 18$ N·s/m. The mass is subjected to a harmonic force which has an amplitude $F_0 = 20$ N and a frequency $\omega_f = 15$ rad/s. The initial conditions are $x_0 = 4$ cm, and $\dot{x}_0 = 0$. Determine the displacement, velocity, and acceleration of the mass after time $t = 0.5$ s.

4.12. Repeat Problem 11 for zero initial conditions, that is, $x_0 = 0$ and $\dot{x}_0 = 0$.

4.13. Repeat Problem 11 for the following two cases of the damping coefficient c:

(a) $c = 180$ N·s/m; (b) $c = 360$ N·s/m.

4.14. Repeat Problem 11 if the frequency of the harmonic force ω_f is equal to the system natural frequency.

4.15. A damped single degree of freedom mass–spring system is excited at resonance by a harmonic forcing function which has an amplitude of 40 N. The system has mass m of 3 kg, a stiffness coefficient k of 2700 N/m, and a damping coefficient c of 20 N·s/m. If the initial conditions are such that $x_0 = 5$ cm, and $\dot{x}_0 = 0$, determine the displacement, velocity, and acceleration of the mass after $t = 0.2$ s.

4.16. A damped single degree of freedom mass–spring system is excited at resonance by a harmonic forcing function which has an amplitude of 80 N. It was observed that the steady state amplitude of the forced vibration is 6 mm. It was also observed that when the frequency of excitation is three times the natural frequency of the system, the amplitude of the steady state vibration becomes 1 mm. Determine the stiffness coefficient k and the damping factor ξ.

4.17. Determine the differential equation of motion and the steady state response of the system shown in Fig. P4, where F_1 and F_2 are harmonic forcing functions given, respectively, by $F_1 = F_{10} \sin \omega_1 t$ and $F_2 = F_{20} \sin \omega_2 t$.

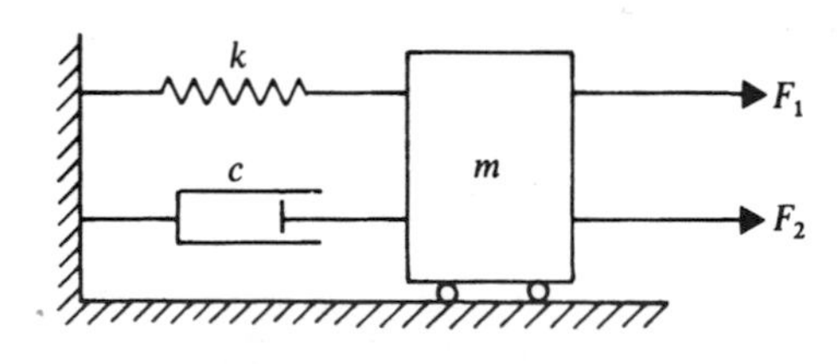

FIG. P4.4

4.18. Assuming small oscillations, derive the differential equation of motion of the system shown in Fig. P5. Determine also the steady state response of this system.

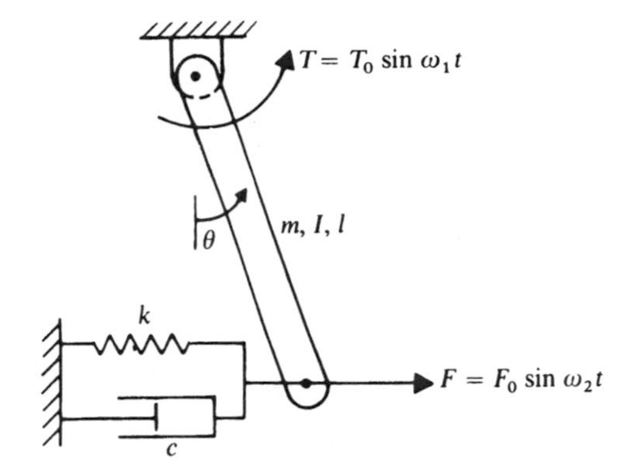

FIG. P4.5

4.19. The data of Problem 18 are such that $m = 0.5$ kg, $l = 0.5$ m, $k = 1000$ N/m, $c = 20$ N·s/m, and the beam is a uniform slender beam; the forcing functions are $F = 40 \sin 20t$ N and $T = 10 \sin 40t$ N·m; and the initial conditions are such that $\theta_0 = 0$ and $\dot{\theta}_0 = 3$ rad/s. Determine the displacement, velocity and acceleration of the center of mass of the rod after time $t = 0.3$ s.

4.20. Repeat Problem 19 for zero initial conditions, that is, $\theta_0 = \dot{\theta}_0 = 0$.

4.21. Derive the differential equation of motion of the system shown in Fig. P6. Obtain the steady state solution of the absolute motion of the mass. Also obtain the displacement of the mass with respect to the moving base.

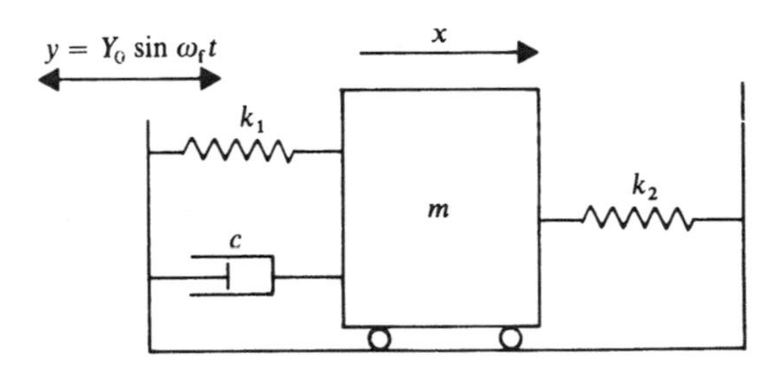

FIG. P4.6

4.22. For the system shown in Fig. P6, let $m = 3$ kg, $k_1 = k_2 = 1350$ N/m, $c = 40$ N·s/m, and $y = 0.04 \sin 15t$. The initial conditions are such that $x_0 = 5$ mm and $\dot{x}_0 = 0$. Determine the displacement, velocity, and acceleration of the mass after time $t = 1$ s.

4.23. Repeat Problem 22 for the following two cases:

(a) $y = 0.04 \sin 30t$; (b) $y = 0.04 \sin 60t$.

4.24. In Problem 22 determine the steady state amplitude of the displacement of the mass with respect to the moving base.

4.25. In Problem 22, determine the steady state amplitude of the force transmitted to the moving base.

4.26. Derive the differential equation of motion of the damped single degree of freedom mass–spring system shown in Fig. P7. Obtain the steady state solution and the amplitude of the force transmitted to the base.

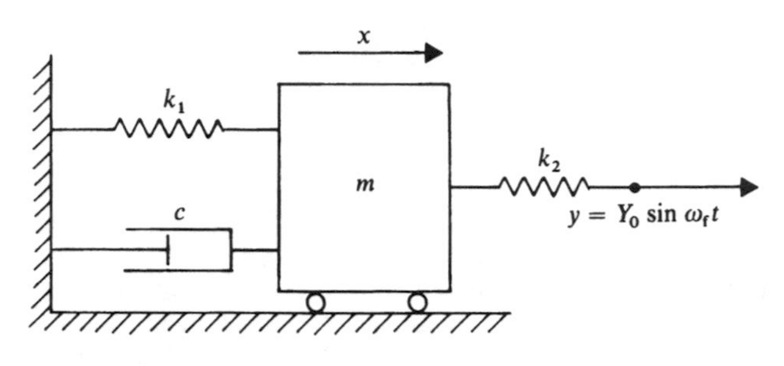

FIG. P4.7

4.27. For the system shown in Fig. P7, let $m = 3$ kg, $k_1 = k_2 = 1350$ N/m, $c = 40$ N·s/m, and $y = 0.04 \sin 15t$. The initial conditions are $x_0 = 5$ mm and $\dot{x}_0 = 0$. Determine the displacement, velocity, and acceleration of the mass after time $t = 1$ s.

4.28. In Problem 27, determine the steady state amplitude of the force transmitted to the base.

4.29. Repeat Problem 27 for the two case:

(a) $k_1 = 0$; (b) $c = 0$.

4.30. Derive the differential equation of motion of the system shown in Fig. P8, assuming small angular oscillations. Determine also the steady state response of this system.

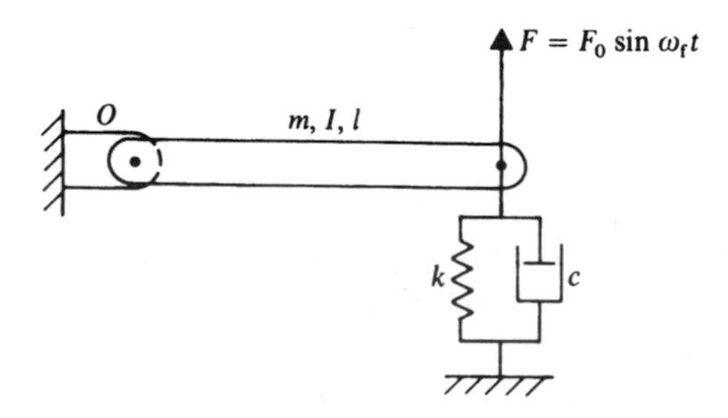

FIG. P4.8

4.31. In Problem 30, let the rod be uniform and slender with mass $m = 0.5$ kg and $l = 0.5$ m. Let $k = 2000$ N/m, $c = 20$ N · s/m, and $F = 10 \sin 10t$ N. The initial conditions are $\theta_0 = 0$ and $\dot{\theta}_0 = 3$ rad/s. Determine the displacement equation of the beam as a function of time.

4.32. Assuming small angular oscillations, derive the differential equation of motion of the uniform slender beam shown in Fig. P9. Obtain an expression for the steady state solution as a function of time.

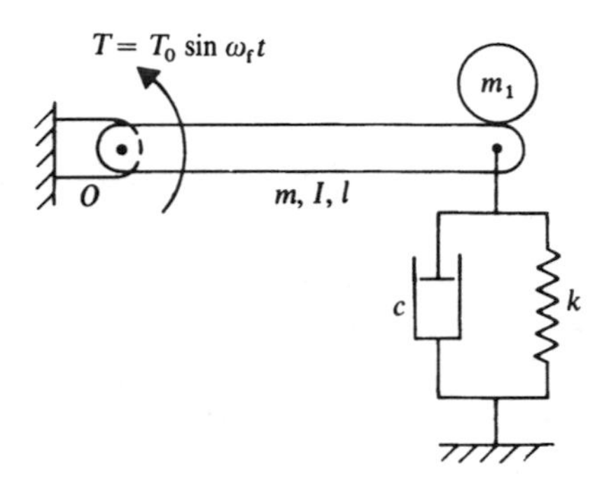

FIG. P4.9

4.33. In Problem 32, let $m = 1$ kg, $m_1 = 0.3$ kg, $l = 0.5$ m, $k = 2000$ N/m, $c = 30$ N · s/m, and $T = 10 \sin 10t$ N · m. The initial conditions are $\theta_0 = 0$ and $\dot{\theta}_0 = 5$ rad/s. Determine the angular velocity of the beam after time $t = 1$ s.

4.34. Assuming small angular oscillations, derive the differential equation of motion of the system shown in Fig. P10. If $m_1 = m_2 = 0.5$ kg, $l_1 = l_2 = 0.5$ m, $k = 10^4$ N/m, $c = 50$ N · s/m, and $T = 5 \sin 5t$ N · m, determine the complete solution as a function of time for the initial conditions $\theta_0 = 0$ and $\dot{\theta}_0 = 0$.

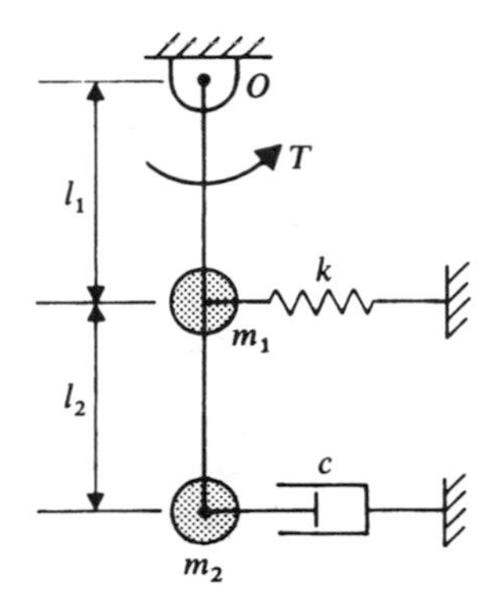

FIG. P4.10

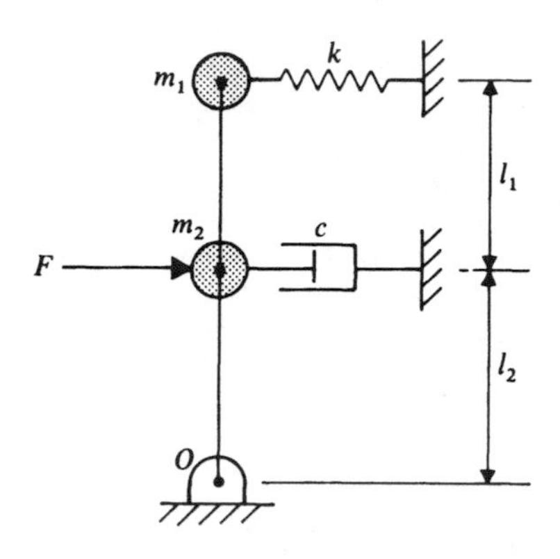

FIG. P4.11

4.35. Assuming small angular oscillations, derive the differential equation of motion of the vibratory system shown in Fig. P11. If $m_1 = m_2 = 0.5$ kg, $l_1 = l_2 = 0.5$ m, $k = 10^4$ N/m, $c = 50$ N·s/m, and $F = 10 \sin 5t$ N, determine the complete solution as a function of time for the initial conditions $\theta_0 = 0$ and $\dot{\theta}_0 = 2$ rad/s.

4.36. The system shown in Fig. P12 consists of a uniform slender rod of mass m and length l. The rod is connected to the ground by a pin joint at O and is supported by a spring and damper at one end, as shown in the figure. The rod is subjected to a harmonic force F that applies at the other end of the rod. Derive the differential equation of motion of this system assuming small angular oscillation. Obtain the steady state solution.

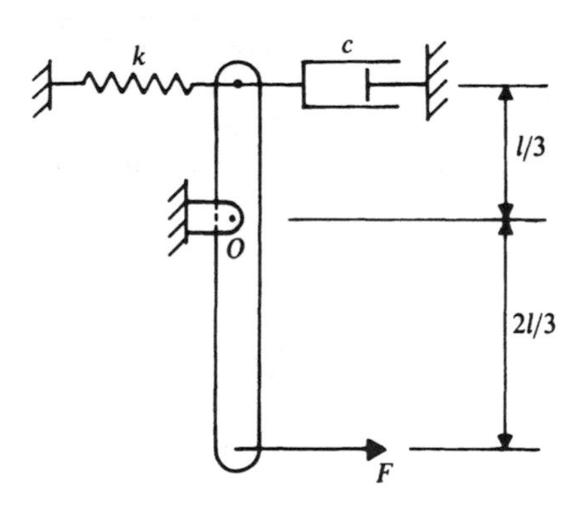

FIG. P4.12

4.37. In Problem 36, let $m = 0.5$ kg, $l = 0.9$ m, $k = 3 \times 10^3$ N/m, $c = 10$ N·s/m, and $F = 5 \sin 10t$ N. Determine the complete solution as a function of time for the following initial conditions $\theta_0 = 0$ and $\dot{\theta}_0 = 3$ rad/s.

4.38. Assuming small angular oscillations, derive the differential equation of motion of the vibratory system shown in Fig. P13.

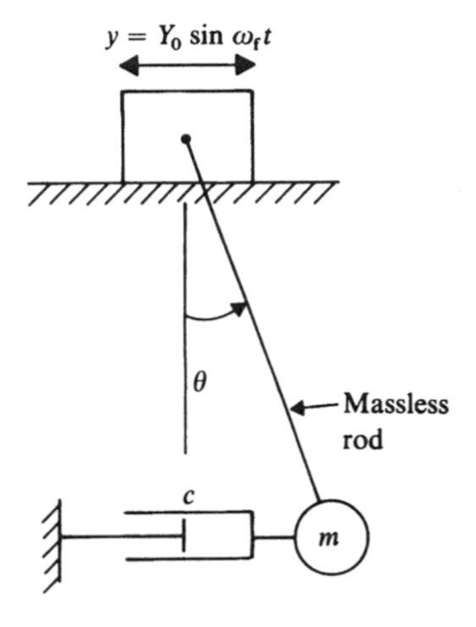

FIG. P4.13

4.39. Assuming small angular oscillations, derive the differential equation of motion of the system shown in Fig. P14 where the rod is assumed to be uniform and slender. The mass and length of the rod are assumed to be m and l, respectively.

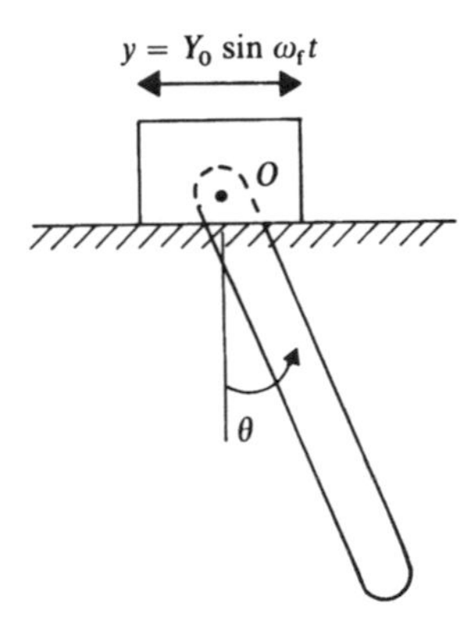

FIG. P4.14

4.40. Derive the differential equation of motion of the simple vehicle model shown in Fig. P15. The vehicle is assumed to travel over the rough surface with a constant vehicle speed v. Obtain also the steady state response of the vehicle.

4.41. For the system shown in Fig. P15, let $m = 10$ kg, $k = 4 \times 10^3$ N/m, $c = 150$ N $\cdot$ s/m, $\lambda = 2$m, and the amplitude $Y_0 = 0.1$ m. Determine the maximum vertical displacement of the mass and the corresponding vehicle speed. Determine also the maximum dynamic force transmitted to the mass at the resonant speed.

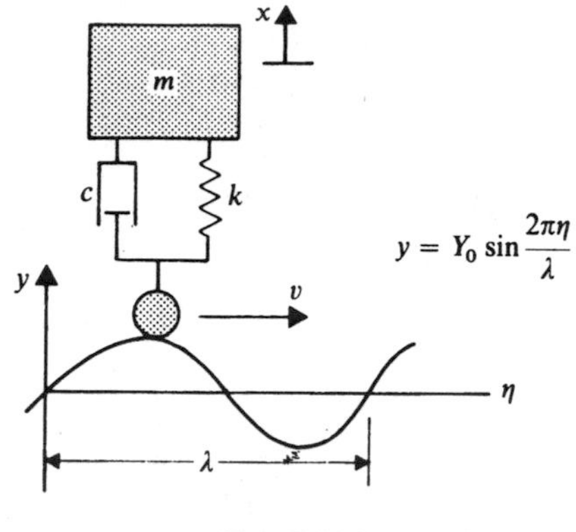

FIG. P4.15

4.42. A vibrometer has a natural frequency of 5.5 Hz and a damping factor ξ of 0.6. This instrument is used to measure the displacement of a machine vibrating at 4 Hz. It was assumed that the amplitude measured by the vibrometer is 0.085 m. Determine the exact value of the amplitude of the machine.

4.43. Design an accelerometer which can be used to measure vibrations in the range 0–30 Hz with a maximum error at 0.8%. Select the mass, stiffness, and damping coefficients for the accelerometer.

5
Response to Nonharmonic Forces

The response of damped and undamped single degree of freedom systems to harmonic forcing functions was discussed in the preceding chapter. It was shown that the steady state response of the system to such excitations is also harmonic, with a phase difference between the force and the displacement which depends on the amount of damping. The analysis presented and the concepts introduced in the preceding chapter are fundamental to the study of the theory of vibration, and the use of these concepts and methods of vibration analysis was demonstrated by several applications.

In this chapter, the response of the single degree of freedom system to more general forcing functions will be discussed. First, we consider the vibration of the single degree of freedom system under a periodic forcing function, which has the property of repeating itself in all details after a certain time interval, called the *period*. In later sections, the vibration of the single degree of freedom system under general forcing functions will be considered.

5.1 PERIODIC FORCING FUNCTIONS

A forcing function $F(t)$ is said to be periodic if there exists a positive real number T_f such that

$$F(t + T_f) = F(t) \tag{5.1}$$

for any given time t. The real number T_f is called the *period* of F. Clearly, if T_f is a period of $F(t)$, then nT_f is also a period, where n is any given positive integer. This can be easily verified, since if we define $t_1 = t + T_f$, then we have

$$F(t_1) = F(t + T_f)$$

Equation 1 then yields

$$F(t_1) = F(t_1 + T_f) = F(t + 2T_f) = F(t)$$

That is, $2T_f$ is also a period. In like manner, one can also verify that nT_f is also a period for any positive integer n. If T_f is the smallest period of F, T_f is called

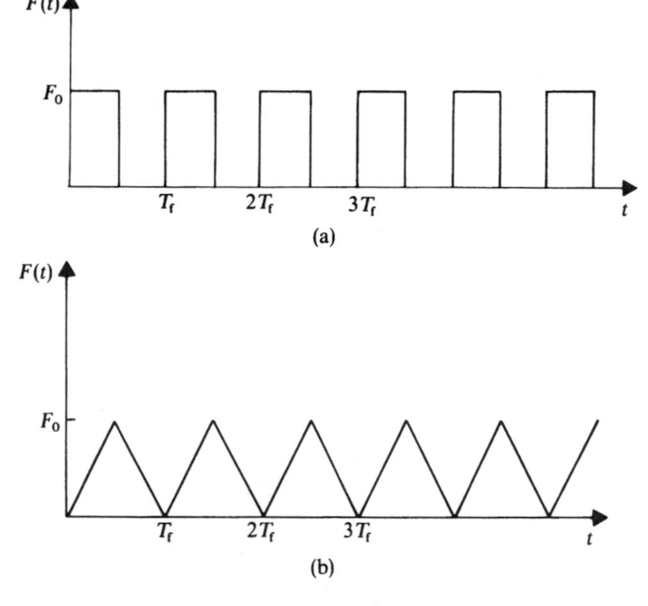

FIG. 5.1. Typical periodic functions.

the *fundamental period* or simply the period. Henceforth, the forcing function F will be said to be periodic of period T_f if and only if T_f is the fundamental period.

Any harmonic function such as the sine and cosine function is periodic. The converse, however, is not true, that is, any periodic function is not necessarily a harmonic function, as demonstrated by the examples of the periodic functions shown in Fig. 1. Periodic functions can be written as the sum of harmonic functions using *Fourier series*.

Fourier Series Let us define the fundamental frequency ω_f to be

$$\omega_f = \frac{2\pi}{T_f} \tag{5.2}$$

For a given periodic forcing function $F(t)$ with a period T_f, the Fourier expansion of this function can be expressed in terms of harmonic functions as

$$F(t) = \frac{a_0}{2} + \sum_{n=1}^{\infty} a_n \cos n\omega_f t + \sum_{n=1}^{\infty} b_n \sin n\omega_f t \tag{5.3}$$

where ω_f is given by Eq. 2, and a_0, a_n, and b_n are constants to be determined in the following section.

Observe that $F(t)$ can also be written in the following form

$$F(t) = F_0 + \sum_{n=1}^{\infty} F_n \sin(\omega_n t + \phi_n) \tag{5.4}$$

where

$$F_0 = \frac{a_0}{2} \tag{5.5}$$

$$F_n = \sqrt{a_n^2 + b_n^2} \tag{5.6}$$

$$\omega_n = n\omega_f \tag{5.7}$$

and

$$\phi_n = \tan^{-1}\left(\frac{a_n}{b_n}\right) \tag{5.8}$$

The amplitudes F_0 and F_n and the phase angle ϕ_n can be determined once the coefficients a_0, a_n, and b_n in the Fourier series are determined.

5.2 DETERMINATION OF THE FOURIER COEFFICIENTS

In this section, methods for the analytical and numerical evaluation of the Fourier coefficients a_0, a_n, and b_n that appear in Eq. 3 are discussed. The numerical technique can be used in the cases in which the function is not described by a simple curve or in which the function is provided in a tabulated form.

Coefficient a_0 By integrating Eq. 3 over the period $(0, T_f)$ or, equivalently, over the period $(-T_f/2, T_f/2)$, one obtains

$$\int_{-T_f/2}^{T_f/2} F(t)\,dt = \int_{-T_f/2}^{T_f/2} \frac{a_0}{2}\,dt + \sum_{n=1}^{\infty} \int_{-T_f/2}^{T_f/2} (a_n \cos n\omega_f t + b_n \sin n\omega_f t)\,dt \tag{5.9}$$

For any integer n, it can be verified that

$$\int_{-T_f/2}^{T_f/2} \cos n\omega_f t\,dt = 0 \tag{5.10}$$

$$\int_{-T_f/2}^{T_f/2} \sin n\omega_f t\,dt = 0 \tag{5.11}$$

Substituting Eqs. 10 and 11 into Eq. 9 yields

$$\int_{-T_f/2}^{T_f/2} F(t)\,dt = \frac{a_0}{2} \int_{-T_f/2}^{T_f/2} dt = \frac{a_0}{2} T_f$$

that is,

$$a_0 = \frac{2}{T_f} \int_{-T_f/2}^{T_f/2} F(t)\,dt \tag{5.12}$$

The integral in Eq. 12 represents the area under the curve $F(t)$ in one period, and therefore, the constant a_0 is this area multiplied by the constant $2/T_f$.

Coefficients a_n, $n = 1, 2, \ldots$ In order to determine the coefficient a_m, for a fixed integer m, in the Fourier series of Eq. 3, we multiply Eq. 3 by $\cos m\omega_f t$ and integrate over the interval $(-T_f/2, T_f/2)$ to obtain

$$\int_{-T_f/2}^{T_f/2} F(t) \cos m\omega_f t \, dt = \int_{-T_f/2}^{T_f/2} \frac{a_0}{2} \cos m\omega_f t \, dt + \sum_{n=1}^{\infty} \int_{-T_f/2}^{T_f/2} a_n \cos n\omega_f t \cos m\omega_f t \, dt + \sum_{n=1}^{\infty} \int_{-T_f/2}^{T_f/2} b_n \sin n\omega_f t \cos m\omega_f t \, dt \tag{5.13}$$

One can verify the following identities

$$\int_{-T_f/2}^{T_f/2} \cos n\omega_f t \cos m\omega_f t \, dt = \begin{cases} 0 & \text{if } m \neq n \\ T_f/2 & \text{if } m = n \end{cases} \tag{5.14}$$

$$\int_{-T_f/2}^{T_f/2} \sin n\omega_f t \cos m\omega_f t \, dt = 0 \tag{5.15}$$

Using these identities and Eq. 10, Eq. 13 can be written as

$$\int_{-T_f/2}^{T_f/2} F(t) \cos m\omega_f t \, dt = a_m T_f/2$$

that is,

$$a_m = \frac{2}{T_f} \int_{-T_f/2}^{T_f/2} F(t) \cos m\omega_f t \, dt \tag{5.16}$$

Coefficients b_n, $n = 1, 2, \ldots$ In order to determine the coefficient b_m in the Fourier expansion of Eq. 3, we multiply Eq. 3 by $\sin m\omega_f t$ and integrate over the interval $(0, T_f)$ or, equivalently, over the interval $(-T_f/2, T_f/2)$. This leads to

$$\int_{-T_f/2}^{T_f/2} F(t) \sin m\omega_f t \, dt = \int_{-T_f/2}^{T_f/2} \frac{a_0}{2} \sin m\omega_f t \, dt + \sum_{n=1}^{\infty} \int_{-T_f/2}^{T_f/2} a_n \cos n\omega_f t \sin m\omega_f t \, dt + \sum_{n=1}^{\infty} \int_{-T_f/2}^{T_f/2} b_n \sin n\omega_f t \sin m\omega_f t \, dt \tag{5.17}$$

The following identity can be verified for any positive integers n and m

$$\int_{-T_f/2}^{T_f/2} \sin n\omega_f t \sin m\omega_f t \, dt = \begin{cases} 0 & \text{if } m \neq n \\ T_f/2 & \text{if } m = n \end{cases} \tag{5.18}$$

Using this identity and the identities of Eqs. 11 and 15, Eq. 17 yields

$$\int_{-T_f/2}^{T_f/2} F(t) \sin m\omega_f t \, dt = b_m T_f/2$$

that is,

$$b_m = \frac{2}{T_f} \int_{-T_f/2}^{T_f/2} F(t) \sin m\omega_f t \, dt \tag{5.19}$$

Equations 12, 16, and 19 are the basic equations that can be used to determine the coefficients that appear in the Fourier expansion of the periodic function $F(t)$. The use of these equations is demonstrated by the following example.

Example 5.1

Find the Fourier series of the periodic function $F(t)$ shown in Fig. 2.

Solution. The function $F(t)$ shown in the figure is defined over the interval $(-T_f/2, T_f/2)$ by

$$F(t) = \begin{cases} 0 & \text{for} \quad -T_f/2 < t < 0 \\ F_0 & \text{for} \quad 0 \leq t \leq T_f/2 \end{cases}$$

Therefore, the coefficients a_0, a_m, and b_m in the Fourier series of Eq. 3 are obtained as follows

$$\begin{aligned} a_0 &= \frac{2}{T_f} \int_{-T_f/2}^{T_f/2} F(t) \, dt = \frac{2}{T_f} \left[\int_{-T_f/2}^{0} (0) \, dt + \int_0^{T_f/2} F_0 \, dt \right] \\ &= \frac{2F_0}{T_f} \frac{T_f}{2} = F_0 \end{aligned}$$

$$\begin{aligned} a_m &= \frac{2}{T_f} \int_{-T_f/2}^{T_f/2} F(t) \cos m\omega_f t \, dt \\ &= \frac{2}{T_f} \left[\int_{-T_f/2}^{0} (0) \cos m\omega_f t \, dt + \int_0^{T_f/2} F_0 \cos m\omega_f t \, dt \right] \\ &= \frac{2F_0}{T_f} \int_0^{T_f/2} \cos m\omega_f t \, dt = \frac{2F_0}{m\omega_f T_f} \sin m\omega_f t \bigg|_0^{T_f/2} \\ &= \frac{2F_0}{m\omega_f T_f} \sin(m\omega_f T_f/2) \end{aligned}$$

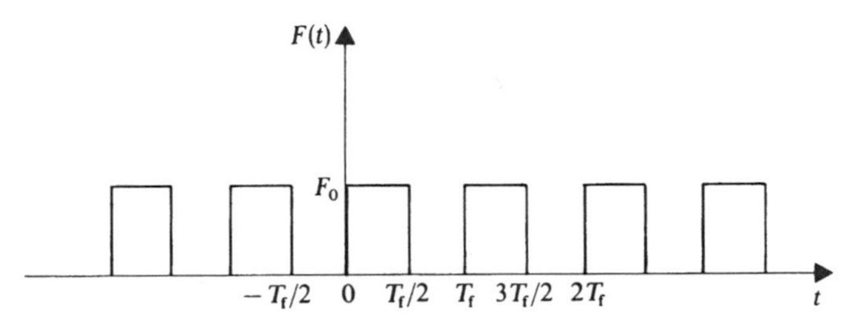

FIG. 5.2. Periodic function $F(t)$.

Using Eq. 2, the coefficient a_m can be written as

$$a_m = \frac{F_0}{\pi m} \sin m\pi = 0$$

$$b_m = \frac{2}{T_f} \int_{-T_f/2}^{T_f/2} F(t) \sin m\omega_f t \, dt = \frac{2}{T_f} \left[\int_{-T_f/2}^{0} (0) \sin m\omega_f t \, dt + \int_{0}^{T_f/2} F_0 \sin m\omega_f t \, dt \right]$$

$$= \frac{2F_0}{T_f} \int_0^{T_f/2} \sin m\omega_f t \, dt = -\frac{2F_0}{m\omega_f T_f} \cos m\omega_f t \Big|_0^{T_f/2} = -\frac{F_0}{\pi m} [\cos m\pi - 1]$$

$$= \begin{cases} \dfrac{2F_0}{\pi m} & \text{if } m \text{ is odd} \\ 0 & \text{if } m \text{ is even} \end{cases}$$

or, equivalently,

$$b_m = \frac{2F_0}{m\pi}, \qquad m = 1, 3, 5, \ldots$$

Therefore, the Fourier series of the periodic forcing function shown in Fig. 2 is given by

$$F(t) = \frac{F_0}{2} + \frac{2F_0}{\pi} \sin \omega_f t + \frac{2F_0}{3\pi} \sin 3\omega_f t + \frac{2F_0}{5\pi} \sin 5\omega_f t + \ldots$$

$$= \frac{F_0}{2} + \sum_{n=1,3,5,}^{\infty} \frac{2F_0}{n\pi} \sin n\omega_f t = F_0 \left[\frac{1}{2} + \sum_{n=1,3,5,}^{\infty} \frac{2}{n\pi} \sin n\omega_f t \right]$$

where $\omega_f = 2\pi/T_f$.

Numerical Solution for Fourier Coefficients In many applications the periodic functions cannot be described by simple curves. These functions may be obtained through experimental measurements and, consequently, their values at different points in time are given in a tabulated form. In such cases, the direct integration of the periodic functions in a closed analytical form may be impossible. One must then resort to numerical calculations in evaluating the Fourier coefficients.

If the period of the function is T_f, one may divide this period into n_p equal intervals. The length of each interval is then given by

$$\Delta t = \frac{T_f}{n_p}$$

Replacing the integrals in Eqs. 12, 16, and 19 by the finite sums, one obtains

$$a_0 = \frac{2}{T_f} \sum_{j=1}^{n_p} F(t_j) \, \Delta t$$

$$a_m = \frac{2}{T_f} \sum_{j=1}^{n_p} \{F(t_j) \cos m\omega_f t_j\} \Delta t$$

$$b_m = \frac{2}{T_f} \sum_{j=1}^{n_p} \{F(t_j) \sin m\omega_f t_j\} \Delta t$$

where

$$\omega_f = \frac{2\pi}{T_f} = \frac{2\pi}{n_p \Delta t}$$

Substituting in the preceding equations for $T_f = n_p \Delta t$, one obtains

$$a_0 = \frac{2}{n_p} \sum_{j=1}^{n_p} F(t_j)$$

$$a_m = \frac{2}{n_p} \sum_{j=1}^{n_p} F(t_j) \cos m\omega_f t_j$$

$$b_m = \frac{2}{n_p} \sum_{j=1}^{n_p} F(t_j) \sin m\omega_f t_j$$

These three equations are the equivalent counterparts of the integrals given by Eqs. 12, 16, and 19 for the evaluation of the Fourier coefficients. The accuracy of the numerical evaluation of Fourier coefficients by using these equations depends on the number of intervals n_p, and this accuracy increases by increasing n_p and decreases by decreasing n_p.

Illustrative Example In order to demonstrate the use of the techniques developed in this section for the evaluation of the coefficients of the Fourier series, we consider the periodic function shown in Fig. 1(b). This function is defined as

$$F(t) = \begin{cases} \dfrac{2F_0}{T_f} t, & 0 \le t < \dfrac{T_f}{2} \\ -\dfrac{2F_0}{T_f} t + 2F_0, & \dfrac{T_f}{2} \le t < T_f \end{cases}$$

The use of Eqs. 12, 16, and 19 show that the coefficient a_0, a_m, and b_m are

$$a_0 = F_0$$

$$a_m = \frac{2F_0}{(m\pi_f)^2}[\cos m\pi - 1] = \begin{cases} \dfrac{-4F_0}{(m\pi)^2} & \text{if } m \text{ is odd} \\ 0 & \text{if } m \text{ is even} \end{cases}$$

$$b_m = 0$$

The coefficient a_0 in this example can be evaluated numerically using the equation

$$a_0 = \frac{2}{n_p} \sum_{j=1}^{n_p} F(t_j)$$

$$= \frac{2}{n_p} \left\{ \sum_{j=1}^{n_p/2} \frac{2F_0}{T_f} t_j + \sum_{j=n_p/2+1}^{n_p} \left[-\frac{2F_0}{T_f} t_j + 2F_0 \right] \right\} = F_0$$

Similarly, the coefficients a_m and b_m can be evaluated numerically as

$$a_m = \frac{2}{n_p}\sum_{j=1}^{n_p} F(t_j)\cos m\omega_f t_j$$

$$= \frac{2}{n_p}\left\{\sum_{j=1}^{n_p/2}\frac{2F_0}{T_f}t_j\cos m\omega_f t_j + \sum_{j=n_p/2+1}^{n_p}\left(-\frac{2F_0}{T_f}t_j + 2F_0\right)\cos m\omega_f t_j\right\}$$

$$b_m = \frac{2}{n_p}\sum_{j=1}^{n_p} F(t_j)\sin m\omega_f t_j$$

$$= \frac{2}{n_p}\left\{\sum_{j=1}^{n_p/2}\frac{2F_0}{T_f}t_j\sin m\omega_f t_j + \sum_{j=n_p/2+1}^{n_p}\left(-\frac{2F_0}{T_f}t_j + 2F_0\right)\sin m\omega_f t_j\right\}$$

Table 1 shows the coefficients a_m and b_m obtained numerically for different harmonics using different numbers of intervals. It is clear from the results presented in this table that the coefficients obtained numerically using a sufficiently large number of intervals are in good agreement with the coefficient obtained by using the analytical methods.

Example 1 shows that the periodic function need not be continuous in order to have a valid Fourier expansion with coefficients determined by the integral of Eqs. 12, 16, and 19. One can show, however, at the points of discontinuity,

TABLE 5.1. Fourier Coefficients ($F_0 = 1$)

Harmonics	Numerical Results						Analytical Results	
	$n_p = 10$		$n_p = 50$		$n_p = 100$			
m	a_m	b_m	a_m	b_m	a_m	b_m	a_m	b_m
1	−0.419	0.000	−0.406	0.000	−0.405	0.000	−0.405	0.000
2	0.000	0.000	0.000	0.000	0.000	0.000	0.000	0.000
3	−0.061	0.000	−0.046	0.000	−0.045	0.000	−0.045	0.000
4	0.000	0.000	0.000	0.000	0.000	0.000	0.000	0.000
5	−0.040	0.000	−0.017	0.000	−0.016	0.000	−0.016	0.000
6	0.000	0.000	0.000	0.000	0.000	0.000	0.000	0.000
7	−0.061	0.000	−0.009	0.000	−0.008	0.000	−0.008	0.000
8	0.000	0.000	0.000	0.000	0.000	0.000	0.000	0.000
9	−0.419	0.000	−0.006	0.000	−0.005	0.000	−0.005	0.000
10	1.000	0.000	0.000	0.000	0.000	0.000	0.000	0.000
11	−0.419	0.000	−0.004	0.000	−0.003	0.000	−0.003	0.000
12	0.000	0.000	0.000	0.000	0.000	0.000	0.000	0.000
13	−0.061	0.000	−0.003	0.000	−0.003	0.000	−0.003	0.000
14	0.000	0.000	0.000	0.000	0.000	0.000	0.000	0.000
15	−0.040	0.000	−0.002	0.000	−0.002	0.000	−0.002	0.000
16	0.000	0.000	0.000	0.000	0.000	0.000	0.000	0.000
17	−0.061	0.000	−0.002	0.000	−0.002	0.000	−0.001	0.000
18	0.000	0.000	0.000	0.000	0.000	0.000	0.000	0.000
19	−0.419	0.000	−0.002	0.000	−0.001	0.000	−0.001	0.000
20	1.000	0.000	0.000	0.000	0.000	0.000	0.000	0.000

the Fourier expansion of the function $F(t)$ converges to the average of the right- and left-hand limits of the function. The questions concerning the convergence of Fourier series are answered by the *theorem of Dirichlet* which states that if $F(t)$ is a bounded periodic function which has a finite number of maximum and minimum points and a finite number of points of discontinuity, then the Fourier expansion of $F(t)$ converges to $F(t)$ at all points where $F(t)$ is continuous and converges to the average of the right- and left-hand limits of $F(t)$ at the points of discontinuity.

In order to illustrate the application of *Dirichlet conditions* when the Fourier coefficients are determined numerically we consider the periodic function given in Example 1. If we assume that n_p is given, and keeping in mind that $F(t) = 0$ half the period and equal F_0 the other half, the coefficient a_0 can be determined as

$$a_0 = \frac{2}{n_p}\sum_{j=1}^{n_p} F(t_j) = \frac{2}{n_p}\left[\sum_{j=1}^{n_p/2}(0) + \sum_{j=(n_p/2+1)}^{n_p} F_0\right] = F_0$$

which is the same result obtained in the preceding example using the integral of Eq. 12. The coefficient a_m can also be determined as

$$\begin{aligned} a_m &= \frac{2}{n_p}\sum_{j=1}^{n_p} F(t_j)\cos m\omega_f t_j \\ &= \frac{2}{n_p}\left[\sum_{j=1}^{n_p/2}(0)\cos m\omega_f t_j + \sum_{j=(n_p/2+1)}^{n_p} F_0\cos m\omega_f t_j\right] \\ &= \frac{2F_0}{n_p}\sum_{j=(n_p/2+1)}^{n_p}\cos m\omega_f t_j = \frac{2F_0}{n_p}S_1 \end{aligned}$$

where S_1 is

$$S_1 = \sum_{j=(n_p/2+1)}^{n_p}\cos m\omega_f t_j$$

Using the fact that

$$t_j = j\Delta t = \frac{jT_f}{n_p} = \frac{2\pi j}{n_p\omega_f}$$

the series S_1 can be written as

$$S_1 = \sum_{j=(n_p/2+1)}^{n_p}\cos\left(\frac{2\pi jm}{n_p}\right)$$

Similarly, one can show that the constant b_m can also be written as

$$\begin{aligned} b_m &= \frac{2}{n_p}\sum_{j=1}^{n_p} F(t_j)\sin m\omega_f t_j \\ &= \frac{2F_0}{n_p}S_2 \end{aligned}$$

TABLE 5.2. Numerical Evaluation of the Fourier Coefficients

Harmonics	S_1				S_2			
m	$n_p = 10$	$n_p = 20$	$n_p = 50$	$n_p = 100$	$n_p = 10$	$n_p = 20$	$n_p = 50$	$n_p = 100$
1	1.000	1.000	1.000	1.000	−3.078	−6.314	−15.895	−31.820
2	0.000	0.000	0.000	0.000	0.000	0.000	0.000	0.000
3	1.000	1.000	1.000	1.000	−0.727	−1.963	−5.242	−10.579
4	0.000	0.000	0.000	0.000	0.000	0.000	0.000	0.000
5	1.000	1.000	1.000	1.000	0.000	−1.000	−3.078	−6.314
6	0.000	0.000	0.000	0.000	0.000	0.000	0.000	0.000
7	1.000	1.000	1.000	1.000	0.727	−0.510	−2.125	−4.474
8	0.000	0.000	0.000	0.000	0.000	0.000	0.000	0.000
9	1.000	1.000	1.000	1.000	3.078	−0.158	−1.576	−3.442
10	5.000	0.000	0.000	0.000	0.000	0.000	0.000	0.000
11	1.000	1.000	1.000	1.000	−3.078	0.158	−1.209	−2.778
12	0.000	0.000	0.000	0.000	0.000	0.000	0.000	0.000
13	1.000	1.000	1.000	1.000	−0.727	0.510	−0.939	−2.311
14	0.000	0.000	0.000	0.000	0.000	0.000	0.000	0.000
15	1.000	1.000	1.000	1.000	0.000	1.000	−0.727	−1.963
16	0.000	0.000	0.000	0.000	0.000	0.000	0.000	0.000
17	1.000	1.000	1.000	1.000	0.727	1.963	−0.550	−1.691
18	0.000	0.000	0.000	0.000	0.000	0.000	0.000	0.000
19	1.000	1.000	1.000	1.000	3.078	6.314	−0.396	−1.471
20	5.000	10.000	0.000	0.000	0.000	0.000	0.000	0.000

$$S_1 = \sum_{k=n_p/2+1}^{n_p} \cos \frac{2\pi km}{n_p}$$

$$S_2 = \sum_{k=n_p/2+1}^{n_p} \sin \frac{2\pi km}{n_p}$$

where S_2 is given by

$$S_2 = \sum_{j=(n_p/2+1)}^{n_p} \sin\left(\frac{2\pi jm}{n_p}\right)$$

The convergence of S_1 and S_2 for different harmonics is shown in Table 2. It is clear from the values of S_1 and S_2 presented in this table that the Fourier coefficients which are determined numerically will not converge to the Fourier coefficients obtained analytically in Example 1. This is mainly due to the fact that the average of the right- and left-hand limits of the function at the points of discontinuity was not used. In order to solve this problem and demonstrate the application of the Dirichlet conditions, the problem is solved again by using the average of the right- and left-hand limits at the points of discontinuity. The numerical results presented in Table 3 show that the Fourier coefficients obtained numerically are in good agreement with the Fourier coefficients obtained analytically using the integrals of Eqs. 12, 16, and 19.

TABLE 5.3. Dirichlet's Condition ($F_0 = 1$)

Harmonics	Numerical Results						Analytical Results	
	$n_p = 10$		$n_p = 50$		$n_p = 100$			
m	a_m	b_m	a_m	b_m	a_m	b_m	a_m	b_m
1	0.000	0.616	0.000	0.636	0.000	0.636	0.000	0.637
2	0.000	0.000	0.000	0.000	0.000	0.000	0.000	0.000
3	0.000	0.145	0.000	0.210	0.000	0.212	0.000	0.212
4	0.000	0.000	0.000	0.000	0.000	0.000	0.000	0.000
5	0.000	0.000	0.000	0.123	0.000	0.126	0.000	0.127
6	0.000	0.000	0.000	0.000	0.000	0.000	0.000	0.000
7	0.000	0.145	0.000	0.085	0.000	0.089	0.000	0.091
8	0.000	0.000	0.000	0.000	0.000	0.000	0.000	0.000
9	0.000	0.616	0.000	0.063	0.000	0.069	0.000	0.071
10	1.000	0.000	0.000	0.000	0.000	0.000	0.000	0.000
11	0.000	0.616	0.000	0.048	0.000	0.056	0.000	0.058
12	0.000	0.000	0.000	0.000	0.000	0.000	0.000	0.000
13	0.000	0.145	0.000	0.038	0.000	0.046	0.000	0.049
14	0.000	0.000	0.000	0.000	0.000	0.000	0.000	0.000
15	0.000	0.000	0.000	0.029	0.000	0.039	0.000	0.042
16	0.000	0.000	0.000	0.000	0.000	0.000	0.000	0.000
17	0.000	0.145	0.000	0.022	0.000	0.034	0.000	0.037
18	0.000	0.000	0.000	0.000	0.000	0.000	0.000	0.000
19	0.000	0.616	0.000	0.016	0.000	0.029	0.000	0.034
20	1.000	0.000	0.000	0.000	0.000	0.000	0.000	0.000

5.3 SPECIAL CASES

In some special cases of the periodic forcing functions, some of the coefficients that appear in the Fourier expansion of the functions are zeros. By considering these special cases, the effort and time spent to obtain the Fourier expansion of many of the periodic functions can be significantly reduced.

Harmonic Functions Harmonic functions are periodic functions in which all the Fourier coefficients are zeros except one coefficient. For example, consider the function $F(t) = F_0 \sin \omega_1 t$ shown in Fig. 3. The period of this function is given by

$$T_f = \frac{2\pi}{\omega_1}$$

and, accordingly,

$$\omega_f = \omega_1$$

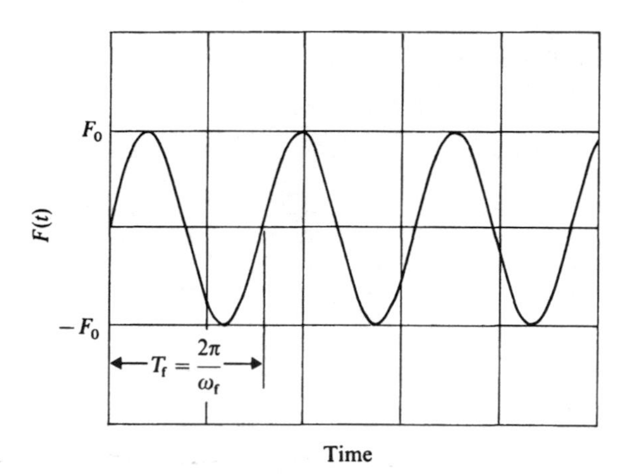

FIG. 5.3. Harmonic function.

By using the identities of Eqs. 11 and 15, one has

$$a_0 = \frac{2}{T_f} \int_{-T_f/2}^{T_f/2} F(t)\, dt = \frac{2}{T_f} \int_{-T_f/2}^{T_f/2} F_0 \sin \omega_1 t\, dt = 0$$

$$a_m = \frac{2}{T_f} \int_{-T_f/2}^{T_f/2} F(t) \cos m\omega_f t\, dt = \frac{2}{T_f} \int_{-T_f/2}^{T_f/2} F_0 \sin \omega_1 t \cos m\omega_f t\, dt = 0$$

By using the identity of Eq. 18, the coefficient b_m is obtained as

$$\begin{aligned} b_m &= \frac{2}{T_f} \int_{-T_f/2}^{T_f/2} F(t) \sin m\omega_f t\, dt = \frac{2}{T_f} \int_{-T_f/2}^{T_f/2} F_0 \sin \omega_1 t \sin m\omega_f t\, dt \\ &= \begin{cases} F_0 & \text{if} \quad m = 1 \\ 0 & \text{if} \quad m \neq 1 \end{cases} \end{aligned}$$

that is,

$$F(t) = F_0 \sin \omega_1 t = b_1 \sin \omega_f t$$

Similar comments apply to the cosine functions.

Even Functions A periodic function $F(t)$ is said to be even if

$$F(t) = F(-t) \tag{5.20}$$

It can be shown that if the function $F(t)$ is an even function, then the coefficients b_m are all zeros, that is,

$$b_m = 0 \qquad \text{for} \quad m = 1, 2, \ldots$$

In this special case, the Fourier series of the function $F(t)$ is given by

$$F(t) = \frac{a_0}{2} + \sum_{n=1}^{\infty} a_n \cos n\omega_f t \tag{5.21}$$

where the coefficients a_0 and a_n are defined by Eqs. 12 and 16. Clearly, the cosine function is an even function, since

$$\cos(\theta) = \cos(-\theta)$$

Odd Functions A periodic function $F(t)$ is said to be odd if

$$F(t) = -F(-t) \tag{5.22}$$

It can be shown that if the function $F(t)$ is an odd function, then the coefficients a_0 and a_m in the Fourier series are identically zero, that is,

$$a_0 = 0, \qquad a_m = 0, \qquad m = 1, 2, \ldots$$

In this special case, the Fourier series of the function $F(t)$ is given by

$$F(t) = \sum_{n=1}^{\infty} b_n \sin n\omega_f t \tag{5.23}$$

where the coefficients b_n, $n = 1, 2, \ldots$, are given by Eq. 19. Observe that the sine function is an odd function, since

$$\sin(-\theta) = -\sin(\theta)$$

Example 5.2

Find the Fourier expansion of the function $F(t)$ shown in Fig. 4.

Solution. The function in the figure is defined as follows

$$F(t) = \begin{cases} 0 & -T_f/2 < t < -T_f/4 \\ F_0 & -T_f/4 \leq t \leq T_f/4 \\ 0 & T_f/4 < t \leq T_f/2 \end{cases}$$

This function is periodic since

$$F(t) = F(t + T_f)$$

Furthermore, the function is an even function since

$$F(t) = F(-t)$$

Therefore,

$$b_m = 0 \qquad \text{for} \quad m = 1, 2, \ldots$$

The coefficient a_0 can be obtained using Eq. 12 as follows

$$a_0 = \frac{2}{T_f} \int_{-T_f/2}^{T_f/2} F(t)\, dt = F_0$$

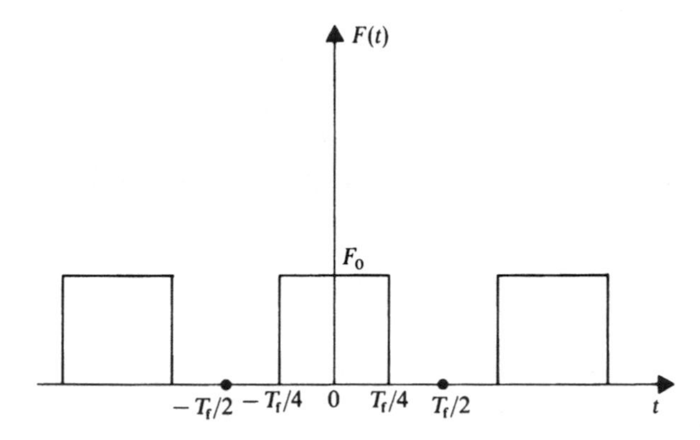

FIG. 5.4. The periodic function $F(t)$.

The coefficient a_m, $m = 1, 2, \ldots$, can be obtained by using Eq. 16 as follows

$$\begin{aligned} a_m &= \frac{2}{T_f}\int_{-T_f/2}^{T_f/2} F(t)\cos m\omega_f t\,dt \\ &= \frac{2}{T_f}\left[\int_{-T_f/2}^{-T_f/4} (0)\cos m\omega_f t\,dt + \int_{-T_f/4}^{T_f/4} F_0\cos m\omega_f t\,dt + \int_{T_f/4}^{T_f/2} (0)\cos m\omega_f t\,dt\right] \\ &= \frac{2F_0}{T_f}\frac{1}{m\omega_f}\sin m\omega_f t\,\Bigg|_{-T_f/4}^{T_f/4} = \frac{F_0}{m\pi}\left[2\sin\frac{m\pi}{2}\right] = \frac{2F_0}{m\pi}\sin\frac{m\pi}{2} \end{aligned}$$

that is,

$$a_m = \begin{cases} 0 & \text{if } m \text{ is even} \\ (-1)^{(m-1)/2}\left(\dfrac{2F_0}{m\pi}\right) & \text{if } m \text{ is odd} \end{cases}$$

5.4 VIBRATION UNDER PERIODIC FORCING FUNCTIONS

The methods for the analytical and numerical evaluation of the Fourier coefficients of periodic functions presented in the preceding sections are used in this section to examine the vibration of the single degree of freedom systems under periodic excitation. Figure 5 shows a single degree of freedom system under the influence of the periodic forcing function $F(t)$. The equation of motion of this system can be written as

$$m\ddot{x} + c\dot{x} + kx = F(t) \tag{5.24}$$

where m is the mass, c is the damping coefficient, and k is the spring coefficient. The periodic force $F(t)$ can be expressed in terms of harmonic functions by using Fourier series as follows

$$F(t) = \frac{a_0}{2} + \sum_{n=1}^{\infty} (a_n \cos n\omega_f t + b_n \sin n\omega_f t) \tag{5.25}$$

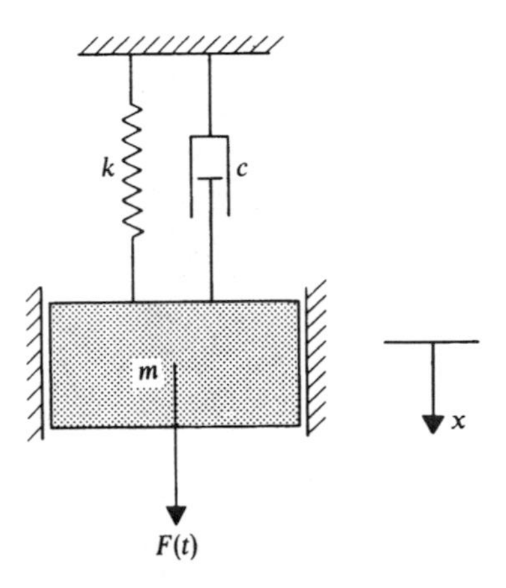

FIG. 5.5. Single degree of freedom system.

where ω_f is the fundamental frequency. Equation 25 can also be written in an alternative form as

$$F(t) = F_0 + \sum_{n=1}^{\infty} F_n \sin(n\omega_f t + \phi_n) \tag{5.26}$$

where

$$F_0 = \frac{a_0}{2}$$

$$F_n = \sqrt{a_n^2 + b_n^2}$$

$$\phi_n = \tan^{-1}\left(\frac{a_n}{b_n}\right)$$

By using Eq. 26, Eq. 24 can be written as

$$m\ddot{x} + c\dot{x} + kx = F_0 + \sum_{n=1}^{\infty} F_n \sin(\omega_n t + \phi_n) \tag{5.27}$$

where

$$\omega_n = n\omega_f$$

The solution of Eq. 27 consists of two parts, the homogeneous function x_h and the particular solution x_p, that is,

$$x = x_h + x_p \tag{5.28a}$$

Methods for obtaining the homogeneous function x_h are discussed in Chapter 3 for the undamped and damped single degree of freedom vibratory systems. Since Eq. 27 is a linear, second-order ordinary differential equation with constant coefficients, the principle of superposition can be applied in order to obtain the particular solution x_p. First, we obtain the particular solution due to the constant term F_0 only. The solution, in this case, is denoted as x_{p0}.

Second, we determine the response of the system due to each of the terms $F_n \sin(\omega_n t + \phi_n)$ in the infinite series in the right-hand side of Eq. 27. The solution in each of these cases wll be denoted as x_{pn}. Applying the principle of superposition, the system response to the forcing function represented by the infinite series in the right-hand side of Eq. 27 will be given by the infinite series $\sum_{n=1}^{\infty} x_{pn}$. Therefore, the complete solution of Eq. 28a can be written as

$$x = x_h + x_{p0} + \sum_{n=1}^{\infty} x_{pn} \tag{5.28b}$$

Response of the System to the Constant Force F_0 Since F_0 is constant, it is clear that x_{p0} is also a constant. Following the procedure discussed in Chapter 2, we assume x_{p0} to be a combination of F_0 and its independent derivatives. In this case, we assume x_{p0} to be

$$x_{p0} = C$$

where C is a constant. It follows that

$$\dot{x}_{p0} = \ddot{x}_{p0} = 0$$

Substituting into the differential equation,

$$m\ddot{x}_{p0} + c\dot{x}_{p0} + kx_{p0} = F_0$$

one obtains

$$kC = F_0$$

that is,

$$x_{p0} = C = \frac{F_0}{k} \tag{5.29}$$

Response of the System to the Force $F_n \sin(\omega_n t + \phi_n)$ The term $F_n \sin(\omega_n t + \phi_n)$ represents a harmonic forcing function. The response of the single degree of freedom system to this type of force was discussed in the preceding chapter, and it was shown therein, that because of the damping of the system, there is a phase angle between the force and the system response. The solution x_{pn}, in this case, can be written as

$$x_{pn} = \frac{F_n/k}{\sqrt{(1 - r_n^2)^2 + (2r_n\xi)^2}} \sin(\omega_n t + \phi_n - \psi_n) \tag{5.30}$$

where

$$r_n = \frac{\omega_n}{\omega} = \frac{n\omega_f}{\omega} = nr_1 \tag{5.31}$$

$$\omega = \sqrt{\frac{k}{m}} \tag{5.32}$$

$$\psi_n = \tan^{-1}\left(\frac{2r_n\xi}{1 - r_n^2}\right) \tag{5.33}$$

and ξ is the damping factor. It follows that

$$\sum_{n=1}^{\infty} x_{pn} = \sum_{n=1}^{\infty} \frac{F_n/k}{\sqrt{(1 - r_n^2)^2 + (2r_n\xi)^2}} \sin(\omega_n t + \phi_n - \psi_n) \tag{5.34}$$

Particular Solution By using Eqs. 29 and 34, the particular solution x_p of Eq. 28 can be written as

$$\begin{aligned} x_p &= x_{p0} + \sum_{n=1}^{\infty} x_{pn} \\ &= \frac{F_0}{k} + \sum_{n=1}^{\infty} \frac{F_n/k}{\sqrt{(1 - r_n^2)^2 + (2r_n\xi)^2}} \sin(\omega_n t + \phi_n - \psi_n) \end{aligned} \tag{5.35}$$

The use of the procedure described in this section, to obtain the particular solution of the vibration equation of the single degree of freedom system under the influence of periodic excitation, is demonstrated by the following examples.

Example 5.3

Determine the steady state response of the single degree of freedom system shown in Fig. 6 to the forcing function $F(t)$, where $F(t)$ is the periodic function given in Example 1.

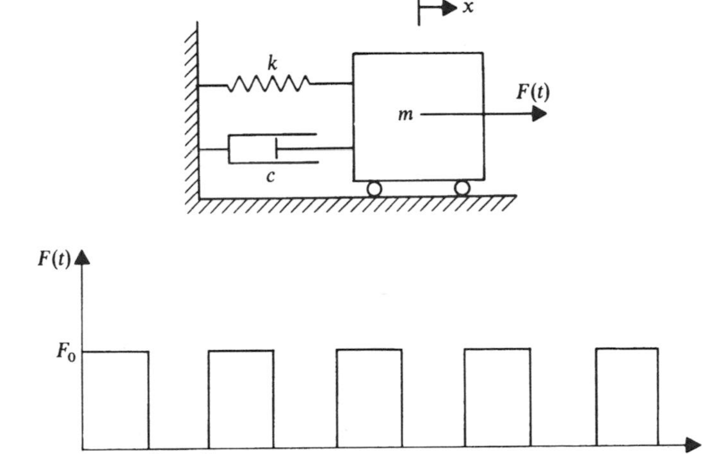

FIG. 5.6. Single degree of freedom mass–spring system.

Solution. The equation of motion of the damped single degree of freedom system due to the periodic excitation is given by

$$m\ddot{x} + c\dot{x} + kx = F(t)$$

It was shown in Example 1 that $F(t)$ can be expressed in terms of the harmonic functions as

$$F(t) = \frac{a_0}{2} + \sum_{n=1}^{\infty} b_n \sin \omega_n t$$

where $\omega_n = n\omega_f = 2\pi n/T_f$

$$a_0 = F_0$$

$$b_n = \begin{cases} 2F_0/n\pi & \text{if } n \text{ is odd} \\ 0 & \text{if } n \text{ is even} \end{cases}$$

The equation of motion can then be written as

$$m\ddot{x} + c\dot{x} + kx = \frac{a_0}{2} + \sum_{n=1}^{\infty} b_n \sin \omega_n t$$

$$= \frac{F_0}{2} + \frac{2F_0}{\pi} \sin \omega_f t + \frac{2F_0}{3\pi} \sin 3\omega_f t + \ldots$$

Clearly, x_{p0} is given by

$$x_{p0} = \frac{F_0}{2k}$$

Since, in this case, $a_n = 0$, $n = 1, 2, \ldots$, it is clear that the phase angle ϕ_n of Eq. 34 is zero, and as such, x_{pn} is defined as

$$x_{pn} = \frac{b_n/k}{\sqrt{(1 - r_n^2)^2 + (2r_n\xi)^2}} \sin(n\omega_f t - \psi_n)$$

$$= \frac{2F_0/n\pi k}{\sqrt{(1 - r_n^2)^2 + (2r_n\xi)^2}} \sin(n\omega_f t - \psi_n), \qquad n = 1, 3, 5, \ldots$$

It follows that

$$x_p = x_{p0} + \sum_{n=1,3,5}^{\infty} x_{pn}$$

$$= \frac{F_0}{2k} + \sum_{n=1,3,5}^{\infty} \frac{2F_0/n\pi k}{\sqrt{(1 - r_n^2)^2 + (2r_n\xi)^2}} \sin(n\omega_f t - \psi_n)$$

$$= \frac{F_0}{k}\left[\frac{1}{2} + \sum_{n=1,3,5}^{\infty} \frac{2}{n\pi\sqrt{(1 - r_n^2)^2 + (2r_n\xi)^2}} \sin(n\omega_f t - \psi_n)\right]$$

Example 5.4

Determine the steady state response of the single degree of freedom system shown in Fig. 7, due to the periodic forcing function $F(t)$ given in Example 2. Neglect the gravity effect.

Solution. The equation of motion of this system is given by

$$m_e\ddot{\theta} + c_e\dot{\theta} + k_e\theta = F(t)\, l$$

where

$$m_e = \frac{ml^2}{3}$$

$$c_e = ca^2$$

$$k_e = ka^2$$

where c and k are, respectively, the damping and spring coefficients. It was shown in Example 2 that the periodic forcing function $F(t)$ can be written as

$$F(t) = \frac{a_0}{2} + \sum_{n=1}^{\infty} a_n \cos n\omega_f t$$

where

$$a_0 = F_0$$

$$a_n = \begin{cases} 0 & \text{if } n \text{ is even} \\ (-1)^{(n-1)/2}(2F_0/n\pi) & \text{if } n \text{ is odd} \end{cases}$$

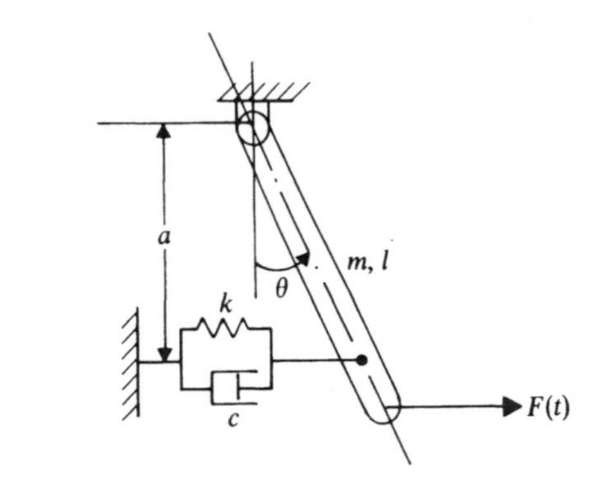

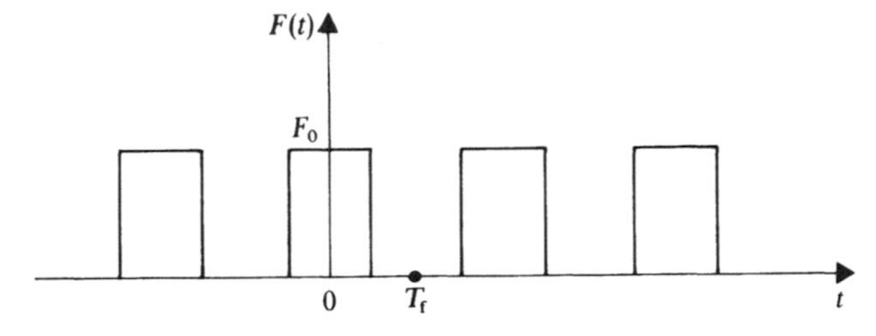

FIG. 5.7. Single degree of freedom pendulum.

Since

$$\cos n\omega_f t = \sin\left(n\omega_f t + \frac{\pi}{2}\right)$$

the periodic force function $F(t)$ can be written as

$$F(t) = \frac{a_0}{2} + \sum_{n=1}^{\infty} a_n \sin\left(n\omega_f t + \frac{\pi}{2}\right)$$

$$= \frac{F_0}{2} + \frac{2F_0}{\pi}\sin\left(\omega_f t + \frac{\pi}{2}\right) - \frac{2F_0}{3\pi}\sin\left(3\omega_f t + \frac{\pi}{2}\right) + \ldots$$

where the angle ϕ_n, $n = 1, 2, \ldots$, of Eq. 26 is given by

$$\phi_n = \frac{\pi}{2}, \qquad n = 1, 2, \ldots$$

The equation of motion of this single degree of freedom system can be written as

$$m_e\ddot{\theta} + c_e\dot{\theta} + k_e\theta = \frac{F_0 l}{2} + \frac{2F_0 l}{\pi}\sin\left(\omega_f t + \frac{\pi}{2}\right) - \frac{2F_0 l}{3\pi}\sin\left(3\omega_f t + \frac{\pi}{2}\right) + \ldots$$

Consequently,

$$x_{p0} = \frac{F_0 l}{2k_e}$$

$$x_{pn} = \begin{cases} 0 & \text{if } n \text{ is even} \\ (-1)^{(n-1)/2} \dfrac{2F_0 l/n\pi k_e}{\sqrt{(1 - r_n^2)^2 + (2r_n\xi)^2}} \sin\left(n\omega_f t + \dfrac{\pi}{2} - \psi_n\right) & \text{if } n \text{ is odd} \end{cases}$$

Therefore,

$$x_p = \frac{F_0 l}{2k_e} + \sum_{n=1,3,5}^{\infty} (-1)^{(n-1)/2} \frac{2F_0 l/n\pi k_e}{\sqrt{(1 - r_n^2)^2 + (2r_n\xi)^2}} \sin\left(n\omega_f t + \frac{\pi}{2} - \psi_n\right)$$

$$= \frac{F_0 l}{2k_e} + \sum_{n=1,3,5}^{\infty} (-1)^{(n-1)/2} \frac{2F_0 l/n\pi k_e}{\sqrt{(1 - r_n^2)^2 + (2r_n\xi)^2}} \cos(n\omega_f t - \psi_n)$$

where the damping factor ξ is given by

$$\xi = \frac{c_e}{C_c} = \frac{ca^2}{2m_e\omega} = \frac{3ca^2}{2ml^2\omega}$$

and ω is the system natural frequency defined by

$$\omega = \sqrt{\frac{k_e}{m_e}} = \sqrt{\frac{3ka^2}{ml^2}}$$

5.5 IMPULSIVE MOTION

An impulsive force is defined as a force which has a large magnitude, and acts during a very short time duration such that the time integral of this force is

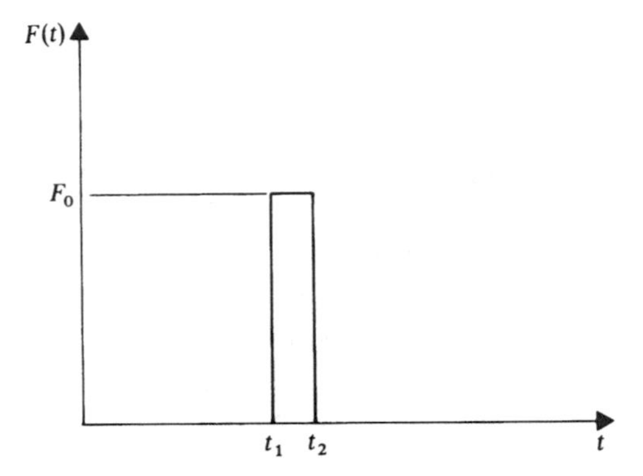

FIG. 5.8. Impulsive force.

finite. Let the impulsive force $F(t)$ shown in Fig. 8 act on the single degree of freedom system shown in Fig. 9. The differential equation of motion of this system can be written as

$$m\ddot{x} + c\dot{x} + kx = F(t)$$

Integrating this equation over the very short interval (t_1, t_2), one obtains

$$\int_{t_1}^{t_2} m\ddot{x}\,dt + \int_{t_1}^{t_2} c\dot{x}\,dt + \int_{t_1}^{t_2} kx\,dt = \int_{t_1}^{t_2} F(t)\,dt \tag{5.36}$$

Since the time interval (t_1, t_2) is assumed to be very small, we assume that x does not change appreciably, and we also assume that the change in the velocity $\dot{x}$ is finite. One, therefore, has

$$\operatorname*{Lim}_{t_1 \to t_2} \int_{t_1}^{t_2} c\dot{x}\,dt = 0$$

$$\operatorname*{Lim}_{t_1 \to t_2} \int_{t_1}^{t_2} kx\,dt = 0$$

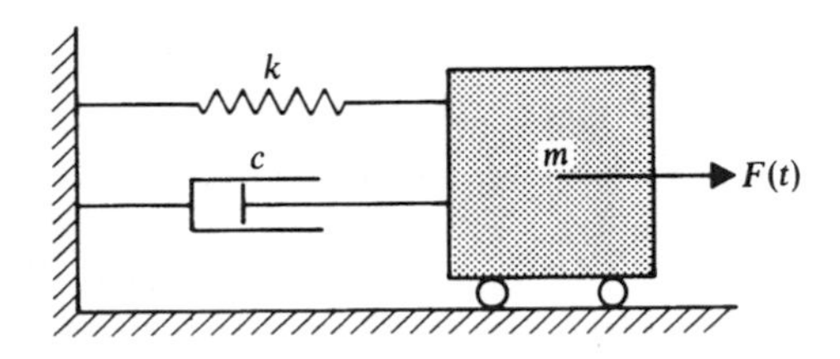

FIG. 5.9. Damped single degree of freedom system under the effect of impulsive force $F(t)$.

Therefore, if t_1 approaches t_2, Eq. 36 yields

$$\int_{t_1}^{t_2} m\ddot{x}\, dt = \int_{t_1}^{t_2} F(t)\, dt \tag{5.37}$$

Since $\ddot{x} = d\dot{x}/dt$, Eq. 37 can be written as

$$\int_{\dot{x}_1}^{\dot{x}_2} m\, d\dot{x} = \int_{t_1}^{t_2} F(t)\, dt \tag{5.38}$$

where $\dot{x}_1$ and $\dot{x}_2$ are, respectively, the velocities at t_1 and t_2. Equation 38 yields

$$m(\dot{x}_2 - \dot{x}_1) = \int_{t_1}^{t_2} F(t)\, dt$$

from which

$$\Delta\dot{x} = \dot{x}_2 - \dot{x}_1 = \frac{1}{m}\int_{t_1}^{t_2} F(t)\, dt \tag{5.39}$$

where $\Delta\dot{x}$ is the jump discontinuity in the velocity of the mass due to the impulsive force. The time integral in Eq. 39 is called the *linear impulse* I and is defined by

$$I = \int_{t_1}^{t_2} F(t)\, dt \tag{5.40}$$

In the particular case in which the linear impulse is equal to one, I is called the *unit impulse*.

Equation 39 can be written as

$$\Delta\dot{x} = \dot{x}_2 - \dot{x}_1 = \frac{I}{m} \tag{5.41}$$

This result indicates that the effect of the impulsive force, which acts over a very short time duration on a system which is initially at rest, can be accounted for by considering the motion of the system with initial velocity I/m and zero initial displacement. That is, in the case of impulsive motion, we consider the system vibrating freely as the result of the initial velocity given by Eq. 41. The free vibration of the underdamped single degree of freedom system shown in Fig. 9 is governed by the equations

$$x(t) = Xe^{-\xi\omega t}\sin(\omega_d t + \phi) \tag{5.42}$$

$$\dot{x}(t) = -\xi\omega Xe^{-\xi\omega t}\sin(\omega_d t + \phi) + \omega_d Xe^{-\xi\omega t}\cos(\omega_d t + \phi) \tag{5.43}$$

where X and ϕ are constants to be determined from the initial conditions, ω is the natural frequency, ξ is the damping factor, and ω_d is the damped natural frequency

$$\omega_d = \omega\sqrt{1 - \xi^2}$$

As the result of applying an impulsive force with a linear impulse I at $t = 0$, the initial conditions are

$$x(t = 0) = 0, \qquad \dot{x}(t = 0) = \frac{I}{m}$$

Since the initial displacement is zero, Eq. 42 yields

$$\phi = 0$$

Using Eq. 43 and the initial velocity, it is an easy matter to verify that

$$x(t) = \frac{I}{m\omega_d} e^{-\xi\omega t} \sin \omega_d t \tag{5.44}$$

which can be written as

$$x(t) = IH(t) \tag{5.45}$$

where $H(t)$ is called the *impulse response function* and is defined as

$$H(t) = \frac{1}{m\omega_d} e^{-\xi\omega t} \sin \omega_d t \tag{5.46}$$

Example 5.5

Find the response of the single degree of freedom system shown in Fig. 9 to the rectangular impulsive force shown in Fig. 8, where $m = 10$ kg, $k = 9{,}000$ N/m, $c = 18$ N·s/m, and $F_0 = 10{,}000$ N. The force is assumed to act at time $t = 0$ and the impact interval is assumed to be 0.005 s.

Solution. The linear impulse I is given by

$$I = \int_{t_1}^{t_2} F(t)\,dt = \int_0^{0.005} 10{,}000\,dt = 10{,}000(0.005) = 50 \text{ N}\cdot\text{s}$$

The natural frequency of the system ω is given by

$$\omega = \sqrt{\frac{k}{m}} = \sqrt{\frac{9000}{10}} = 30 \text{ rad/s}$$

The critical damping coefficient C_c is

$$C_c = 2m\omega = 2(10)(30) = 600$$

The damping factor ξ is

$$\xi = \frac{c}{C_c} = \frac{18}{600} = 0.03$$

The damped natural frequency ω_d is

$$\omega_d = \omega\sqrt{1 - \xi^2} = 30\sqrt{1 - (0.03)^2} = 29.986 \text{ rad/s}$$

The system response to the impulsive force is then given by

$$\begin{aligned} x(t) &= \frac{I}{m\omega_d} e^{-\xi\omega t} \sin \omega_d t \\ &= \frac{50}{(10)(29.986)} e^{-(0.03)(30)t} \sin 29.986t \\ &= 0.1667 e^{-0.9t} \sin 29.986t \end{aligned}$$

5.6 RESPONSE TO AN ARBITRARY FORCING FUNCTION

In this section, we consider the response of a damped single degree of freedom system to an arbitrary forcing fucntion $F(t)$, shown in Fig. 10. The procedure described in the preceding section for obtaining the impulse response can be used as a basis for developing a general expression for the response of the system to an arbitrary forcing function. The arbitrary forcing function $F(t)$ can be regarded as a series of impulsive forces $F(\tau)$ acting over a very short-lived interval $d\tau$. The force $F(\tau)$ then produces the short duration impulse $F(\tau)\,d\tau$, and the response of the system to this impulse for all $t > \tau$ is given by

$$dx = F(\tau)\,d\tau\,H(t-\tau) \tag{5.47}$$

where $H(t)$ is the impulse response function defined by Eq. 46. That is,

$$dx = F(\tau)\,d\tau \cdot \frac{1}{m\omega_d} e^{-\xi\omega(t-\tau)} \sin \omega_d(t-\tau) \tag{5.48}$$

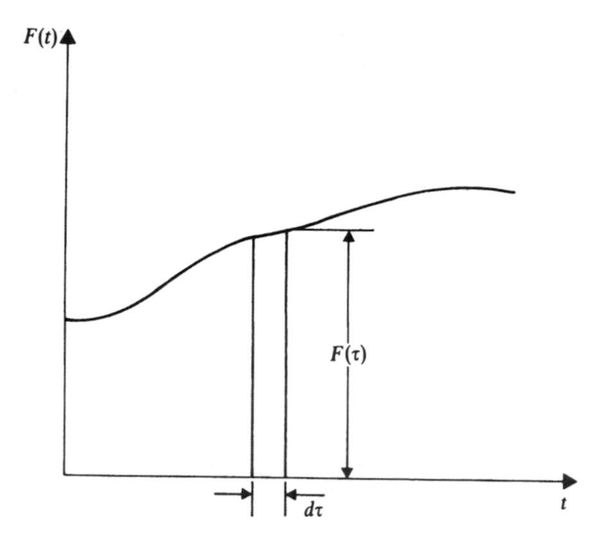

FIG. 5.10. Arbitrary forcing function $F(t)$.

In this equation, dx represents the incremental response of the damped single degree of freedom system to the incremental impulse $F(\tau)\,d\tau$ for $t > \tau$. In order to determine the total response, we integrate Eq. 48 over the entire interval

$$x(t) = \int_0^t F(\tau)H(t - \tau)\,d\tau \tag{5.49}$$

or

$$x(t) = \frac{1}{m\omega_d}\int_0^t F(\tau)e^{-\xi\omega(t-\tau)} \sin \omega_d(t - \tau)\,d\tau \tag{5.50}$$

Equation 49 or Eq. 50 is called the *Duhamel integral* or the *convolution integral*. It is important to emphasize, however, that in obtaining the convolution integral we made use of the principle of superposition which is valid only for linear systems. Furthermore, in deriving the convolution integral, no mention was given to the initial conditions and, accordingly, the integral of Eq. 49, or Eq. 50, provides only the forced response. If the initial conditions are not equal to zero, that is,

$$x_0 = x(t = 0) \neq 0 \qquad \text{and/or} \qquad \dot{x}_0 = \dot{x}(t = 0) \neq 0$$

then Eq. 50 must be modified to include the effect of the initial conditions. To this end, we first define the homogeneous solution and determine the arbitrary constants in the case of free vibration as the result of these initial conditions, and then use the principle of superposition to add the homogeneous function to the forced response.

Special Case A special case of the preceding development is the case of an *undamped single degree of freedom system*. In this case, $\omega_d = \omega$ and $\xi = 0$ and the impulse response function of Eq. 46 reduces to

$$H(t) = \frac{1}{m\omega} \sin \omega t \tag{5.51}$$

The forced response, in this special case, is given by

$$x(t) = \frac{1}{m\omega}\int_0^t F(\tau) \sin \omega(t - \tau)\,d\tau \tag{5.52a}$$

If the effect of the initial conditions is considered, the general solution is given by

$$x(t) = \frac{\dot{x}_0}{\omega} \sin \omega t + x_0 \cos \omega t + \frac{1}{m\omega}\int_0^t F(\tau) \sin \omega(t - \tau)\,d\tau \tag{5.52b}$$

Example 5.6

Find the forced response of the damped single degree of freedom system to the step function shown in Fig. 11.

FIG. 5.11. Step function.

Solution. The forced response of the damped single degree of freedom system to an arbitrary forcing function is

$$x(t) = \frac{1}{m\omega_d} \int_0^t F(\tau) e^{-\xi\omega(t-\tau)} \sin \omega_d(t - \tau)\, d\tau$$

In the case of a step function, the forcing function $F(t)$ is defined as

$$F(t) = F_0; \qquad t > 0$$

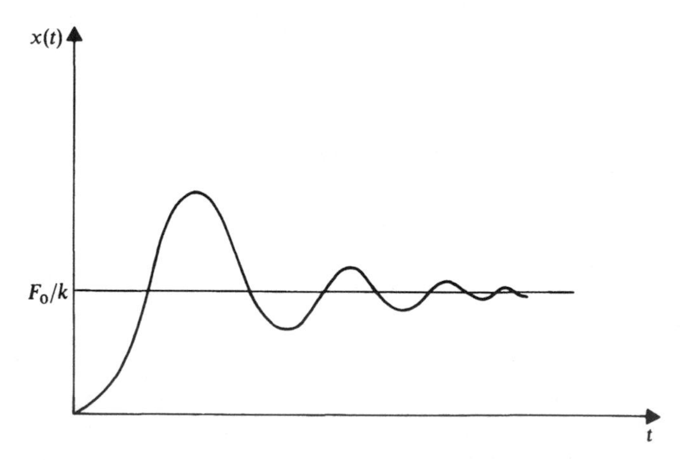

FIG. 5.12. Response of the damped single degree of freedom system to the step forcing function.

that is,

$$x(t) = \frac{1}{m\omega_d} \int_0^t F_0 e^{-\xi\omega(t-\tau)} \sin \omega_d(t-\tau)\, d\tau$$

$$= \frac{F_0}{m\omega_d} \int_0^t e^{-\xi\omega(t-\tau)} \sin \omega_d(t-\tau)\, d\tau$$

$$= \frac{F_0}{k}\left[1 - \frac{e^{-\xi\omega t}}{\sqrt{1-\xi^2}} \cos(\omega_d t - \psi)\right]$$

where the angle ψ is defined as

$$\psi = \tan^{-1}\left(\frac{\xi}{\sqrt{1-\xi^2}}\right)$$

The response of this system is shown in Fig. 12.

Example 5.7

Find the forced response of the undamped single degree of freedom system to the forcing function shown in Fig. 13.

Solution. The forced response of the undamped single degree of freedom system to an arbitrary forcing function is given by

$$x(t) = \frac{1}{m\omega} \int_0^t F(\tau) \sin \omega(t-\tau)\, d\tau$$

The forcing function $F(t)$ shown in Fig. 13 is defined as

$$\begin{aligned} F(t) &= F_0 t/t_1, \qquad & 0 \le t \le t_1 \\ &= F_0, & t > t_1 \end{aligned}$$

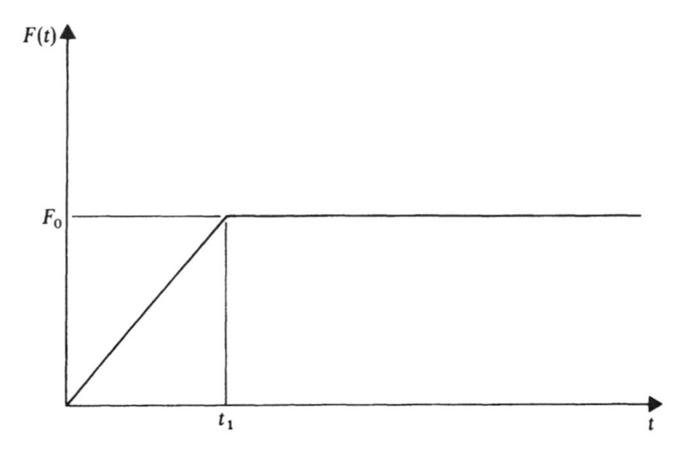

FIG. 5.13. Forcing function $F(t)$.

Therefore, the forced response is given by

$$x(t) = \frac{1}{m\omega}\left[\int_0^{t_1} \frac{F_0\tau}{t_1} \sin\omega(t-\tau)\,d\tau + \int_{t_1}^{t} F_0 \sin\omega(t-\tau)\,d\tau\right]$$

Using integration by parts, the response $x(t)$ is given by

$$\begin{aligned} x(t) &= \frac{F_0}{m\omega}\left[\frac{\tau\cos\omega(t-\tau)}{\omega t_1}\bigg|_0^{t_1} - \int_0^{t_1}\frac{\cos\omega(t-\tau)}{\omega t_1}\,d\tau + \frac{\cos\omega(t-\tau)}{\omega}\bigg|_{t_1}^{t}\right] \\ &= \frac{F_0}{m\omega}\left\{\frac{\cos\omega(t-t_1)}{\omega} + \frac{\sin\omega(t-\tau)}{\omega^2 t_1}\bigg|_0^{t_1} + \frac{1}{\omega} - \frac{\cos\omega(t-t_1)}{\omega}\right\} \\ &= \frac{F_0}{m\omega}\left(\frac{1}{\omega} + \frac{\sin\omega(t-t_1)}{\omega^2 t_1} - \frac{\sin\omega t}{\omega^2 t_1}\right) \end{aligned}$$

Numerical Evaluation of the Duhamel Integral In the preceding examples, where the external forces are given by simple, integrable functions, we were able to determine the dynamic response in a closed form by the use of the Duhamel integral. In many practical applications, however, the forcing function may be obtained from experimental data, or may be given in a complex form such that the analytical evaluation of the Duhamel integral is difficult. In these cases, one must then resort to numerical methods in order to determine the response of the system by using incremental summation instead of the integration of Eq. 49 or Eq. 52a. Observe that Eq. 52a can be obtained as a special case of Eq. 49 in which the damping factor ξ is equal to zero. Therefore, in the following discussion, we will consider only the numerical evaluation of the Duhamel integral of Eq. 49. For convenience, we write the exponential and harmonic functions in Eq. 49 in the following forms

$$e^{-\xi\omega(t-\tau)} = \frac{e^{\xi\omega\tau}}{e^{\xi\omega t}} \tag{5.53a}$$

$$\sin\omega_d(t-\tau) = \sin\omega_d t\cos\omega_d\tau - \cos\omega_d t\sin\omega_d\tau \tag{5.53b}$$

Substituting these two equations into Eq. 50, one obtains

$$x(t) = \frac{e^{-\xi\omega t}}{m\omega_d}\left[\int_0^{t} F(\tau)e^{\xi\omega\tau}\sin\omega_d t\cos\omega_d\tau\,d\tau - \int_0^{t} F(\tau)e^{\xi\omega\tau}\cos\omega_d t\sin\omega_d\tau\,d\tau\right]$$

Since the integration is with respect to τ, the terms which are functions of t

can be factored out of the integral. This leads to

$$x(t) = \frac{e^{-\xi\omega t}}{m\omega_d}\left[\sin \omega_d t \int_0^t F(\tau)e^{\xi\omega\tau} \cos \omega_d \tau \, d\tau - \cos \omega_d t \int_0^t F(\tau)e^{\xi\omega\tau} \sin \omega_d \tau \, d\tau\right] \tag{5.54}$$

This equation can be written compactly as

$$x(t) = \frac{e^{-\xi\omega t}}{m\omega_d}[I_1 \sin \omega_d t - I_2 \cos \omega_d t] \tag{5.55}$$

where

$$I_1(t) = \int_0^t F(\tau)e^{\xi\omega\tau} \cos \omega_d \tau \, d\tau \tag{5.56a}$$

$$I_2(t) = \int_0^t F(\tau)e^{\xi\omega\tau} \sin \omega_d \tau \, d\tau \tag{5.56b}$$

The integrals I_1 and I_2 are the ones which will be evaluated numerically. We, therefore, rewrite them in the following simple form

$$I_1(t) = \int_0^t y_1(\tau) \, d\tau \tag{5.57a}$$

$$I_2(t) = \int_0^t y_2(\tau) \, d\tau \tag{5.57b}$$

where

$$y_1(\tau) = F(\tau)e^{\xi\omega\tau} \cos \omega_d \tau \tag{5.58a}$$

$$y_2(\tau) = F(\tau)e^{\xi\omega\tau} \sin \omega_d \tau \tag{5.58b}$$

Figure 14 shows an arbitrary forcing function $F(\tau)$. The time domain of the function up to point t is divided into n equal intervals with length $\Delta\tau$. Note that

$$t = n\Delta\tau, \qquad \Delta\tau = \frac{t}{n} \tag{5.59}$$

$$\tau_j = j\Delta\tau = \frac{j}{n}t \tag{5.60}$$

Using simple summations, one can then write the integrals of Eqs. 57a and 57b as

$$I_1(t) = \sum_{j=0}^{n} y_1(\tau_j)\Delta\tau$$

$$I_2(t) = \sum_{j=0}^{n} y_2(\tau_j)\Delta\tau$$

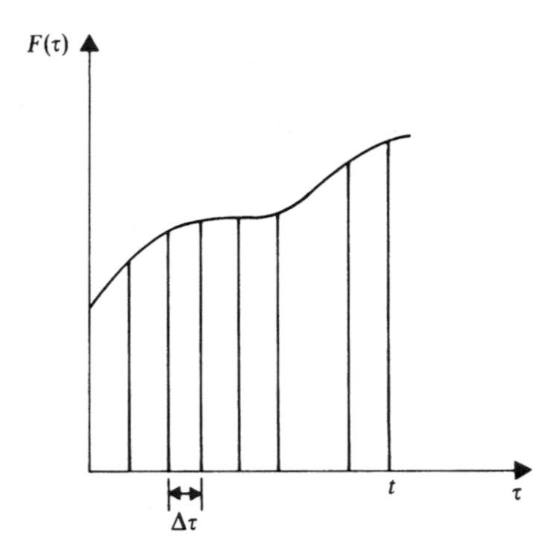

FIG. 5.14. Numerical evaluation of the Duhamel integral.

which, upon using Eqs. 59 and 60, leads to

$$I_1(t) = \sum_{j=0}^{n} y_1\left(\frac{j}{n}t\right)\frac{t}{n} = \frac{t}{n}\sum_{j=0}^{n} y_1\left(\frac{j}{n}t\right) \tag{5.61}$$

$$I_2(t) = \sum_{j=0}^{n} y_2\left(\frac{j}{n}t\right)\frac{t}{n} = \frac{t}{n}\sum_{j=0}^{n} y_2\left(\frac{j}{n}t\right) \tag{5.62}$$

Substituting Eqs. 61 and 62 into Eq. 55, one obtains $x(t)$ as

$$x(t) = \frac{te^{-\xi\omega t}}{nm\omega_d}[D_1 \sin \omega_d t - D_2 \cos \omega_d t] \tag{5.63}$$

where

$$D_1 = \sum_{j=0}^{n} y_1(\tau_j) \tag{5.64a}$$

$$D_2 = \sum_{j=0}^{n} y_2(\tau_j) \tag{5.64b}$$

in which the functions y_1 and y_2 are defined by Eqs. 58a and 58b.

In the numerical evaluation of the integrals of Eqs. 61 and 62, a simple summation is used. More accurate methods for the numerical evaluation of the integrals such as the trapezoidal rule and Simpson's rule can also be used (Carnahan et al., 1969). By increasing the number of intervals, however, the convergence of the simple summation procedure used in this section is acceptable in most practical applications.

In the special case of undamped vibration, Eq. 63 reduces to

$$x(t) = \frac{t}{nm\omega}[D_1 \sin \omega t - D_2 \cos \omega t] \tag{5.65}$$

and the functions y_1 and y_2 of Eq. 58 reduce to

$$y_1 = F(\tau)\cos\omega\tau \tag{5.66a}$$

$$y_2 = F(\tau)\sin\omega\tau \tag{5.66b}$$

The numerical procedure described in this section is demonstrated by the following example.

Example 5.8

Use the method of the numerical evaluation of the Duhamel integral discussed in this section to obtain the dynamic response of the single degree of freedom system of Example 6. Assume that $m = 1$ kg, $k = 2500$ N/m, $c = 10$ N·s/m, $F_0 = 10$ N.

Solution. The natural frequency of the system ω

$$\omega = \sqrt{\frac{k}{m}} = \sqrt{\frac{2500}{1}} = 50 \text{ rad/s}$$

The damping factor ξ is given by

$$\xi = \frac{c}{C_c} = \frac{c}{2m\omega} = \frac{10}{2(1)(50)} = 0.1$$

and

$$\omega_d = \omega\sqrt{1-\xi^2} = 50\sqrt{1-(0.1)^2} = 49.749 \text{ rad/s}$$

Therefore,

$$\frac{e^{-\xi\omega t}}{m\omega_d} = \frac{e^{-(0.1)(50)t}}{(1)(49.749)} = 0.0201e^{-5t}$$

Equation 63 can be written for this example as

$$x(t) = \frac{0.0201te^{-5t}}{n}[D_1\sin\omega_d t - D_2\cos\omega_d t] \tag{5.67}$$

where D_1 and D_2 are defined by Eqs. 64a and 64b. Since the external force is given in this example by the simple step function, one has

$$y_1 = F_0e^{\xi\omega\tau}\cos\omega_d\tau = 10e^{5\tau}\cos 49.749\tau$$

$$y_2 = F_0e^{\xi\omega\tau}\sin\omega_d\tau = 10e^{5\tau}\sin 49.749\tau$$

That is,

$$y_1(\tau_j) = 10e^{5\tau_j}\cos 49.749\tau_j$$

$$y_2(\tau_j) = 10e^{5\tau_j}\sin 49.749\tau_j$$

where τ_j is defined by Eq. 60. Hence

$$D_1 = \sum_{j=0}^{n} y_1(\tau_j) = \sum_{j=0}^{n} 10e^{5\tau_j}\cos 49.749\tau_j$$

$$D_2 = \sum_{j=0}^{n} y_2(\tau_j) = \sum_{j=0}^{n} 10e^{5\tau_j}\sin 49.749\tau_j$$

TABLE 5.4. Coefficients D_1 and D_2

	D_1				D_2			
t	$n = 5$	$n = 50$	$n = 100$	$n = 200$	$n = 5$	$n = 50$	$n = 100$	$n = 200$
0.1	−7.98	−156.88	−321.09	−649.43	−4.48	32.99	74.01	156.03
0.2	−13.49	−91.14	−175.76	−344.90	2.96	152.57	312.81	632.86
0.3	−12.80	79.19	170.60	352.75	19.78	163.89	313.41	611.71
0.4	12.91	187.47	354.67	687.38	45.26	−17.40	−68.50	−169.88
0.5	113.55	25.28	−15.95	−97.10	35.06	−231.91	−454.81	−897.33
0.6	125.19	−319.18	−658.81	−1330.51	−379.51	−120.22	−137.78	−174.02

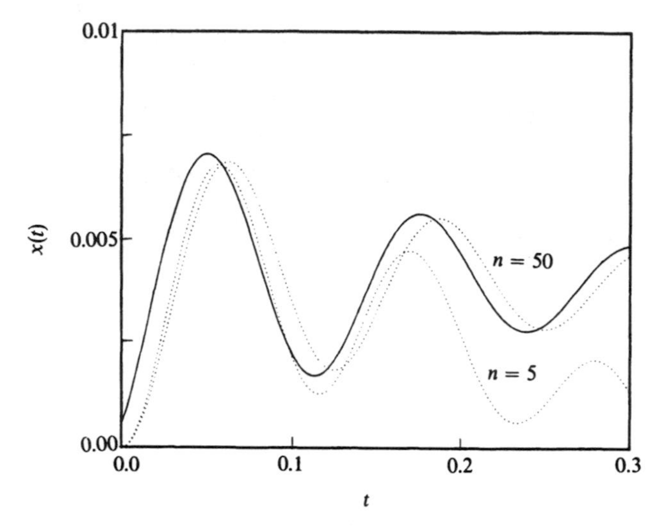

FIG. 5.15. Comparison between the exact and numerical solutions.

Table 4 shows the values of D_1 and D_2 for different numbers of intervals. The numerical values of D_1 and D_2 can be substituted into Eq. 67 in order to obtain $x(t)$. In Example 6 it was shown that the closed form solution of this problem is given by

$$x(t) = \frac{F_0}{k}\left[1 - \frac{e^{-\xi\omega t}}{\sqrt{1-\xi^2}}\cos(\omega_d t - \psi)\right]$$
$$= 4 \times 10^{-3}[1 - 1.005e^{-5t}\cos(49.749t - 5.7392°)]$$

A comparison between this exact solution and the numerical solution obtained by using Eq. 67 is presented in Fig. 15 for different values of n. Observe that a better accuracy is obtained by inceasing n.

5.7 FREQUENCY CONTENTS IN ARBITRARY FORCING FUNCTIONS

The frequency content of a periodic forcing function as determined by Fourier series expansion includes frequencies that are multiples of the fundamental frequency ω_f. Therefore, the frequency spectrum of a periodic forcing function is defined at only discrete points in the frequency domain, and the magnitudes of the coefficients F_n in the Fourier series expansions determine which frequencies are significant. A similar procedure, which is based on the *Fourier transform method*, also can be used to determine the frequency contents of an arbitrary forcing function that is not periodic. In this procedure, the arbitrary forcing function is assumed to be a periodic function that has a fundamental periodic time $T_f = \infty$. In this case, we can define the Fourier transform $F(\omega_f)$ of the function $F(t)$ as

$$F(\omega_f) = \int_0^\infty F(t)e^{-i\omega_f t}\, dt$$

where i is the complex operator. The preceding equation can be written, using Euler's formula, as

$$F(\omega_f) = \int_0^\infty F(t) \cos \omega_f t\, dt - i \int_0^\infty F(t) \sin \omega_f t\, dt$$

This equation also can be written as

$$F(\omega_f) = a - ib$$

where a and b are coefficients that resemble the Fourier coefficients and are defined as

$$a = \int_0^\infty F(t) \cos \omega_f t\, dt$$

$$b = \int_0^\infty F(t) \sin \omega_f t\, dt$$

These equations define the amplitude and phase of the Fourier transform as

$$|F(\omega_f)| = \sqrt{a^2 + b^2}$$

$$\phi = \tan^{-1} \frac{-b}{a}$$

Note that the definition of the phase is slightly different from the phase angles defined using the Fourier coefficients since it represents here the phase angle associated with a vector whose components are defined by the real and imaginary parts of the Fourier transform. It is also important to point out that, in the case of complex functions or functions defined in tabulated forms, the integrals in the Fourier transform can be evaluated numerically by using a procedure similar to the one used in the numerical evaluation of the Fourier coefficients and the Duhamel integral.

In order to demonstrate the use of the Fourier transform method to examine the frequency contents in an arbitrary forcing function, we use as an example the function $F(t)$ shown in Fig. 16. The coefficients a and b of the Fourier

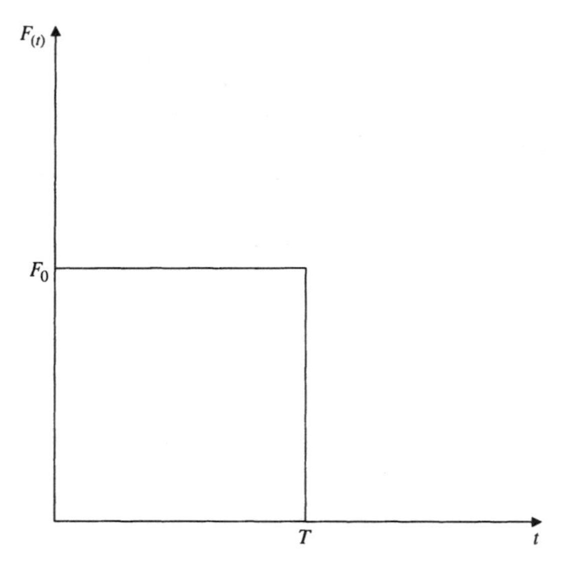

FIG. 5.16. Nonperiodic functions.

transform of this function are defined as

$$a = \int_0^\infty F(t) \cos \omega_f t \, dt = \int_0^T F_0 \cos \omega_f t \, dt = \frac{F_0 \sin \omega_f T}{\omega_f}$$

$$b = \int_0^\infty F(t) \sin \omega_f t \, dt = \int_0^T F_0 \sin \omega_f t \, dt = \frac{F_0(1 - \cos \omega_f T)}{\omega_f}$$

It follows that

$$|F(\omega_f)| = \frac{F_0}{\omega_f}\sqrt{2 - 2 \cos \omega_f T}$$

$$\phi = \tan^{-1}\left(\frac{\cos \omega_f T - 1}{\sin \omega_f T}\right)$$

Figure 17 shows the amplitude of the Fourier transform of the function shown in Fig. 16. The results presented in this figure show the frequency contents in the transient nonperiodic function $F(t)$, just as the Fourier coefficients show the frequency contents of the periodic functions. Using similar plots, it can be shown that as T decreases, $F(\omega)$ has large amplitude at higher frequencies. For this reason, rapidly varying forces are known to have higher frequency contents as compared to slowly varying forces. Note also that for transient nonperiodic functions, the frequency spectrum is a continuous function of the frequency ω_f, whereas in the case of a periodic function, the frequency spectrum is defined at discrete points in the frequency domain.

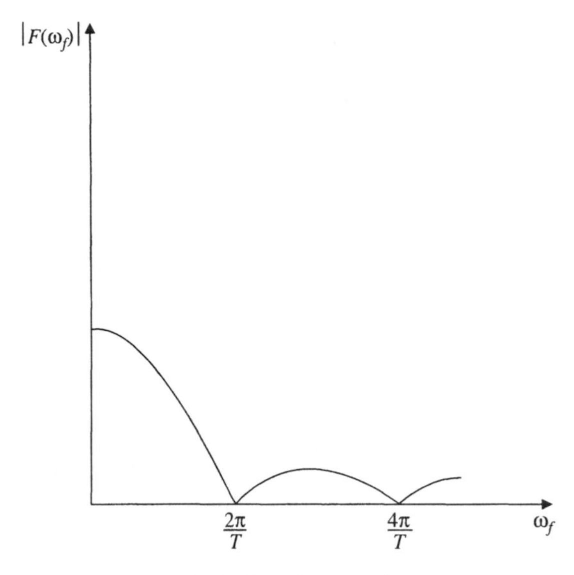

FIG. 5.17. Fourier transform.

5.8 COMPUTER METHODS IN NONLINEAR VIBRATION

Thus far, we have considered only the solutions of the vibration equations of systems in which the differential equations of motion are linear. As pointed out in Chapter 2, a differential equation is said to be linear if the equation contains only the first power of the dependent variable or its derivatives. When the equations are linear the principle of superposition can be applied, and consequently, the response of the system to a set of forces can be determined by adding the responses obtained as the result of application of each force separately. In fact, this is the principle which enabled us to derive the Duhamel integral, and therefore, the Duhamel integral can be used only when the system is linear.

In many applications, the differential equation of motion contains quadratic, cubic, or even trigonometric functions of the dependent variable. In these cases, the equation is said to be *nonlinear*. Unlike linear equations, where a closed-form solution can always be obtained, the solution of most nonlinear equations can only be obtained numerically. It is a common practice to try to linearize nonlinear equations so that the techniques for solving linear systems can be used. This approach can be used only when the nonlinear effect can be neglected. For example, the free oscillations of the pendulum in Fig. 18 are governed by the nonlinear equation

$$\frac{ml^2}{3}\ddot{\theta} + mg\frac{l}{2}\sin\theta = 0 \tag{5.68}$$

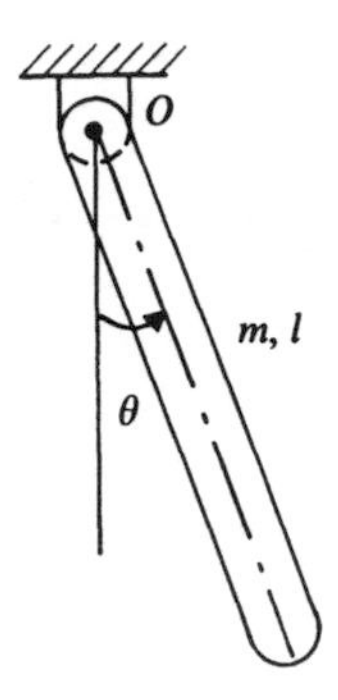

FIG. 5.18. Nonlinear free oscillations.

If the oscillations are small ($\theta \leq 10°$), linearization techniques may be used, and one can write

$$\sin \theta \approx \theta \tag{5.69}$$

Using this approximation, Eq. 68 can be linearized, leading to

$$\frac{ml^2}{3}\ddot{\theta} + mg\frac{l}{2}\theta = 0 \tag{5.70}$$

This is a linear equation which can be solved using the techniques described in this chapter and the preceding chapters. If the assumption of small oscillations cannot be made, the nonlinear equation given by Eq. 68 must be solved using computer and numerical methods.

The nonlinearity that appears in Eq. 68 results mainly from the large rotation of the pendulum. Another type of nonlinearity may arise, when the elastic or damping forces are nonlinear functions of the displacement and its time derivatives. If the spring force is quadratic function in the displacement, one has

$$F_s = kx^2 \tag{5.71}$$

The equation of motion of a single degree of freedom mass–spring system is given, in this case, by

$$m\ddot{x} + c\dot{x} + kx^2 = F(t) \tag{5.72}$$

This is again a nonlinear equation. While elastically linear systems have one equilibrium position only, elastically nonlinear systems may have more than one equilibrium configuration. For example, consider the following nonlinear differential equation of motion of the single degree of freedom system:

$$m\ddot{x} + c\dot{x} + k\left(x^3 - \frac{x}{6}\right) = 0$$

At equilibrium, $\dot{x} = \ddot{x} = 0$, which, upon substitution into the nonlinear differential equation, yields

$$k\left(x^3 - \frac{x}{6}\right) = 0$$

This equation has the following three roots:

$$x_1 = 0, \qquad x_2 = \frac{1}{\sqrt{6}}, \qquad x_3 = -\frac{1}{\sqrt{6}}$$

which define three different static equilibrium configurations.

State Space Representation Unlike linear systems, there is no standard simple form which all nonlinear equations can be assumed to take, since any power of the dependent variable and its derivatives may appear. Furthermore, a closed-form solution for many of the nonlinear equations cannot be obtained. One, in these cases, must resort to computer and numerical methods, by which the solutions of the nonlinear equations are obtained by direct numerical integration.

The application of Newton's second law leads to a linear equation in the acceleration. This equation, for linear and nonlinear systems, can be written in the following general form

$$\ddot{x} = G(x, \dot{x}, t) \tag{5.73}$$

where $G(x, \dot{x}, t)$ can be a nonlinear function in its arguments x, $\dot{x}$, and t. For example, Eq. 68, which describes the nonlinear vibration of the pendulum, can be written in the form of Eq. 73 as

$$\ddot{\theta} = -\frac{3g}{2l}\sin\theta$$

where the function G of Eq. 73 is recognized, in this case, as

$$G = -\frac{3g}{2l}\sin\theta$$

Similarly, Eq. 72, which describes the nonlinear vibration of the damped single degree of freedom mass–spring system, can be written in the form of Eq. 73 as

$$\ddot{x} = \frac{1}{m}[F(t) - c\dot{x} - kx^2]$$

where the function G of Eq. 73 is recognized, in this case, as

$$G = \frac{1}{m}[F(t) - c\dot{x} - kx^2]$$

Equation 73 is a second-order ordinary differential equation which is equivalent to two first-order ordinary differential equations. In order to determine these two differential equations, let us define the following *state variables*

$$y_1 = x \tag{5.74a}$$

$$y_2 = \dot{y}_1 = \dot{x} \tag{5.74b}$$

It is clear that

$$\dot{y}_2 = \ddot{x} = G(y_1, y_2, t) \tag{5.75}$$

Equations 74b and 75 can then be written as

$$\dot{y}_1 = y_2 \tag{5.76a}$$

$$\dot{y}_2 = G(y_1, y_2, t) \tag{5.76b}$$

which can be written in a vector form as

$$\dot{\mathbf{y}} = \mathbf{f}(\mathbf{y}, t) \tag{5.77}$$

where

$$\mathbf{y} = \begin{bmatrix} y_1 \\ y_2 \end{bmatrix}, \qquad \mathbf{f}(\mathbf{y}, t) = \begin{bmatrix} y_2 \\ G(y_1, y_2, t) \end{bmatrix} \tag{5.78}$$

Equation 77 is the vector of *state equations* of the system which can be integrated numerically to determine y_1 and y_2. Once y_1 and y_2 are determined, Eq. 74 can be used to determine x and $\dot{x}$.

Numerical Integration There are several numerical integration methods for solving Eq. 77. The simplest method is called *Euler's method.* In order to understand Euler's method, Eq. 77 can be written as

$$\dot{\mathbf{y}} = \frac{d\mathbf{y}}{dt} = \mathbf{f}(\mathbf{y}, t)$$

or

$$d\mathbf{y} = \mathbf{f}(\mathbf{y}, t)\, dt$$

which leads to

$$\int_{\mathbf{y}_0}^{\mathbf{y}_1} d\mathbf{y} = \int_{t_0}^{t_1} \mathbf{f}(\mathbf{y}, t)\, dt \tag{5.79}$$

Define

$$\mathbf{y}_0 = \begin{bmatrix} y_1(t_0) \\ y_2(t_0) \end{bmatrix}, \qquad \mathbf{y}_1 = \begin{bmatrix} y_1(t_1) \\ y_2(t_1) \end{bmatrix} \tag{5.80}$$

Clearly

$$\int_{\mathbf{y}_0}^{\mathbf{y}_1} d\mathbf{y} = \mathbf{y}_1 - \mathbf{y}_0 \tag{5.81}$$

If we assume that t_1 is selected such that $t_1 - t_0 = \Delta t$ is very small, the integral on the right-hand side of Eq. 79 can be approximated as

$$\int_{t_0}^{t_1} \mathbf{f}(\mathbf{y}, t)\, dt \approx \mathbf{f}(\mathbf{y}_0, t_0)\Delta t \tag{5.82}$$

Substituting Eqs. 81 and 82 into Eq. 79, one obtains

$$\mathbf{y}_1 = \mathbf{y}_0 + \mathbf{f}(\mathbf{y}_0, t_0)\Delta t \tag{5.83}$$

Observe that if the initial conditions x_0 and $\dot{x}_0$ are given, the vector $\mathbf{y}_0$ can be evaluated, using Eqs. 74 and 80, as

$$\mathbf{y}_0 = \begin{bmatrix} y_1(t_0) \\ y_2(t_0) \end{bmatrix} = \begin{bmatrix} x_0 \\ \dot{x}_0 \end{bmatrix} \tag{5.84a}$$

This vector can also be substituted into Eq. 78 in order to determine the function $\mathbf{f}(y_0, t_0)$ as

$$\mathbf{f}(\mathbf{y}_0, t_0) = \begin{bmatrix} y_2(t_0) \\ G(y_1(t_0), y_2(t_0), t_0) \end{bmatrix} = \begin{bmatrix} \dot{x}_0 \\ G(x_0, \dot{x}_0, t_0) \end{bmatrix} \tag{5.84b}$$

By substituting Eqs. 84a and 84b into Eq. 83, the state vector $\mathbf{y}_1$ can be defined. One can then use this vector to advance the numerical integration one step. To this end, we write

$$\mathbf{y}_2 = \mathbf{y}_1 + \mathbf{f}(\mathbf{y}_1, t_1)\Delta t \tag{5.85}$$

In general, one has the following recursive formula based on Euler's method

$$\mathbf{y}_n = \mathbf{y}_{n-1} + \mathbf{f}(\mathbf{y}_{n-1}, t_{n-1})\Delta t \tag{5.86}$$

The use of Euler's method is demonstrated by the following simple example.

Example 5.9

In order to examine the accuracy of Euler's method, we apply the numerical procedure discussed in this section to a linear system whose solution is known in a closed form. The results obtained from the numerical solution are then compared with the exact solution. We consider, for this purpose, the free undamped vibration of the system given in Example 8. If the initial displacement is assumed to be 0.01 m and the initial velocity is assumed to be zero, the exact solution for the equation of free vibration is

$$\begin{aligned} x(t) &= x_0 \cos \omega t = 0.01 \cos 50t \\ \dot{x}(t) &= -0.5 \sin 50t \end{aligned} \tag{5.87}$$

In order to demonstrate the use of the numerical procedure discussed in this section, we write the equation of free vibration as

$$m\ddot{x} + kx = 0$$

or

$$\ddot{x} = -\omega^2 x$$

TABLE 5.5. Euler's Method

Time	Approximate Values					Exact Values	
	y_1	y_2	$G = -\omega^2 y_1$	$y_2 \Delta t$	$G \Delta t$	y_1	y_2
0.00	0.010	0.000	−25.00	0.000	−0.250	0.010	0.000
0.01	0.010	−0.250	−25.00	-2.5×10^{-3}	−0.25	8.776×10^{-3}	−0.2397
0.02	7.5×10^{-3}	−0.5	−18.75	$-5. \times 10^{-3}$	−0.1875	5.403×10^{-3}	−0.42074
0.03	2.5×10^{-3}	−0.6875	−6.250	-6.875×10^{-3}	−0.0625	7.0737×10^{-4}	−0.49875
0.04	-4.375×10^{-3}	−0.750	10.9375	-7.5×10^{-3}	0.109375	-4.1615×10^{-3}	−0.45465
0.05	−0.011875	−0.6406	29.6875	-6.406×10^{-3}	0.29675	-8.0114×10^{-3}	−0.29924
0.06	−0.01828	−0.34375	45.703	-3.438×10^{-3}	0.45703	-9.899×10^{-3}	−0.07056
0.07	−0.02172	0.1133	54.297	1.133×10^{-3}	0.54297	-9.365×10^{-3}	0.1754
0.08	−0.02059	0.65625	51.4648	6.5625×10^{-3}	0.51465	-6.536×10^{-3}	0.3784
0.09	−0.01402	1.170898	35.0585	0.0117089	0.35059	-2.1079×10^{-3}	0.48877
0.10	-2.3144×10^{-3}	1.52148	5.78604	0.015215	0.05786	2.837×10^{-3}	0.4795
0.11	0.0129	1.5793	−32.25109	0.015793	−0.322511	7.0867×10^{-3}	0.3528
0.12	0.02869	1.2568	−71.7336	0.012568	−0.71734	9.602×10^{-3}	0.1397
0.13	0.041262	0.5395	−103.1544	5.395×10^{-3}	−1.03154	9.766×10^{-3}	−0.10756

This equation is in the same form as Eq. 73. Using Eq. 74, one can then define the state vector of Eq. 78 as

$$\mathbf{y} = \begin{bmatrix} y_1 \\ y_2 \end{bmatrix} = \begin{bmatrix} x \\ \dot{x} \end{bmatrix}$$

and the state equation of Eq. 77 as

$$\begin{bmatrix} \dot{y}_1 \\ \dot{y}_2 \end{bmatrix} = \begin{bmatrix} y_2 \\ -\omega^2 y_1 \end{bmatrix}$$

One may select the step size Δt to be 0.01 which is less than one-tenth of the period of oscillation. Equation 83 can be used to predict the state vector $\mathbf{y}_1$, at time $t_1 = t_0 + \Delta t = 0 + 0.01 = 0.01$, as

$$\begin{aligned} \mathbf{y}_1 &= \begin{bmatrix} y_1(t_1) \\ y_2(t_1) \end{bmatrix} = \mathbf{y}_0 + \mathbf{f}(\mathbf{y}_0, t_0)\Delta t \\ &= \begin{bmatrix} y_1(t_0) \\ y_2(t_0) \end{bmatrix} + \begin{bmatrix} y_2(t_0) \\ -\omega^2 y_1(t_0) \end{bmatrix} \Delta t \\ &= \begin{bmatrix} 0.01 \\ 0 \end{bmatrix} + 0.01 \begin{bmatrix} 0 \\ -(50)^2(0.01) \end{bmatrix} = \begin{bmatrix} 0.01 \\ -0.25 \end{bmatrix} \end{aligned}$$

The exact solution obtained from using Eq. 87 is

$$y_1(t_1) = x(t_1) = 0.0087758 \text{ m}$$

$$\dot{y}_1(t_1) = \dot{x}(t_1) = -0.23971 \text{ m/s}$$

Similarly, at time $t_2 = t_1 + \Delta t = 0.020$

$$\begin{aligned} \mathbf{y}_2 &= \mathbf{y}_1 + \mathbf{f}(\mathbf{y}_1, t_1)\Delta t \\ &= \begin{bmatrix} y_1(t_1) \\ y_2(t_1) \end{bmatrix} + \begin{bmatrix} y_2(t_1) \\ -\omega^2 y_1(t_1) \end{bmatrix} \Delta t \\ &= \begin{bmatrix} 0.01 \\ -0.25 \end{bmatrix} + 0.01 \begin{bmatrix} -0.25 \\ -(50)^2(0.01) \end{bmatrix} = \begin{bmatrix} 0.0075 \\ -0.5 \end{bmatrix} = \begin{bmatrix} y_1(t_2) \\ y_2(t_2) \end{bmatrix} \end{aligned}$$

The exact solution obtained from using Eq. 87 is

$$y_1(t_2) = 0.005403 \qquad \text{and} \qquad y_2(t_2) = -0.420735$$

Table 5 shows the approximate results obtained by using Euler's method. These results are compared with the exact solutions obtained by using Eq. 87.

Remarks It is clear from the results presented in the preceding example, and the comparison between the exact solution and the approximate solution obtained by using Euler's method, that Euler's method is not a very accurate method. One, however, can show that better results can be obtained using this method by reducing the step size Δt. Nonetheless, the results obtained using this method will continue to diverge, especially in the cases of highly nonlinear systems. In order to understand the approximation used in Euler's method, we use Taylor's series to write the solution, at time $t + \Delta t$, as

$$\mathbf{y}(t + \Delta t) = \mathbf{y}(t) + \dot{\mathbf{y}}(t)\,\Delta t + \frac{(\Delta t)^2}{2}\ddot{\mathbf{y}}(t) + \cdots \tag{5.88}$$

Euler's method can be obtained from Taylor's series by truncating terms higher than the first order, that is,

$$\mathbf{y}(t + \Delta t) = \mathbf{y}(t) + \dot{\mathbf{y}}(t)\,\Delta t \tag{5.89}$$

In this case, the error of integration can be evaluated as

$$\mathbf{E} = \frac{(\Delta t)^2}{2}\ddot{\mathbf{y}}(t) + \frac{(\Delta t)^3}{3!}\dddot{\mathbf{y}}(t) + \cdots \tag{5.90}$$

Observe that in the preceding example $\ddot{\mathbf{y}}(t) = -\omega^2\mathbf{y}(t)$, and if the frequency ω is very high, the error $\mathbf{E}$ given by Eq. 90 can be very large. In fact, the higher the frequency content of the function is, the less accurate Euler's method is going to be. Therefore, with rapidly varying functions the use of Euler's method is not recommended. In fact, Euler's method is rarely used in practical applications because of its low order of integration. Higher-order methods such as the *Runge–Kutta* and *Adams methods* (Atkinson, 1978) are often used.

The Runge–Kutta method is closely related to Taylor's series expansion, but no differentiation of $\dot{\mathbf{y}}$ is required. In the Runge–Kutta method, higher-order terms in Taylor's series are ignored and the first derivative is replaced by an average slope. In this case, the method can be written as

$$\mathbf{y}(t + \Delta t) = \mathbf{y}(t) + \Delta t\,\mathbf{f}_a \tag{5.91}$$

where $\mathbf{f}_a$ is the average slope which can be obtained using Simpson's rule (Carnahan et al., 1969; and Atkinson, 1978) as

$$\mathbf{f}_a = \frac{1}{6}\left[\dot{\mathbf{y}}(t) + 4\dot{\mathbf{y}}\left(t + \frac{\Delta t}{2}\right) + \dot{\mathbf{y}}(t + \Delta t)\right]$$

In the Runge–Kutta method, the central term in this equation is split into two terms, that is,

$$\mathbf{f}_a = \tfrac{1}{6}[\mathbf{f}_1 + 2\mathbf{f}_2 + 2\mathbf{f}_3 + \mathbf{f}_4] \tag{5.92}$$

where

$$\left.\begin{aligned} \mathbf{f}_1 &= \dot{\mathbf{y}}(t, \mathbf{y}(t)) \\ \mathbf{f}_2 &= \dot{\mathbf{y}}\left(t + \frac{\Delta t}{2}, \mathbf{y}(t) + \frac{\Delta t}{2}\mathbf{f}_1\right) \\ \mathbf{f}_3 &= \dot{\mathbf{y}}\left(t + \frac{\Delta t}{2}, \mathbf{y}(t) + \frac{\Delta t}{2}\mathbf{f}_2\right) \\ \mathbf{f}_4 &= \dot{\mathbf{y}}(t + \Delta t, \mathbf{y}(t) + \Delta t\,\mathbf{f}_3) \end{aligned}\right\} \tag{5.93}$$

Substituting Eq. 92 into Eq. 91, one obtains

$$\mathbf{y}(t + \Delta t) = \mathbf{y}(t) + \frac{\Delta t}{6}[\mathbf{f}_1 + 2\mathbf{f}_2 + 2\mathbf{f}_3 + \mathbf{f}_4] \tag{5.94}$$

which leads to the following recursive relations

$$\mathbf{y}_{n+1} = \mathbf{y}_n + \frac{\Delta t}{6}[\mathbf{f}_1 + 2\mathbf{f}_2 + 2\mathbf{f}_3 + \mathbf{f}_4]_n \tag{5.95}$$

where

$$\mathbf{y}_n = \mathbf{y}(t_n) \tag{5.96}$$

In order to see the improved accuracy obtained using the Runge–Kutta method, we will use the data of the system of the preceding example, and we will compare the results with the results obtained using the first-order Euler method. To this end, we first summarize the main steps of the numerical algorithm. For the purpose of illustration, we carry out the calculations for one fixed time step based on the data of the preceding example.

The Runge–Kutta Algorithm Based on the Runge–Kutta numerical technique presented in this section, a computer algorithm can be developed. The main steps of this numerical algorithm are as follows:

(1) The initial velocity and displacement of the system must be specified.
(2) By using the equation of motion, one can define the state equations of Eq. 77 and the function $\mathbf{f}(\mathbf{y}, t)$ of Eq. 78.
(3) A relatively small time step Δt is selected. Set $t = t_0$ as the initial time.
(4) Evaluate the function $\mathbf{f}_1 = \dot{\mathbf{y}}(t, \mathbf{y}(t))$ of Eq. 93 at this point in time.
(5) Use the function $\mathbf{f}_1$ evaluted in the preceding step to calculate $\bar{\mathbf{y}}_1 = \mathbf{y}(t) + (\Delta t/2)\,\mathbf{f}_1$. The vector $\bar{\mathbf{y}}_1$ can be used to evaluate the vector function $\mathbf{f}_2$.
(6) Use the vector function $\mathbf{f}_2$ evaluated in the preceding step to calculate $\bar{\mathbf{y}}_2 = \mathbf{y}(t) + (\Delta t/2)\,\mathbf{f}_2$. The vector $\mathbf{y}_2$ can be used to evaluate $\mathbf{f}_3$ of Eq. 93.
(7) The function $\mathbf{f}_3$ can be used to calculate the vector $\bar{\mathbf{y}}_3 = \mathbf{y}(t) + \Delta t\mathbf{f}_3$. This vector can be used to evaluate numerically the vector function $\mathbf{f}_4$ of Eq. 93.
(8) Having evaluated $\mathbf{f}_1$, $\mathbf{f}_2$, $\mathbf{f}_3$, and $\mathbf{f}_4$, Eq. 94 or, equivalently, Eq. 95 can bc used to evaluate $\mathbf{y}(t + \Delta t)$.
(9) If $t < t_e$ where t_e is the final time, set $t = t + \Delta t$ and repeat steps (4)–(8), otherwise stop.

In the following, the use of the Runge–Kutta algorithm is demonstrated for one step starting with $t = 0$. We use the data of the preceding example and proceed as follows:

(1) The initial conditions of the system are $x_0 = 0.01$ m and $\dot{x}_0 = 0$.
(2) The function $\dot{\mathbf{y}} = \mathbf{f}(\mathbf{y}, t)$ of Eq. 78 is defined for this example as

$$\dot{\mathbf{y}} = \mathbf{f}(\mathbf{y}, t) = \begin{bmatrix} y_2 \\ -\omega^2 y_1 \end{bmatrix} = \begin{bmatrix} \dot{x} \\ -\omega^2 x \end{bmatrix}$$

(3) We select the step size to be $\Delta t = 0.01$.

(4) We first evaluate the function $\mathbf{f}_1$ as

$$\mathbf{f}_1 = \begin{bmatrix} y_2 \\ -\omega^2 y_1 \end{bmatrix} = \begin{bmatrix} 0.000 \\ -25.000 \end{bmatrix}$$

(5) Using $\mathbf{f}_1$ we can evaluate $\bar{\mathbf{y}}_1$ as

$$\bar{\mathbf{y}}_1 = \mathbf{y}(0) + \frac{\Delta t}{2}\mathbf{f}_1 = \begin{bmatrix} 0.010 \\ 0.000 \end{bmatrix} + \frac{0.01}{2}\begin{bmatrix} 0.000 \\ -25.00 \end{bmatrix} = \begin{bmatrix} 0.010 \\ -0.125 \end{bmatrix}$$

Therefore,

$$\mathbf{f}_2 = \begin{bmatrix} -0.125 \\ -(50)^2(0.01) \end{bmatrix} = \begin{bmatrix} -0.125 \\ -25.0 \end{bmatrix}$$

(6) The vector $\mathbf{f}_2$ can then be used to evaluate $\bar{\mathbf{y}}_2$ as

$$\bar{\mathbf{y}}_2 = \mathbf{y}(0) + \frac{\Delta t}{2}\mathbf{f}_2 = \begin{bmatrix} 0.010 \\ 0.000 \end{bmatrix} + \frac{0.01}{2}\begin{bmatrix} -0.125 \\ -25.0 \end{bmatrix} = \begin{bmatrix} 9.375 \times 10^{-3} \\ -0.125 \end{bmatrix}$$

Therefore,

$$\mathbf{f}_3 = \begin{bmatrix} -0.125 \\ -(50)^2(9.375 \times 10^{-3}) \end{bmatrix} = \begin{bmatrix} -0.125 \\ -23.4375 \end{bmatrix}$$

(7) The vector function $\mathbf{f}_3$ can then be used to evaluate $\bar{\mathbf{y}}_3$ as

$$\bar{\mathbf{y}}_3 = \mathbf{y}(0) + \Delta t\mathbf{f}_3 = \begin{bmatrix} 0.01 \\ 0.00 \end{bmatrix} + 0.01\begin{bmatrix} -0.125 \\ -23.4375 \end{bmatrix} = \begin{bmatrix} 8.75 \times 10^{-3} \\ -0.234375 \end{bmatrix}$$

Therefore, $\mathbf{f}_4$ of Eq. 93 can be evaluated as

$$\mathbf{f}_4 = \begin{bmatrix} -0.234375 \\ -(50)^2(8.75 \times 10^{-3}) \end{bmatrix} = \begin{bmatrix} -0.234375 \\ -21.875 \end{bmatrix}$$

(8) Finally, the response of the system at time $t + \Delta t$, which at this step $(0 + \Delta t) = \Delta t = 0.01$, is given by direct susbtitution of $\mathbf{f}_1, \mathbf{f}_2, \mathbf{f}_3$, and $\mathbf{f}_4$ into Eq. 94 or, equivalently, Eq. 95. This yields

$$\begin{aligned}
\mathbf{y}(0.01) &= \mathbf{y}(0) + \frac{\Delta t}{6}\left[\mathbf{f}_1 + 2\mathbf{f}_2 + 2\mathbf{f}_3 + \mathbf{f}_4\right] \\
&= \begin{bmatrix} 0.01 \\ 0.00 \end{bmatrix} + \frac{0.01}{6}\begin{bmatrix} 0.000 - 2 \times 0.125 - 2 \times 0.125 - 0.234375 \\ -25.00 - 2 \times 25 - 2 \times 23.4375 - 21.875 \end{bmatrix} \\
&= \begin{bmatrix} 0.01 \\ 0.00 \end{bmatrix} + \begin{bmatrix} -1.224 \times 10^{-3} \\ -0.239583 \end{bmatrix} \\
&= \begin{bmatrix} 8.77604 \times 10^{-3} \\ -0.239583 \end{bmatrix} = \begin{bmatrix} y_1(0.01) \\ y_2(0.01) \end{bmatrix}
\end{aligned}$$

Clearly these results are in a very good agreement with the exact solution presented in Table 3 and they are much more accurate than the results obtained by using Euler's method.

Based on the Runge–Kutta algorithm presented in this section, a simple Fortran computer program can be developed to numerically solve nonlinear differential equations. An example of such a program is shown below:

```
  DIMENSION Y(20),YP(20),YOUT(20),YT(20), YP2(20),YP3(20)
  READ(5,800)NEQN,NSTP,XI,XE
  READ(5,900)(Y(I),I=1,NEQN)
  X = XI
  H = (XE-XI)/NSTP
  DO 60 I=1,NSTP
  CALL RUNGK0(Y,YP,NEQN,X,H,YOUT,YT,YP2,YP3)
  DO 20 I=1,NEQN
  Y(I) = YOUT(I)
20 WRITE(6,1000)X,I,YOUT(I)
60 X=X+H
800 FORMAT(2I5,2F10.0)
900 FORMAT(6F10.0)
1000 FORMAT(' X',3X,E12.5,5X,'Y(',I5,')',3X,E12.5)
  STOP
  END
C
C
  SUBROUTINE RUNGK0(Y,YP,NEQN,X,H,YOUT,YT,YP2,YP3)
  DIMENSION Y(1),YP(1),YOUT(1),YT(1),YP2(1),YP3(1)
  H5   = 0.5*H
  H6   = H/6.0
  XU   = X+H5
  XPH  = X+H
  CALL F(X,Y,YP)
  DO 100 I=1,NEQN
  YT(I) = Y(I)+H5*YP(I)
100 CONTINUE
  CALL F(XU,YT,YP2)
  DO 200 I=1,NEQN
  YT(I) = Y(I)+H5*YP2(I)
200 CONTINUE
  CALL F(XU,YT,YP3)
  DO 300 I=1,NEQN
  YT(I)  = Y(I)+H*YP3(I)
  YP2(I) = YP2(I)+YP3(I)
300 CONTINUE
  CALL F(XPH,YT,YP3)
  DO 400 I=1,NEQN
  YOUT(I) = Y(I)+H6*(YP(I)+2.*YP2(I)+YP3(I))
```

```
400 CONTINUE
    RETURN
    END
C
C
    SUBROUTINE F(X,Y,YP)
    DIMENSION Y(1),YP(1)
C
C      Insert the first order ordinary differential equations here.
C
    RETURN
    END
```

This computer program for solving first-order differential equations consists of three parts. The first part is the main program. The second part is Subroutine **RUNGK0**, which is a numerical integrator that uses the Runge–Kutta method to solve a system of first-order differential equations. The user of this program must define the ordinary differential equations to be integrated in the third part, Subroutine **F**. The arguments that appear in this program are as follows:

X	Independent variable that represents time.
Y	State vector in which the user must store the initial conditions.
YP	Function $\mathbf{F}(\mathbf{y}, t)$ in the differential equation.
YOUT	The integrated state vector at the current **X**.
NEQN	Number of state equations.
NSTP	Number of integration steps.
XI	Initial time of the simulation.
XE	End time of the simulation.
H	Step size.

All other vectors are working arrays used in the numerical integration. The maximum number of equations that can be solved by this program is 20. If the user has more than 20 equations, the dimensions of the arrays must be increased.

Problems

5.1. Determine analytically the Fourier series expansion for the periodic forcing function $F(t)$ shown in Fig. P1. Also determine numerically the Fourier coefficients assuming that $F_0 = 10$ N. Compare the results obtained numerically with the analytical results.

5.2. Find the forced response of the undamped single degree of freedom system to the periodic forcing function shown in Fig. P1.

5.3. Find the forced response of the single degree of freedom system with viscous damping to the periodic forcing function shown in Fig. P1.

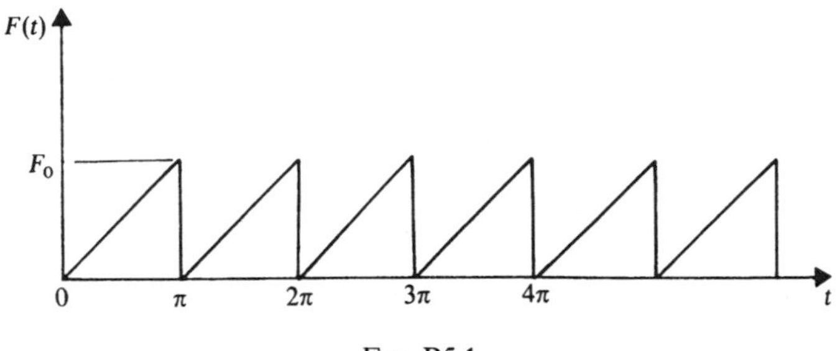

FIG. P5.1

5.4. Determine the forced response of the single degree of freedom system shown in Fig. P2 to the periodic forcing function shown in Fig. P1.

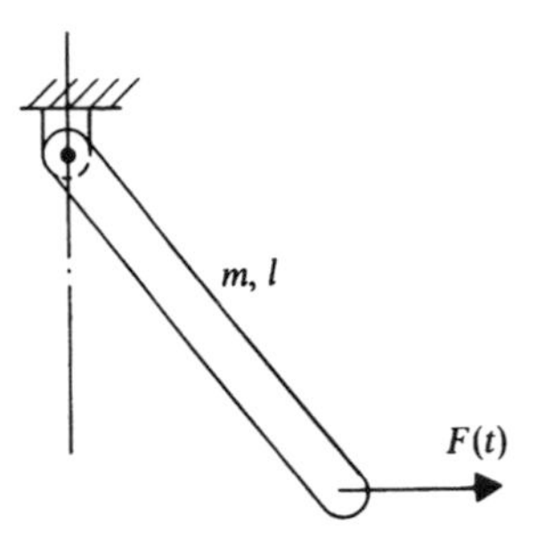

FIG. P5.2

5.5. Obtain the forced response of the damped single degree of freedom system shown in Fig. P3 to the periodic forcing function shown in Fig. P1.

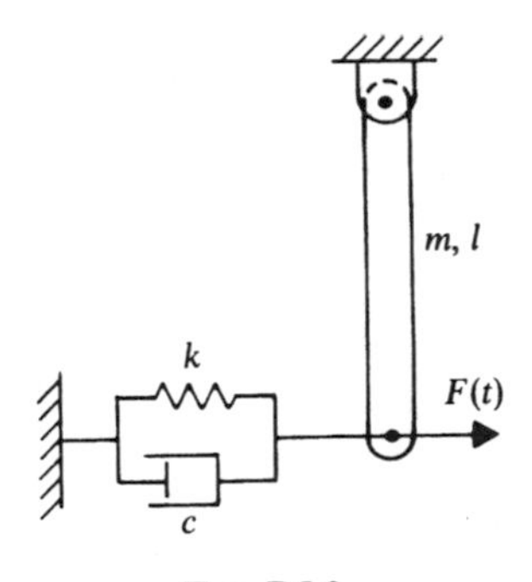

FIG. P5.3

5.6. Find the Fourier series expansion of the function $F(t)$ shown in Fig. P4.

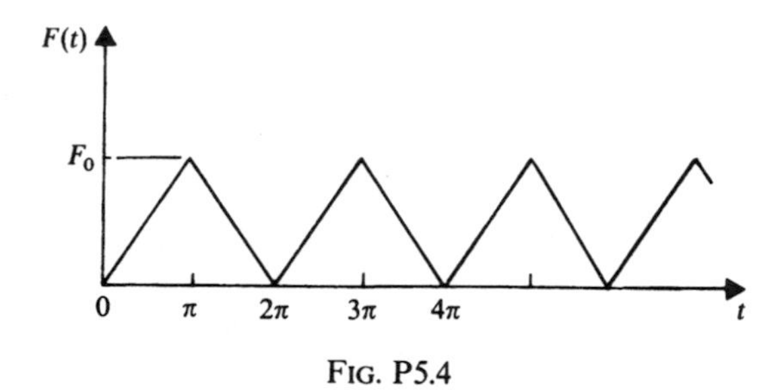

FIG. P5.4

5.7. Find the response of the undamped single degree of freedom system to the periodic forcing function shown in Fig. P4.

5.8. Determine the response of the damped single degree of freedom system to the periodic forcing function shown in Fig. P4.

5.9. Find the response of the single degree of freedom system shown in Fig. P2 to the periodic forcing function $F(t)$ shown in Fig. P4.

5.10. Find the response of the single degree of freedom system shown in Fig. P3 to the periodic forcing function $F(t)$ shown in Fig. P4.

5.11. Determine the response of the single degree of freedom system shown in Fig. P5 to the periodic forcing function $F(t)$ shown in Fig. P1.

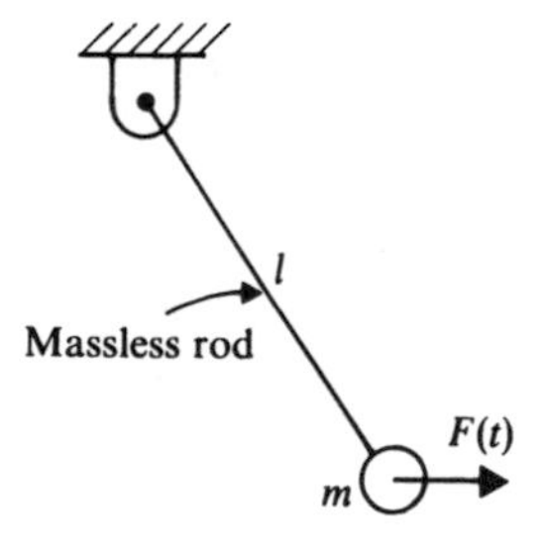

FIG. P5.5

5.12. Find the forced response of the single degree of freedom system shown in Fig. P5 to the periodic forcing function $F(t)$ shown in Fig. P4.

5.13. Obtain the Fourier series expansion of the periodic function $F(t)$ shown in Fig. P6.

5.14. Determine the Fourier series expansion of the periodic function $F(t)$ shown in Fig. P6, assuming that $T = \pi$.

5.15. Determine the forced response of the single degree of freedom system shown in Fig. P3 to the periodic forcing function $F(t)$ shown in Fig. P6.

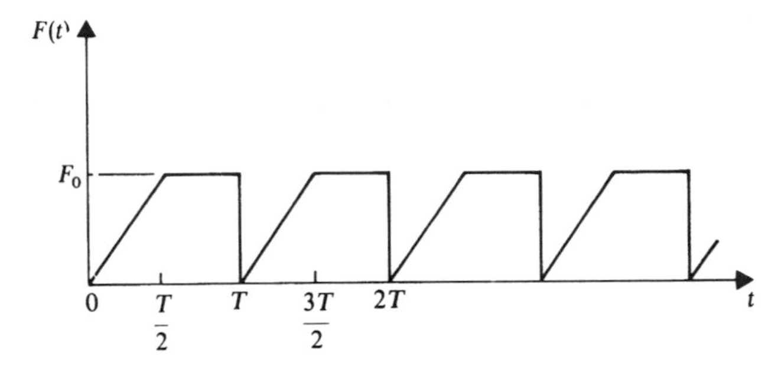

FIG. P5.6

5.16. Determine the forced response of the single degree of freedom system shown in Fig. P5 to the periodic forcing function $F(t)$ shown in Fig. P6.

5.17. Using analytical methods, find the Fourier expansion of the periodic function shown in Fig. P7. Also determine numerically the Fourier coefficients assuming that $M_0 = 10$ N·m and $T = 4$ s. Compare the results obtained numerically with the analytical solution.

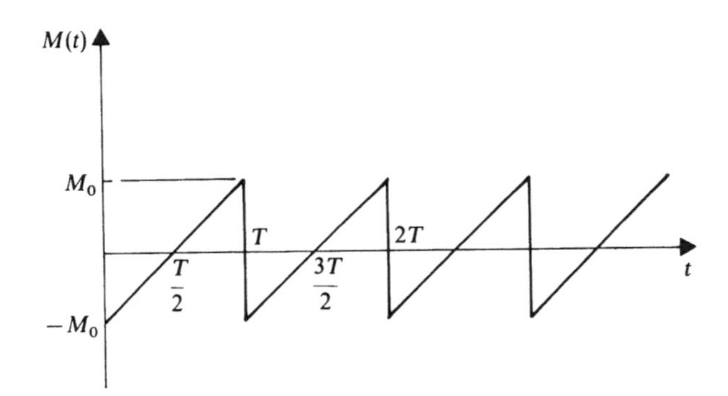

FIG. P5.7

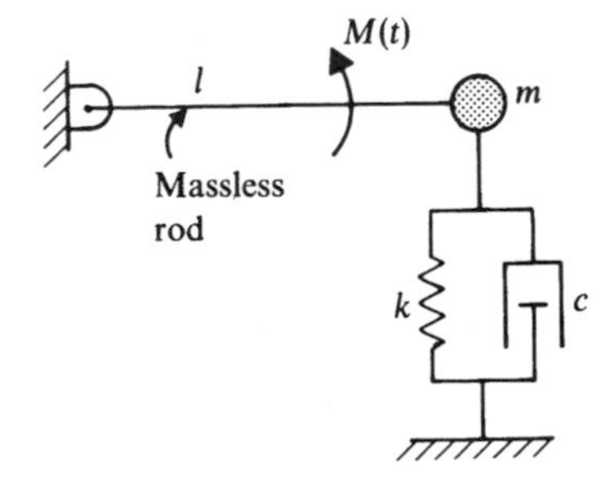

FIG. P5.8

5.18. Determine the forced response of the single degree of freedom system shown in Fig. P8 to the periodic function $M(t)$ shown in Fig. P7.

5.19. Determine the response of the single degree of freedom system shown in Fig. P9 to the periodic function $M(t)$ shown in Fig. P7.

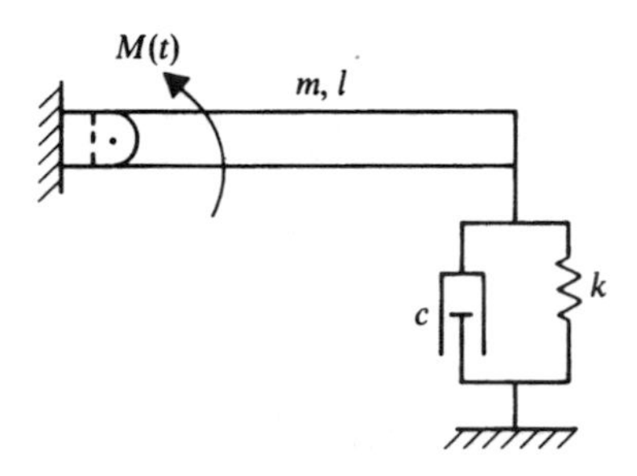

FIG. P5.9

5.20. Determine the response of the single degree of freedom system shown in Fig. P10, where $F(t)$ is the periodic forcing function given in Fig. P6 and $M(t)$ is the periodic function shown in Fig. P7.

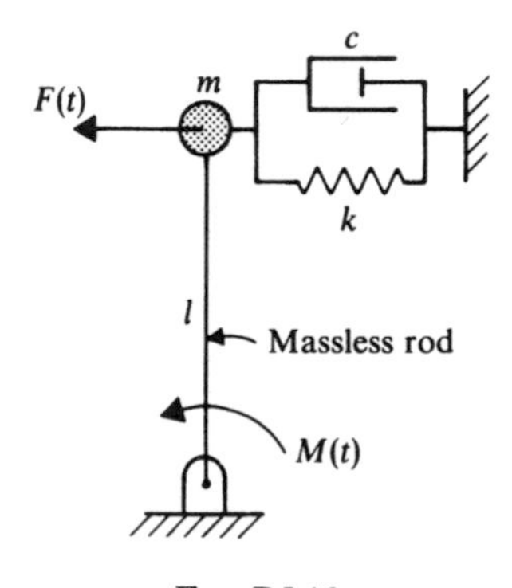

FIG. P5.10

5.21. Show that the time t_m corresponding to the peak response of a damped single degree of freedom system due to an impulsive force is given by the equation

$$\tan \omega_d t_m = \frac{\sqrt{1-\xi^2}}{\xi}$$

where ω_d is the damped natural frequency and ξ is the damping factor.

5.22. Determine the maximum displacement of a damped spring–mass system due to the excitation of an impulsive force whose linear impulse is I.

5.23. If an arbitrary force $F(t)$ is applied to an undamped single degree of freedom mass–spring system with nonzero initial conditions, show that the response of the system must be written in the form

$$x(t) = x_0 \cos \omega t + \frac{\dot{x}_0}{\omega} \sin \omega t + \frac{1}{m\omega} \int_0^t F(\tau) \sin \omega(t - \tau)\, d\tau$$

where x_0 is the initial displacement and $\dot{x}_0$ is the initial velocity.

5.24. Determine the forced response of the undamped single degree of freedom spring–mass system to the forcing function shown in Fig. P11.

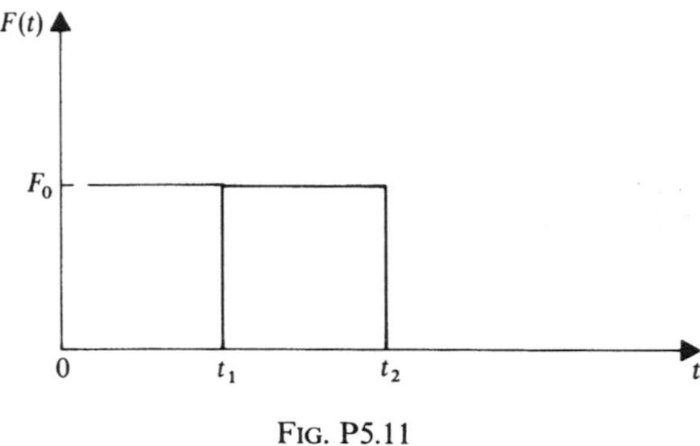

FIG. P5.11

5.25. Determine the forced response of the damped single degree of freedom mass–spring system to the forcing function shown in Fig. P11.

5.26. Find the response of the single degree of freedom system shown in Fig. P3 to the forcing function shown in Fig. P11.

5.27. Determine the response of the damped single degree of freedom mass–spring system to the forcing function shown in Fig. P12.

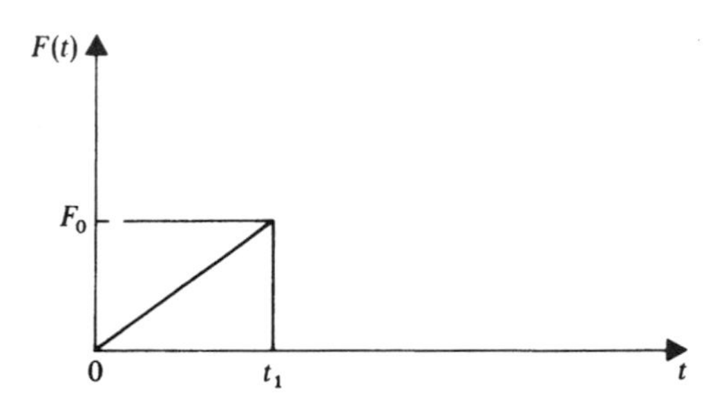

FIG. P5.12

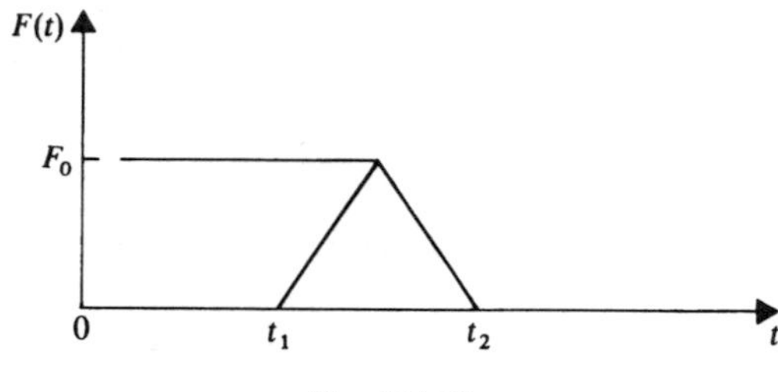

FIG. P5.13

5.28. Obtain the response of the damped mass–spring system to the forcing function shown in Fig. P13.

5.29. Determine the forced response of the damped single degree of freedom system shown in Fig. P3 to the forcing function shown in Fig. P12.

5.30. A single degree of freedom mass–spring system is subjected to the force shown in Fig. P11. Let $m = 1$ kg, $k = 2 \times 10^3$ N/m, $F_0 = 10$ N, $t_1 = 1$ s, and $t_2 = 3$ s. Use the numerical evaluation of the Duhamel integral to obtain the solution assuming zero initial conditions.

5.31. Repeat Problem 30 assuming that the system is damped with $c = 10$ N·s/m.

5.32. Repeat Probelm 30 assuming that the forcing function is replaced by the force given in Fig. P12, with $F_0 = 10$ N, and $t_1 = 2$ s.

5.33. Repeat Problem 30 assuming that the forcing function is replaced by the force given in Fig. P13, with $F_0 = 10$ N, $t_1 = 1$ s, and $t_2 = 3$ s.

5.34. The equation of motion of a single degree of freedom system is given by

$$m\ddot{x} + c\dot{x} + kx^2 = 0$$

where $m = 1$ kg, $c = 10$ N·s/m, $k = 2500$ N/m, $x_0 = 0$, and $\dot{x}_0 = 1$ m/s. Use Euler's method to find the solution of the vibration equation for one cycle assuming a step size $\Delta t = 0.01$.

5.35. Repeat Probelm 34 assuming that $\Delta t = 0.001$.

6
Systems with More Than One Degree of Freedom

Thus far, the theory of vibration of damped and undamped single degree of freedom systems was considered. Both free and forced motions of such systems were discussed and the governing differential equations and their solutions were obtained. Basic concepts and definitions, which are fundamental in understanding the vibration of single degree of freedom systems, were introduced. It is the purpose of this chapter to generalize the analytical development presented in the preceding chapters to the case in which the systems have more than one degree of freedom. We will start with the free and forced vibrations of both damped and undamped two degree of freedom systems.

A system is said to be a two degree of freedom system if only two independent coordinates are required in order to define completely the system configuration. It is important, however, to emphasize that the set of system degrees of freedom is not unique, since any two coordinates can be considered as degrees of freedom as long as they are independent. Examples of two degree of freedom systems are shown in Fig. 1. In Fig. 1(a), the displacements x_1 and x_2 are the system degrees of freedom. For the two degree of freedom system shown in Fig. 1(b), one may select θ_1 and θ_2 to be the system degrees of freedom since these two angular displacements are sufficient to determine the displacements of the two masses. The relationships between the displacements of the masses and the independent coordinates θ_1 and θ_2 are

$$x_1 = l_1 \sin \theta_1$$

$$x_2 = l_1 \sin \theta_1 + l_2 \sin \theta_2$$

The coordinates x_1 and x_2 can also be used as the system degrees of freedom. Similarly, in Fig. 1(c), if we assume that the horizontal motion of the beam is not allowed, the system configuration can be identified by using the coordinates y and θ, since the location of an arbitrary point p at a distance a from the center of mass of the beam can be written as

$$y_p = y + a \sin \theta$$

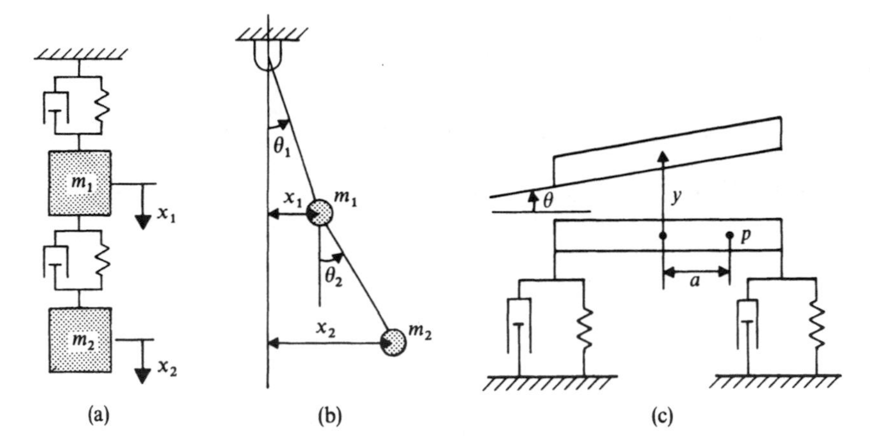

FIG. 6.1. Examples of two degree of freedom systems.

In the following sections we will study the linear theory of vibration of the two degree of freedom systems, and develop that theory to approximately the same level as was reached with single degree of freedom systems. The last section of this chapter will be devoted to a brief introduction to the analysis of multi-degree of freedom systems.

6.1 FREE UNDAMPED VIBRATION

In this section the differential equations of the free vibration of undamped two degree of freedom systems are developed.

Differential Equations of Motion Figure 2(a) shows an example of a two degree of freedom system. The masses m_1 and m_2 have only the freedom

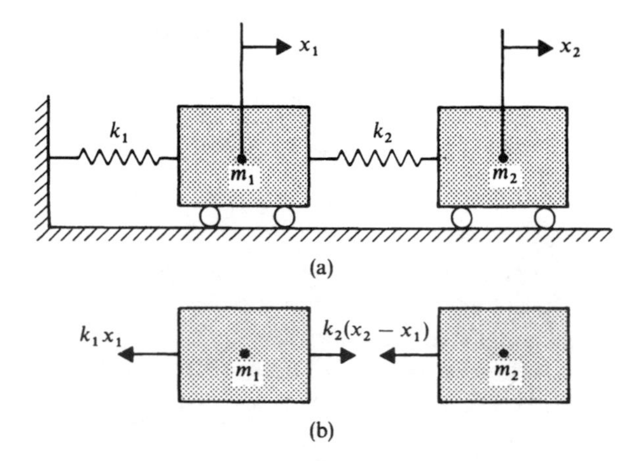

FIG. 6.2. Two degree of freedom system.

to translate in the horizontal direction, and therefore, one coordinate is required to define the position of each mass. The mass m_1 is connected to the ground by a linear spring which has a stiffness coefficient k_1, while the two masses are connected elastically through the linear spring k_2. Figure 2(b) shows a free body diagram for each mass. Using this free body diagram one can easily verify that the equations of motion of the mass m_1 is given by

$$m_1 \ddot{x}_1 = -k_1 x_1 + k_2(x_2 - x_1) \tag{6.1}$$

Similarly, the equation of motion of the mass m_2 is given by

$$m_2 \ddot{x}_2 = -k_2(x_2 - x_1) \tag{6.2}$$

where x_1 and x_2 are the coordinates of the masses m_1 and m_2 measured from the static equilibrium position. Equations 1 and 2 can be written as

$$m_1 \ddot{x}_1 + (k_1 + k_2) x_1 - k_2 x_2 = 0 \tag{6.3}$$

$$m_2 \ddot{x}_2 + k_2 x_2 - k_2 x_1 = 0 \tag{6.4}$$

These equations, which represent the differential equations of motion of the two degree of freedom system shown in Fig. 2(a), are coupled homogeneous second-order differential equations with constant coefficients. Equations 3 and 4 must be solved simultaneously in order to define the displacement coordinates x_1 and x_2 of the two masses m_1 and m_2.

Solution of the Equations of Motion In order to obtain a solution for Eqs. 3 and 4, we follow a procedure similar to the one used for a system with one degree of freedom, and assume a solution in the form

$$x_1 = X_1 \sin(\omega t + \phi) \tag{6.5}$$

$$x_2 = X_2 \sin(\omega t + \phi) \tag{6.6}$$

where X_1 and X_2 are the amplitudes of vibration, ω is the circular frequency, and ϕ is the phase angle. Differentiating Eqs. 5 and 6 twice with respect to time yields

$$\ddot{x}_1 = -\omega^2 X_1 \sin(\omega t + \phi) \tag{6.7}$$

$$\ddot{x}_2 = -\omega^2 X_2 \sin(\omega t + \phi) \tag{6.8}$$

Substituting Eqs. 5–8 into Eqs. 3 and 4 yields

$$(k_1 + k_2 - \omega^2 m_1) X_1 \sin(\omega t + \phi) - k_2 X_2 \sin(\omega t + \phi) = 0$$

$$(k_2 - \omega^2 m_2) X_2 \sin(\omega t + \phi) - k_2 X_1 \sin(\omega t + \phi) = 0$$

These two equations yield

$$(k_1 + k_2 - \omega^2 m_1) X_1 - k_2 X_2 = 0 \tag{6.9}$$

$$(k_2 - \omega^2 m_2) X_2 - k_2 X_1 = 0 \tag{6.10}$$

One possible solution of these two algebraic equations is the trivial solution $X_1 = X_2 = 0$. Equations 9 and 10 have nontrivial solutions if and only if the determinant of the coefficients of X_1 and X_2 in these equations is equal to zero, that is,

$$\begin{vmatrix} k_1 + k_2 - \omega^2 m_1 & -k_2 \\ -k_2 & k_2 - \omega^2 m_2 \end{vmatrix} = 0 \tag{6.11}$$

which yields

$$(k_1 + k_2 - \omega^2 m_1)(k_2 - \omega^2 m_2) - k_2^2 = 0$$

or

$$m_1 m_2 \omega^4 - [m_1 k_2 + m_2(k_1 + k_2)]\,\omega^2 + k_1 k_2 = 0 \tag{6.12}$$

This equation, which is called the *characteristic equation*, is a quadratic function in ω^2, and has the following two roots

$$\omega_1^2 = \frac{-b + \sqrt{b^2 - 4ac}}{2a} \tag{6.13}$$

$$\omega_2^2 = \frac{-b - \sqrt{b^2 - 4ac}}{2a} \tag{6.14}$$

where

$$a = m_1 m_2, \qquad b = -[m_1 k_2 + m_2(k_1 + k_2)], \qquad c = k_1 k_2$$

Thus, in the case of two degree of freedom systems, the characteristic equation yields two natural frequencies, ω_1 and ω_2, that depend on the masses and spring constants in the system. Therefore, there are two solutions, one associated with the first natural frequency ω_1 and the second associated with the second natural frequency ω_2. For ω_1, Eqs. 9 and 10 yield

$$(k_1 + k_2 - \omega_1^2 m_1)\,X_1 - k_2 X_2 = 0$$

$$(k_2 - \omega_1^2 m_2)\,X_2 - k_2 X_1 = 0$$

Since ω_1 is obtained from Eq. 11, the above two equations provide the same ratio between X_1 and X_2, that is,

$$\beta_1 = \left(\frac{X_1}{X_2}\right)_{\omega=\omega_1} = \frac{X_{11}}{X_{21}} = \frac{k_2}{k_1 + k_2 - \omega_1^2 m_1} = \frac{k_2 - \omega_1^2 m_2}{k_2} \tag{6.15}$$

where X_{11} and X_{21} are, respectively, the amplitudes of the masses m_1 and m_2, if the system vibrates at its first natural frequency ω_1. Therefore, the solutions corresponding to ω_1 can be written as

$$x_{11} = X_{11} \sin(\omega_1 t + \phi_1)$$

$$x_{21} = X_{21} \sin(\omega_1 t + \phi_1)$$

Using Eq. 15, one obtains

$$x_{11} = \beta_1 X_{21} \sin(\omega_1 t + \phi_1) \tag{6.16}$$

$$x_{21} = X_{21} \sin(\omega_1 t + \phi_1) \tag{6.17}$$

Similarly, for ω_2, Eqs. 9 and 10 yield

$$\beta_2 = \left(\frac{X_1}{X_2}\right)_{\omega=\omega_2} = \frac{X_{12}}{X_{22}} = \frac{k_2}{k_1 + k_2 - \omega_2^2 m_1} = \frac{k_2 - \omega_2^2 m_2}{k_2} \tag{6.18}$$

that is,

$$x_{12} = \beta_2 X_{22} \sin(\omega_2 t + \phi_2) \tag{6.19}$$

$$x_{22} = X_{22} \sin(\omega_2 t + \phi_2) \tag{6.20}$$

The amplitude ratios β_1 and β_2 are called the *mode shapes* or the *principal modes of vibration*.

The complete solution, x_1 and x_2, can then be obtained by summing the two solutions given by Eqs. 16 and 17 and Eqs. 19 and 20, that is,

$$x_1(t) = x_{11} + x_{12} = \beta_1 X_{21} \sin(\omega_1 t + \phi_1) + \beta_2 X_{22} \sin(\omega_2 t + \phi_2) \tag{6.21}$$

$$x_2(t) = x_{21} + x_{22} = X_{21} \sin(\omega_1 t + \phi_1) + X_{22} \sin(\omega_2 t + \phi_2) \tag{6.22}$$

In these two equations, there are four constants X_{21}, X_{22}, ϕ_1, and ϕ_2 which can be determined using the initial conditions.

Initial Conditions Differentiating Eqs. 21 and 22 with respect to time, one obtains

$$\dot{x}_1(t) = \omega_1 \beta_1 X_{21} \cos(\omega_1 t + \phi_1) + \omega_2 \beta_2 X_{22} \cos(\omega_2 t + \phi_2) \tag{6.23}$$

$$\dot{x}_2(t) = \omega_1 X_{21} \cos(\omega_1 t + \phi_1) + \omega_2 X_{22} \cos(\omega_2 t + \phi_2) \tag{6.24}$$

We consider the case in which the initial conditions are given by the initial displacements and velocities of the two masses. These initial conditions are given by

$$x_1(t = 0) = x_{10}, \qquad \dot{x}_1(t = 0) = \dot{x}_{10} \tag{6.25}$$

$$x_2(t = 0) = x_{20}, \qquad \dot{x}_2(t = 0) = \dot{x}_{20} \tag{6.26}$$

Substituting these equations into Eqs. 21–24, the following four algebraic equations can be obtained

$$x_{10} = \beta_1 X_{21} \sin \phi_1 + \beta_2 X_{22} \sin \phi_2 \tag{6.27}$$

$$x_{20} = X_{21} \sin \phi_1 + X_{22} \sin \phi_2 \tag{6.28}$$

$$\dot{x}_{10} = \omega_1 \beta_1 X_{21} \cos \phi_1 + \omega_2 \beta_2 X_{22} \cos \phi_2 \tag{6.29}$$

$$\dot{x}_{20} = \omega_1 X_{21} \cos \phi_1 + \omega_2 X_{22} \cos \phi_2 \tag{6.30}$$

These four algebraic equations can be solved for the four unknowns X_{21}, X_{22}, ϕ_1, and ϕ_2.

Example 6.1

For the two degree of freedom system shown in Fig. 2, let $m_1 = m_2 = 5$ kg and $k_1 = k_2 = 2000$ N/m and let $x_{10} = x_{20} = \dot{x}_{10} = 0$ and $\dot{x}_{20} = 0.3$ m/s. Determine the system response as a function of time.

Solution. The natural frequencies ω_1 and ω_2 can be calculated by using Eqs. 13 and 14, where the constants in these equations are given by

$$a = m_1 m_2 = (5)(5) = 25$$

$$b = -[m_1 k_2 + m_2(k_1 + k_2)] = -[5(2000) + 5(2000 + 2000)]$$
$$= -30{,}000$$

$$c = k_1 k_2 = (2000)(2000) = 4 \times 10^6$$

Thus

$$\omega_1^2 = \frac{30{,}000 + \sqrt{(30{,}000)^2 - 4(25)(4 \times 10^6)}}{2(25)} = 1047.21$$

$$\omega_2^2 = \frac{30{,}000 - \sqrt{(30{,}000)^2 - 4(25)(4 \times 10^6)}}{2(25)} = 152.786$$

that is,

$$\omega_1 = 32.361 \text{ rad/s}, \qquad \omega_2 = 12.361 \text{ rad/s}$$

The amplitude ratios β_1 and β_2 can be determined from Eqs. 15 and 18 as follows

$$\beta_1 = \frac{k_2}{k_1 + k_2 - \omega_1^2 m_1} = \frac{2000}{4000 - (1047.21)\,5} = -1.6181$$

$$\beta_2 = \frac{k_2}{k_1 + k_2 - \omega_2^2 m_1} = \frac{2000}{4000 - (152.786)\,5} = 0.618$$

Applying the initial conditions, Eqs. 27–30 yield

$$0 = \beta_1 X_{21} \sin\phi_1 + \beta_2 X_{22} \sin\phi_2$$
$$0 = X_{21} \sin\phi_1 + X_{22} \sin\phi_2$$
$$0 = \omega_1 \beta_1 X_{21} \cos\phi_1 + \omega_2 \beta_2 X_{22} \cos\phi_2$$
$$0.3 = \omega_1 X_{21} \cos\phi_1 + \omega_2 X_{22} \cos\phi_2$$

Given the values of ω_1, ω_2, β_1, and β_2, these four equations can be written as

$$-1.6181 X_{21} \sin\phi_1 + 0.618 X_{22} \sin\phi_2 = 0$$
$$X_{21} \sin\phi_1 + X_{22} \sin\phi_2 = 0$$
$$-52.363 X_{21} \cos\phi_1 + 7.639\, X_{22} \cos\phi_2 = 0$$
$$32.361 X_{21} \cos\phi_1 + 12.361 X_{22} \cos\phi_2 = 0.3$$

These algebraic equations can be solved for the four unknowns X_{21}, X_{22}, ϕ_1, and ϕ_2. It is clear from the first two equations that $X_{21} \sin\phi_1 = X_{22} \sin\phi_2 = 0$. The third and fourth equations can be written in the following matrix form

$$\begin{bmatrix} -52.363 & 7.639 \\ 32.361 & 12.361 \end{bmatrix} \begin{bmatrix} X_{21} \cos\phi_1 \\ X_{22} \cos\phi_2 \end{bmatrix} = \begin{bmatrix} 0 \\ 0.3 \end{bmatrix}$$

The solution of this matrix equation can be obtained by using Cramer's rule or matrix methods as

$$\begin{bmatrix} X_{21} \cos \phi_1 \\ X_{22} \cos \phi_2 \end{bmatrix} = \begin{bmatrix} 2.562 \times 10^{-3} \\ 1.756 \times 10^{-2} \end{bmatrix}$$

One, therefore, has $X_{21} = 2.562 \times 10^{-3}$, $X_{22} = 1.756 \times 10^{-2}$, and $\phi_1 = \phi_2 = 0$.

6.2 MATRIX EQUATIONS

It is more convenient to use matrix notations to write the differential equations of motion of systems which have more than one degree of freedom. In this section, the general matrix equations of the two degree of freedom systems are presented and the solution procedure is outlined using vector and matrix notations.

Differential Equations The differential equations of motion given by Eqs. 3 and 4, that govern the vibration of the system shown in Fig. 2, can be written in a matrix form as

$$\begin{bmatrix} m_1 & 0 \\ 0 & m_2 \end{bmatrix} \begin{bmatrix} \ddot{x}_1 \\ \ddot{x}_2 \end{bmatrix} + \begin{bmatrix} k_1 + k_2 & -k_2 \\ -k_2 & k_2 \end{bmatrix} \begin{bmatrix} x_1 \\ x_2 \end{bmatrix} = \begin{bmatrix} 0 \\ 0 \end{bmatrix} \tag{6.31}$$

which can be written in a compact matrix form as

$$\mathbf{M}\ddot{\mathbf{x}} + \mathbf{K}\mathbf{x} = \mathbf{0} \tag{6.32}$$

where $\mathbf{x}$ and $\ddot{\mathbf{x}}$ are, respectively, the vectors of displacements and accelerations defined as

$$\mathbf{x} = \begin{bmatrix} x_1 \\ x_2 \end{bmatrix}, \quad \ddot{\mathbf{x}} = \begin{bmatrix} \ddot{x}_1 \\ \ddot{x}_2 \end{bmatrix} \tag{6.33}$$

and $\mathbf{M}$ and $\mathbf{K}$ are, respectively, the *symmetric mass* and the *stiffness matrices* of the two degree of freedom system shown in Fig. 2 and are given by

$$\mathbf{M} = \begin{bmatrix} m_1 & 0 \\ 0 & m_2 \end{bmatrix}, \quad \mathbf{K} = \begin{bmatrix} k_1 + k_2 & -k_2 \\ -k_2 & k_2 \end{bmatrix} \tag{6.34}$$

In general, the equations of motion of the free undamped vibration of any two degree of freedom system can be written in the matrix form of Eq. 32. These equations can be written more explicitly as

$$\begin{bmatrix} m_{11} & m_{12} \\ m_{21} & m_{22} \end{bmatrix} \begin{bmatrix} \ddot{x}_1 \\ \ddot{x}_2 \end{bmatrix} + \begin{bmatrix} k_{11} & k_{12} \\ k_{21} & k_{22} \end{bmatrix} \begin{bmatrix} x_1 \\ x_2 \end{bmatrix} = \begin{bmatrix} 0 \\ 0 \end{bmatrix} \tag{6.35}$$

where the mass matrix $\mathbf{M}$ and the stiffness matrix $\mathbf{K}$ of Eq. 32 are recognized as

$$\mathbf{M} = \begin{bmatrix} m_{11} & m_{12} \\ m_{21} & m_{22} \end{bmatrix}, \qquad \mathbf{K} = \begin{bmatrix} k_{11} & k_{12} \\ k_{21} & k_{22} \end{bmatrix} \tag{6.36}$$

and the vector $\mathbf{x}$ is the vector of coordinates which describes any type of translational or rotational motion. The coefficients m_{ij} $(i, j = 1, 2)$ in the mass matrix are called the *mass coefficients* or the *inertia coefficients.* If $m_{12} = m_{21} = 0$, there is no inertia coupling between the system coordinates, and the two coordinates x_1 and x_2 are said to be *dynamically decoupled.* If the coefficients m_{12} and m_{21} are not equal to zero, the coordinates x_1 and x_2 are said to be *dynamically coupled.*

The coefficients k_{ij} $(i, j = 1, 2)$ in the stiffness matrix of Eq. 36 are called the *stiffness coefficients* or the *elastic coefficients.* If the coefficients k_{12} and k_{21} are equal to zero, the coordinates x_1 and x_2 are said to be *elastically decoupled*; otherwise, they are said to be *elastically coupled.* The coordinates in a two degree of freedom system may be dynamically and/or elastically coupled. For instance, in the two degree of freedom system shown in Fig. 2, the coordinates x_1 and x_2 are dynamically decoupled since from Eq. 34 we have

$$m_{11} = m_1, \qquad m_{22} = m_2, \qquad \text{and} \qquad m_{12} = m_{21} = 0$$

On the other hand, the coordinates x_1 and x_2 are elastically coupled since from Eq. 34 we have

$$k_{11} = k_1 + k_2, \qquad k_{22} = k_2, \qquad k_{12} = k_{21} = -k_2$$

Solution Procedure Equation 32 is similar to the differential equation of motion of an undamped single degree of freedom system except that scalars are replaced by matrices and vectors. In order to solve this matrix equation, we follow a procedure similar to the one used for single degree of freedom systems and assume a solution in the form

$$\mathbf{x} = \mathbf{X} \sin(\omega t + \phi) \tag{6.37}$$

where ω is the circular frequency, ϕ is the phase angle, and $\mathbf{X}$ is the vector of amplitudes given by

$$\mathbf{X} = \begin{bmatrix} X_1 \\ X_2 \end{bmatrix} \tag{6.38}$$

Differentiating Eq. 37 twice with respect to time yields

$$\ddot{\mathbf{x}} = -\omega^2 \mathbf{X} \sin(\omega t + \phi) \tag{6.39}$$

Substituting Eqs. 37 and 39 into Eq. 32 yields

$$-\omega^2 \mathbf{M}\mathbf{X} \sin(\omega t + \phi) + \mathbf{K}\mathbf{X} \sin(\omega t + \phi) = \mathbf{0}$$

which can be written as

$$[\mathbf{K} - \omega^2\mathbf{M}]\mathbf{X} \sin(\omega t + \phi) = \mathbf{0} \tag{6.40}$$

Since this equation must be valid at every point in time, one has

$$[\mathbf{K} - \omega^2\mathbf{M}]\mathbf{X} = \mathbf{0} \tag{6.41}$$

This is a system of homogeneous algebraic equations in the vector of amplitudes $\mathbf{X}$, which has a nontrivial solution if and only if the determinant of the coefficient matrix is equal to zero, that is,

$$|\mathbf{K} - \omega^2\mathbf{M}| = 0 \tag{6.42}$$

By using the general definition of the mass matrix $\mathbf{M}$ and the stiffness matrix $\mathbf{K}$ of Eq. 36, the coefficient matrix $[\mathbf{K} - \omega^2\mathbf{M}]$ of Eq. 41 can be written as

$$\mathbf{K} - \omega^2\mathbf{M} = \begin{bmatrix} k_{11} & k_{12} \\ k_{21} & k_{22} \end{bmatrix} - \omega^2 \begin{bmatrix} m_{11} & m_{12} \\ m_{21} & m_{22} \end{bmatrix}$$
$$= \begin{bmatrix} k_{11} - \omega^2 m_{11} & k_{12} - \omega^2 m_{12} \\ k_{21} - \omega^2 m_{21} & k_{22} - \omega^2 m_{22} \end{bmatrix}$$

and, accordingly, the condition of Eq. 42 implies

$$\begin{aligned} |\mathbf{K} - \omega^2\mathbf{M}| &= \begin{vmatrix} k_{11} - \omega^2 m_{11} & k_{12} - \omega^2 m_{12} \\ k_{21} - \omega^2 m_{21} & k_{22} - \omega^2 m_{22} \end{vmatrix} \\ &= (k_{11} - \omega^2 m_{11})(k_{22} - \omega^2 m_{22}) - (k_{12} - \omega^2 m_{12})(k_{21} - \omega^2 m_{21}) \\ &= 0 \end{aligned} \tag{6.43}$$

which also can be written as

$$\omega^4(m_{11}m_{22} - m_{12}m_{21}) + \omega^2(m_{12}k_{21} + m_{21}k_{12} - m_{11}k_{22} - m_{22}k_{11}) + k_{11}k_{22} - k_{12}k_{21} = 0 \tag{6.44}$$

This is the characteristic equation which is a quadratic function of ω^2. This equation can be solved to determine the roots ω_1^2 and ω_2^2.

If the mass and stiffness matrices are symmetric, that is,

$$m_{12} = m_{21} \qquad \text{and} \qquad k_{12} = k_{21},$$

the characteristic equation of Eq. 44 reduces to

$$\omega^4(m_{11}m_{22} - m_{12}^2) + \omega^2(2m_{12}k_{12} - m_{11}k_{22} - m_{22}k_{11}) + k_{11}k_{22} - k_{12}^2 = 0$$

Having determined the natural frequencies ω_1 and ω_2, the solution can be written in a vector form as

$$\mathbf{x} = \begin{bmatrix} x_1 \\ x_2 \end{bmatrix} = \mathbf{X}_1 \sin(\omega_1 t + \phi_1) + \mathbf{X}_2 \sin(\omega_2 t + \phi_2) \tag{6.45}$$

where $\mathbf{X}_1$ and $\mathbf{X}_2$ are the vectors of amplitudes given by

$$\mathbf{X}_1 = \begin{bmatrix} X_{11} \\ X_{21} \end{bmatrix}, \qquad \mathbf{X}_2 = \begin{bmatrix} X_{12} \\ X_{22} \end{bmatrix} \tag{6.46}$$

As described in the preceding section, Eq. 41 can be used to determine the amplitude ratios β_1 and β_2 corresponding to ω_1 and ω_2, where β_1 and β_2 are defined in Eqs. 15 and 18 as

$$\beta_1 = \frac{X_{11}}{X_{21}}, \qquad \beta_2 = \frac{X_{12}}{X_{22}}$$

or

$$X_{11} = \beta_1 X_{21} \qquad \text{and} \qquad X_{12} = \beta_2 X_{22}$$

Therefore, the vectors $\mathbf{X}_1$ and $\mathbf{X}_2$ can be written in terms of the amplitude ratios β_1 and β_2 as

$$\mathbf{X}_1 = \begin{bmatrix} \beta_1 X_{21} \\ X_{21} \end{bmatrix} = \begin{bmatrix} \beta_1 \\ 1 \end{bmatrix} X_{21} \tag{6.47a}$$

$$\mathbf{X}_2 = \begin{bmatrix} \beta_2 X_{22} \\ X_{22} \end{bmatrix} = \begin{bmatrix} \beta_2 \\ 1 \end{bmatrix} X_{22} \tag{6.47b}$$

Equation 45 can then be written as

$$\mathbf{x} = \begin{bmatrix} x_1 \\ x_2 \end{bmatrix} = \begin{bmatrix} \beta_1 \\ 1 \end{bmatrix} X_{21} \sin(\omega_1 t + \phi_1) + \begin{bmatrix} \beta_2 \\ 1 \end{bmatrix} X_{22} \sin(\omega_2 t + \phi_2)$$

$$= \begin{bmatrix} \beta_1 & \beta_2 \\ 1 & 1 \end{bmatrix} \begin{bmatrix} X_{21} \sin(\omega_1 t + \phi_1) \\ X_{22} \sin(\omega_2 t + \phi_2) \end{bmatrix} \tag{6.48}$$

As indicated earlier, the solution $\mathbf{x}$ contains four arbitrary constants X_{21}, X_{22}, ϕ_1, and ϕ_2 which can be determined from the initial conditions. Given the following initial conditions

$$\mathbf{x}_0 = \begin{bmatrix} x_{10} \\ x_{20} \end{bmatrix}, \qquad \dot{\mathbf{x}}_0 = \begin{bmatrix} \dot{x}_{10} \\ \dot{x}_{20} \end{bmatrix}, \tag{6.49}$$

one has the following algebraic equations which can be solved for the constants X_{21}, X_{22}, ϕ_1, and ϕ_2

$$\mathbf{x}_0 = \begin{bmatrix} x_{10} \\ x_{20} \end{bmatrix} = \begin{bmatrix} \beta_1 & \beta_2 \\ 1 & 1 \end{bmatrix} \begin{bmatrix} X_{21} \sin \phi_1 \\ X_{22} \sin \phi_2 \end{bmatrix} \tag{6.50}$$

$$\dot{\mathbf{x}}_0 = \begin{bmatrix} \dot{x}_{10} \\ \dot{x}_{20} \end{bmatrix} = \begin{bmatrix} \beta_1 & \beta_2 \\ 1 & 1 \end{bmatrix} \begin{bmatrix} \omega_1 X_{21} \cos \phi_1 \\ \omega_2 X_{22} \cos \phi_2 \end{bmatrix} \tag{6.51}$$

Example 6.2

Derive the differential equations of motion of the two degree of freedom system shown in Fig. 3.

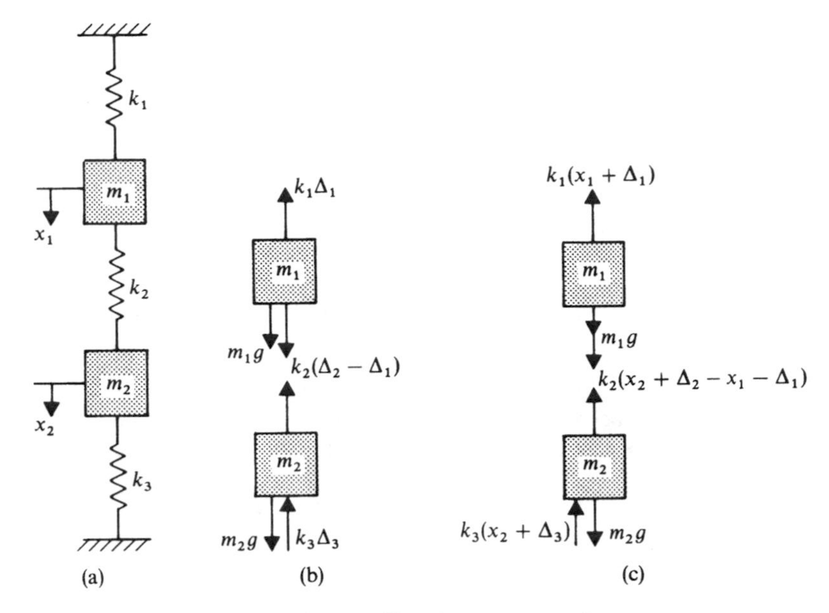

FIG. 6.3. Two degree of freedom mass–spring system.

Solution. Since the spring forces balance the weights at the static equilibrium position, one can show that the effect of the weights is equal to the spring forces due to the static deflection. In the static equilibrium position we have, from Fig. 3(b),

$$m_1g - k_1\Delta_1 + k_2(\Delta_2 - \Delta_1) = 0 \tag{6.52}$$

$$m_2g - k_2(\Delta_2 - \Delta_1) - k_3\Delta_3 = 0 \tag{6.53}$$

where m_1 and m_2 are the masses, g is the gravitational constant, k_1, k_2, and k_3 are the spring constants, and Δ_1, Δ_2, and Δ_3 are the static deflections of the springs in the static equilibrium position. Without loss of generality, we assume that $x_2 > x_1$. From the free body diagram shown in Fig. 3(c), the dynamic equations of the vibratory motion of the two degree of freedom system are given by

$$m_1\ddot{x}_1 = m_1g - k_1(x_1 + \Delta_1) + k_2(x_2 + \Delta_2 - x_1 - \Delta_1)$$

$$m_2\ddot{x}_2 = m_2g - k_2(x_2 + \Delta_2 - x_1 - \Delta_1) - k_3(x_2 + \Delta_3)$$

Using the static equilibrium conditions given by Eqs. 52 and 53, the above differential equations reduce to

$$m_1\ddot{x}_1 + (k_1 + k_2)x_1 - k_2x_2 = 0$$

$$m_2\ddot{x}_2 + (k_2 + k_3)x_2 - k_2x_1 = 0$$

which can be written in a matrix form as

$$\begin{bmatrix} m_1 & 0 \\ 0 & m_2 \end{bmatrix}\begin{bmatrix} \ddot{x}_1 \\ \ddot{x}_2 \end{bmatrix} + \begin{bmatrix} k_1 + k_2 & -k_2 \\ -k_2 & k_2 + k_3 \end{bmatrix}\begin{bmatrix} x_1 \\ x_2 \end{bmatrix} = \begin{bmatrix} 0 \\ 0 \end{bmatrix}$$

or

$$\mathbf{M}\ddot{\mathbf{x}} + \mathbf{K}\mathbf{x} = \mathbf{0}$$

where **M** and **K** are the symmetric mass and stiffness matrices defined by

$$\mathbf{M} = \begin{bmatrix} m_1 & 0 \\ 0 & m_2 \end{bmatrix}, \qquad \mathbf{K} = \begin{bmatrix} k_1 + k_2 & -k_2 \\ -k_2 & k_2 + k_3 \end{bmatrix}$$

Example 6.3

Determine the differential equations of motion of the two degree of freedom torsional system shown in Fig. 4.

Solution. Let θ_1 and θ_2 be, respectively, the angles of rotations of the discs I_1 and I_2. From the free body diagram shown in Fig. 4, and assuming that $\theta_2 > \theta_1$, one can verify that the differential equations of motion are

$$I_1\ddot{\theta}_1 = -k_1\theta_1 + k_2(\theta_2 - \theta_1)$$

$$I_2\ddot{\theta}_2 = -k_2(\theta_2 - \theta_1) - k_3\theta_2$$

which yield

$$I_1\ddot{\theta}_1 + (k_1 + k_2)\theta_1 - k_2\theta_2 = 0$$

$$I_2\ddot{\theta}_2 + (k_2 + k_3)\theta_2 - k_2\theta_1 = 0$$

where k_i is the torsional stiffness of shaft i, $i = 1, 2, 3$, defined as

$$k_i = \frac{G_i J_i}{l_i}$$

where G_i is the modulus of rigidity, J_i is the polar moment of inertia, and l_i is the length of the shaft. By using the matrix notation, the differential equations of motion can be written as

$$\begin{bmatrix} I_1 & 0 \\ 0 & I_2 \end{bmatrix}\begin{bmatrix} \ddot{\theta}_1 \\ \ddot{\theta}_2 \end{bmatrix} + \begin{bmatrix} k_1 + k_2 & -k_2 \\ -k_2 & k_2 + k_3 \end{bmatrix}\begin{bmatrix} \theta_1 \\ \theta_2 \end{bmatrix} = \begin{bmatrix} 0 \\ 0 \end{bmatrix}$$

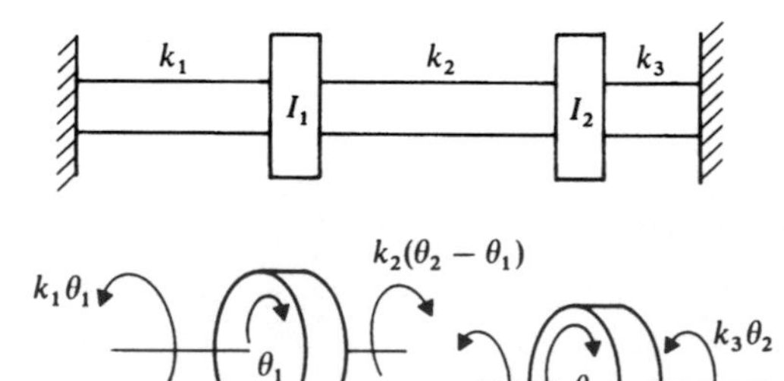

FIG. 6.4. Torsional system.

or, equivalently,

$$\mathbf{M}\ddot{\boldsymbol{\theta}} + \mathbf{K}\boldsymbol{\theta} = \mathbf{0}$$

where **M** and **K** are the symmetric mass and stiffness matrices defined as

$$\mathbf{M} = \begin{bmatrix} I_1 & 0 \\ 0 & I_2 \end{bmatrix}, \qquad \mathbf{K} = \begin{bmatrix} k_1 + k_2 & -k_2 \\ -k_2 & k_2 + k_3 \end{bmatrix}$$

Example 6.4

The bar AB shown in Fig. 5, which represents a simplified model for the chassis of a vehicle, has length l, mass m, and mass moment of inertia I about its mass center C. The bar is supported by two linear springs which have constants k_1 and k_2. Determine the differential equations of motion assuming that the motion of the bar in the horizontal direction is small and can be neglected.

Solution. Since the displacement of the bar in the x direction is neglected, the system configuration can be identified using the two variables y and θ, where y is the vertical displacement of the center of mass and θ is the angular rotation of the bar. It is left to the reader as an exercise to show that the weight of the bar cancels with the deflection of the springs at the static equilibrium position. Therefore, the differential equations of motion are given by

$$m\ddot{y} = -k_1\left(y - \frac{l}{2}\theta\right) - k_2\left(y + \frac{l}{2}\theta\right)$$

$$I\ddot{\theta} = k_1\left(y - \frac{l}{2}\theta\right)\frac{l}{2}\cos\theta - k_2\left(y + \frac{l}{2}\theta\right)\frac{l}{2}\cos\theta$$

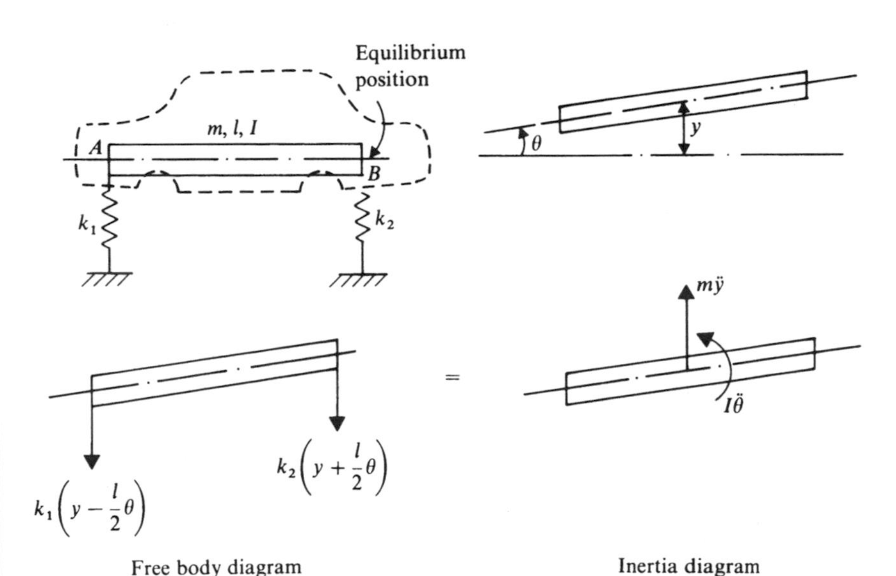

FIG. 6.5. Vibration of the rigid bar.

For small angular oscillations, $\cos\theta \approx 1$, and the differential equations reduce to

$$m\ddot{y} + (k_1 + k_2)y + (k_2 - k_1)\frac{l}{2}\theta = 0$$

$$I\ddot{\theta} + (k_2 + k_1)\frac{l^2}{4}\theta + (k_2 - k_1)\frac{l}{2}y = 0$$

which can be written in matrix form as

$$\begin{bmatrix} m & 0 \\ 0 & I \end{bmatrix}\begin{bmatrix} \ddot{y} \\ \ddot{\theta} \end{bmatrix} + \begin{bmatrix} k_1 + k_2 & (k_2 - k_1)\frac{l}{2} \\ (k_2 - k_1)\frac{l}{2} & (k_2 + k_1)\frac{l^2}{4} \end{bmatrix}\begin{bmatrix} y \\ \theta \end{bmatrix} = \begin{bmatrix} 0 \\ 0 \end{bmatrix}$$

which can be written in a more compact form as

$$\mathbf{M}\ddot{\mathbf{x}} + \mathbf{K}\mathbf{x} = \mathbf{0}$$

where $\mathbf{M}$ and $\mathbf{K}$ are the symmetric mass and stiffness matrices given by

$$\mathbf{M} = \begin{bmatrix} m & 0 \\ 0 & I \end{bmatrix}, \qquad \mathbf{K} = \begin{bmatrix} k_1 + k_2 & (k_2 - k_1)\frac{l}{2} \\ (k_2 - k_1)\frac{l}{2} & (k_2 + k_1)\frac{l^2}{4} \end{bmatrix},$$

and the vectors $\mathbf{x}$ and $\ddot{\mathbf{x}}$ are

$$\mathbf{x} = \begin{bmatrix} y \\ \theta \end{bmatrix}, \qquad \ddot{\mathbf{x}} = \begin{bmatrix} \ddot{y} \\ \ddot{\theta} \end{bmatrix}$$

One can then assume a solution in the form

$$\mathbf{x} = \mathbf{X}\sin(\omega t + \phi)$$

where $\mathbf{X}$ is the vector of amplitudes defined as

$$\mathbf{X} = \begin{bmatrix} Y \\ \Theta \end{bmatrix}$$

The acceleration vector $\ddot{\mathbf{x}}$ is given by

$$\ddot{\mathbf{x}} = -\omega^2\mathbf{X}\sin(\omega t + \phi)$$

Substituting $\mathbf{x}$ and $\ddot{\mathbf{x}}$ into the differential equation one obtains

$$[\mathbf{K} - \omega^2\mathbf{M}]\,\mathbf{X} = \mathbf{0}$$

For a nontrivial solution, the determinant of the coefficient matrix must be equal to zero, and hence

$$|\mathbf{K} - \omega^2\mathbf{M}| = 0$$

which can be written in a more explicit form as

$$\begin{vmatrix} k_{11} - \omega^2 m & k_{12} \\ k_{21} & k_{22} - \omega^2 I \end{vmatrix} = 0$$

where $k_{11} = k_1 + k_2$, $k_{12} = k_{21} = (k_2 - k_1)(l/2)$, and $k_{22} = (k_1 + k_2)(l^2/4)$.

The characteristic equation is then given by

$$(k_{11} - \omega^2 m)(k_{22} - \omega^2 I) - k_{12}^2 = 0$$

or

$$\omega^4 Im - \omega^2[mk_{22} + Ik_{11}] + k_{11}k_{22} - k_{12}^2 = 0$$

which yields the following equation

$$\omega^4 Im - \omega^2(k_1 + k_2)\left[m\frac{l^2}{4} + I\right] + 4k_1k_2\frac{l^2}{4} = 0$$

The roots of this equation are given by

$$\omega_1^2 = \frac{-b + \sqrt{b^2 - 4ac}}{2a}, \qquad \omega_2^2 = \frac{-b - \sqrt{b^2 - 4ac}}{2a}$$

where

$$a = mI$$

$$b = -(k_1 + k_2)\left(m\frac{l^2}{4} + I\right)$$

$$c = 4k_1k_2\frac{l^2}{4}$$

The amplitude ratios β_1 and β_2 are given by

$$\beta_1 = \frac{X_{11}}{X_{21}} = -\frac{(k_2 - k_1)\dfrac{l}{2}}{k_1 + k_2 - \omega_1^2 m} = -\frac{(k_1 + k_2)\dfrac{l^2}{4} - \omega_1^2 I}{(k_2 - k_1)\dfrac{l}{2}}$$

$$\beta_2 = \frac{X_{12}}{X_{22}} = -\frac{(k_2 - k_1)\dfrac{l}{2}}{k_1 + k_2 - \omega_2^2 m} = -\frac{(k_1 + k_2)\dfrac{l^2}{4} - \omega_2^2 I}{(k_2 - k_1)\dfrac{l}{2}}$$

The solution is then given by

$$x_1 = \beta_1 X_{21} \sin(\omega_1 t + \phi_1) + \beta_2 X_{22} \sin(\omega_2 t + \phi_2)$$

$$x_2 = X_{21} \sin(\omega_1 t + \phi_1) + X_{22} \sin(\omega_2 t + \phi_2)$$

Selection of Coordinates In the preceding example, the use of the vertical displacement of the center of mass y and the angular orientation of the bar θ as the system coordinates leads to a dynamically decoupled system of equations in which $m_{12} = m_{21} = 0$. The selection of other sets of coordinates may lead to a simpler or more complex mathematical model. For instance, let us select the vertical displacement of point A as the translational coordinate and θ as the rotational coordinate of the bar of the preceding example. In terms of this new set of coordinates, the displacement and acceleration of the center of mass, in the case of small oscillations, is given by

$$y = y_a + \frac{l}{2}\theta$$

$$\ddot{y} = \ddot{y}_a + \frac{l}{2}\ddot{\theta}$$

where y_a is the vertical displacement of the end point A of the rod. Using these equations with the force diagrams shown in Fig. 5, one can show that the equations of motion of the bar AB, in the case of small oscillations, is given by

$$m\ddot{y}_a + m\frac{l}{2}\ddot{\theta} + (k_1 + k_2)y_a + k_2 l\theta = 0$$

$$I\ddot{\theta} + k_2\frac{l^2}{2}\theta + (k_2 - k_1)\frac{l}{2}y_a = 0$$

which can be written in matrix form as

$$\begin{bmatrix} m & m\frac{l}{2} \\ 0 & I \end{bmatrix}\begin{bmatrix} \ddot{y}_a \\ \ddot{\theta} \end{bmatrix} + \begin{bmatrix} k_1 + k_2 & k_2 l \\ (k_2 - k_1)\frac{l}{2} & k_2\frac{l^2}{2} \end{bmatrix}\begin{bmatrix} y_a \\ \theta \end{bmatrix} = \begin{bmatrix} 0 \\ 0 \end{bmatrix}$$

which contains both dynamic and elastic coupling. Furthermore, the mass and stiffness matrices that appear in this equation are not symmetric. This is mainly because the moment equilibrium condition is obtained by taking the moments of the inertia and external forces with respect to the center of mass of the bar. If the moment equation is defined at point A instead of the center of mass, the use of the coordinates y_a and θ leads to symmetric mass and stiffness matrices. In fact, the symmetric form of the mass and stiffness matrices can be simply obtained from the preceding matrix equation by multiplying the first equation by $l/2$ and adding the resulting equation to the second moment equation, leading to

$$m\frac{l}{2}\ddot{y}_a + \left(I + \frac{ml^2}{4}\right)\ddot{\theta} + k_2 l y_a + k_2 l^2\theta = 0$$

This equation may be combined with the equation that defines the equilibrium in the vertical direction, leading to

$$\begin{bmatrix} m & m\dfrac{l}{2} \\ m\dfrac{l}{2} & I + \dfrac{ml^2}{4} \end{bmatrix} \begin{bmatrix} \ddot{y}_a \\ \ddot{\theta} \end{bmatrix} + \begin{bmatrix} k_1 + k_2 & k_2 l \\ k_2 l & k_2 l^2 \end{bmatrix} \begin{bmatrix} y_a \\ \theta \end{bmatrix} = \begin{bmatrix} 0 \\ 0 \end{bmatrix}$$

This equation, in which the resulting mass and stiffness matrices are symmetric, contains both dynamic and elastic coupling between the coordinates y_a and θ. It is also clear that in terms of one set of coordinates, the form of the resulting equations is not unique, since multiplying an equation by a constant or adding and/or subtracting two equations remains a valid operation.

As a third choice of coordinates, one may take y_b and θ, where y_b is the vertical displacement of point B. By following a similar procedure as described before, one can verify that the matrix differential equation of motion of the system in terms of these coordinates is given by

$$\begin{bmatrix} m & m\dfrac{l}{2} \\ m\dfrac{l}{2} & I + m\dfrac{l^2}{4} \end{bmatrix} \begin{bmatrix} \ddot{y}_b \\ \ddot{\theta} \end{bmatrix} + \begin{bmatrix} k_1 + k_2 & -k_1 l \\ -k_1 l & k_1 l^2 \end{bmatrix} \begin{bmatrix} y_b \\ \theta \end{bmatrix} = \begin{bmatrix} 0 \\ 0 \end{bmatrix}$$

which also exhibits both dynamic and elastic coupling.

Principal Coordinates It is clear that the choice of coordinates has a significant effect on both the inertia (dynamic) and stiffness (elastic) coupling that appear in the differential equations. Therefore, a natural question is to ask if there exists a set of coordinates that totally eliminates both the inertia and stiffness coupling. The answer to this question is positive. In fact, most of the steps required to define these coordinates have already been discussed. Consider the coefficient matrix formed by the amplitude ratios β_1 and β_2, and shown in Eq. 48. Let us denote this matrix as $\mathbf{\Phi}$, that is,

$$\mathbf{\Phi} = \begin{bmatrix} \beta_1 & \beta_2 \\ 1 & 1 \end{bmatrix}$$

The inertia and stiffness coupling can be eliminated by using the following coordinate transformation

$$\mathbf{x} = \mathbf{\Phi q}$$

where $\mathbf{q} = [q_1 \quad q_2]^{\mathrm{T}}$ are called the *principal* or *modal coordinates.*

In order to describe the procedure for obtaining the uncoupled differential equations, the results of the preceding example are used. In terms of the coordinates y and θ, it was shown that the matrix differential equation is given by

$$\begin{bmatrix} m & 0 \\ 0 & I \end{bmatrix}\begin{bmatrix} \ddot{y} \\ \ddot{\theta} \end{bmatrix} + \begin{bmatrix} k_{11} & k_{12} \\ k_{12} & k_{22} \end{bmatrix}\begin{bmatrix} y \\ \theta \end{bmatrix} = \begin{bmatrix} 0 \\ 0 \end{bmatrix}$$

where $k_{11} = k_1 + k_2$, $k_{12} = (k_2 - k_1)(l/2)$, and $k_{22} = (k_2 + k_1)(l^2/4)$.

In order to decouple the preceding two differential equations, the following coordinate transformation is used

$$\begin{bmatrix} y \\ \theta \end{bmatrix} = \begin{bmatrix} \beta_1 & \beta_2 \\ 1 & 1 \end{bmatrix}\begin{bmatrix} q_1 \\ q_2 \end{bmatrix}$$

where q_1 and q_2 are called the *principal* or *modal coordinates*, and β_1 and β_2 are defined in the preceding example. Substituting this transformation into the differential equation and premultiplying by the transpose of the matrix $\mathbf{\Phi}$, one obtains

$$\begin{bmatrix} \beta_1 & 1 \\ \beta_2 & 1 \end{bmatrix}\begin{bmatrix} m & 0 \\ 0 & I \end{bmatrix}\begin{bmatrix} \beta_1 & \beta_2 \\ 1 & 1 \end{bmatrix}\begin{bmatrix} \ddot{q}_1 \\ \ddot{q}_2 \end{bmatrix} + \begin{bmatrix} \beta_1 & 1 \\ \beta_2 & 1 \end{bmatrix}\begin{bmatrix} k_{11} & k_{12} \\ k_{12} & k_{22} \end{bmatrix}\begin{bmatrix} \beta_1 & \beta_2 \\ 1 & 1 \end{bmatrix}\begin{bmatrix} q_1 \\ q_2 \end{bmatrix} = 0$$

which yields

$$\begin{bmatrix} \beta_1^2 m + I & \beta_1\beta_2 m + I \\ \beta_1\beta_2 m + I & \beta_2^2 m + I \end{bmatrix}\begin{bmatrix} \ddot{q}_1 \\ \ddot{q}_2 \end{bmatrix}$$

$$+ \begin{bmatrix} \beta_1^2 k_{11} + 2\beta_1 k_{12} + k_{22} & \beta_1\beta_2 k_{11} + (\beta_1 + \beta_2)k_{12} + k_{22} \\ \beta_1\beta_2 k_{11} + (\beta_1 + \beta_2)k_{12} + k_{22} & \beta_2^2 k_{11} + 2\beta_2 k_{12} + k_{22} \end{bmatrix}\begin{bmatrix} q_1 \\ q_2 \end{bmatrix}$$

$$= \begin{bmatrix} 0 \\ 0 \end{bmatrix}$$

Using the definition of β_1 and β_2 and the definition of ω_1 and ω_2 given in the preceding example, one can verify the following

$$\beta_1\beta_2 m + I = 0$$

$$\beta_1\beta_2 k_{11} + (\beta_1 + \beta_2)k_{12} + k_{22} = 0$$

Using these two identities the matrix differential equation reduces to

$$\begin{bmatrix} \beta_1^2 m + I & 0 \\ 0 & \beta_2^2 m + I \end{bmatrix}\begin{bmatrix} \ddot{q}_1 \\ \ddot{q}_2 \end{bmatrix}$$

$$+ \begin{bmatrix} \beta_1^2 k_{11} + 2\beta_1 k_{12} + k_{22} & 0 \\ 0 & \beta_2^2 k_{11} + 2\beta_2 k_{12} + k_{22} \end{bmatrix}\begin{bmatrix} q_1 \\ q_2 \end{bmatrix} = \begin{bmatrix} 0 \\ 0 \end{bmatrix}$$

which shows that the resulting equations expressed in terms of the principal coordinates q_1 and q_2 are uncoupled.

Example 6.5

In the preceding example, let $m = 1000$ kg, $l = 4$ m, $I = 1300$ kg·m², $k_1 = 50 \times 10^3$ N/m, and $k_2 = 70 \times 10^3$ N/m. Determine the system response as a function of time if the initial conditions are given by

$$y(t = 0) = y_0 = 0.03, \qquad \dot{y}(t = 0) = \dot{y}_0 = 0$$

$$\theta(t = 0) = \theta_0 = 0, \qquad \dot{\theta}(t = 0) = \dot{\theta}_0 = 0$$

Solution. The constants a, b, and c are given by

$$a = mI = (1000)(1300) = 1.3 \times 10^6$$

$$b = -\left[m(k_1 + k_2)\frac{l^2}{4} + I(k_1 + k_2)\right]$$

$$= -\left[(1000)(12 \times 10^4)\frac{(4)^2}{4} + 1300(12 \times 10^4)\right]$$

$$= -[4.8 \times 10^8 + 1.56 \times 10^8] = -6.36 \times 10^8$$

$$c = 4k_1k_2\frac{l^2}{4} = 4(50 \times 10^3)(70 \times 10^3)\frac{(4)^2}{4}$$

$$= 5.6 \times 10^{10}$$

Thus

$$\omega_1^2 = \frac{6.36 \times 10^8 - \sqrt{40.45 \times 10^{16} - 4(1.3 \times 10^6)(5.6 \times 10^{10})}}{2(1.3 \times 10^6)}$$

$$= \frac{6.36 \times 10^8 - 3.366 \times 10^8}{2.6 \times 10^6} = 1.152 \times 10^2$$

$$\omega_2^2 = \frac{6.36 \times 10^8 + \sqrt{40.45 \times 10^{16} - 4(1.3 \times 10^6)(5.6 \times 10^{10})}}{2(1.3 \times 10^6)}$$

$$= \frac{6.36 \times 10^8 + 3.366 \times 10^8}{2.6 \times 10^6} = 3.741 \times 10^2$$

which yield

$$\omega_1 = 10.731 \text{ rad/s}, \qquad \omega_2 = 19.341 \text{ rad/s}$$

The frequency ratios β_1 and β_2 are given by

$$\beta_1 = \frac{X_{11}}{X_{21}} = -\frac{(k_2 - k_1)\dfrac{l}{2}}{k_1 + k_2 - \omega_1^2 m} = -\frac{20 \times 10^3\left(\dfrac{4}{2}\right)}{120 \times 10^3 - (1.152 \times 10^2)(10^3)}$$

$$= -8.333$$

$$\beta_2 = \frac{X_{12}}{X_{22}} = -\frac{(k_2 - k_1)\dfrac{l}{2}}{k_1 + k_2 - \omega_2^2 m} = -\frac{20 \times 10^3\left(\dfrac{4}{2}\right)}{120 \times 10^3 - (3.741 \times 10^2)(10^3)}$$

$$= 0.157$$

The solution can then be written as

$$y(t) = \beta_1 X_{21} \sin(\omega_1 t + \phi_1) + \beta_2 X_{22} \sin(\omega_2 t + \phi_2)$$

$$\theta(t) = X_{21} \sin(\omega_1 t + \phi_1) + X_{22} \sin(\omega_2 t + \phi_2)$$

It follows that

$$y(t) = -8.333X_{21} \sin(10.731t + \phi_1) + 0.157X_{22} \sin(19.341t + \phi_2)$$

$$\theta(t) = X_{21} \sin(10.731t + \phi_1) + X_{22} \sin(19.341t + \phi_2)$$

The velocities can be obtained by differentiating these two equations with respect to time to yield

$$\dot{y}(t) = -89.421X_{21} \cos(10.731t + \phi_1) + 3.037X_{22} \cos(19.341t + \phi_2)$$

$$\dot{\theta}(t) = 10.731X_{21} \cos(10.731t + \phi_1) + 19.341X_{22} \cos(19.341t + \phi_2)$$

The constants X_{21}, X_{22}, ϕ_1, and ϕ_2 can be determined using the initial conditions

$$y_0 = 0.03 = -8.333X_{21} \sin \phi_1 + 0.157X_{22} \sin \phi_2$$

$$\theta_0 = 0 = X_{21} \sin \phi_1 + X_{22} \sin \phi_2$$

$$\dot{y}_0 = 0 = -89.421X_{21} \cos \phi_1 + 3.037X_{22} \cos \phi_2$$

$$\dot{\theta}_0 = 0 = 10.731X_{21} \cos \phi_1 + 19.341X_{22} \cos \phi_2$$

6.3 DAMPED FREE VIBRATION

In the preceding sections, we considered the free undamped vibration of two degree of freedom systems. In this section, the effect of viscous damping on the free vibration of the two degree of freedom system is discussed.

Differential Equations of Motion The two degree of freedom system shown in Fig. 6 consists of two masses m_1 and m_2 connected to each other

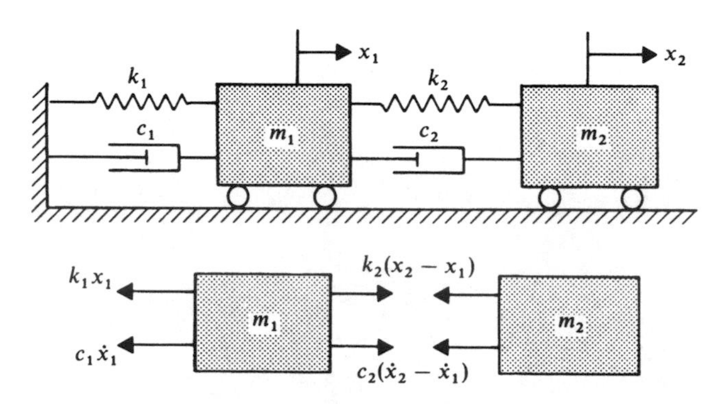

FIG. 6.6. Two degree of freedom system with viscous damping.

and to the ground by springs which have coefficients k_1 and k_2 and viscous dampers which have coefficients c_1 and c_2. Let $x_1(t)$ and $x_2(t)$ denote the displacements of the two masses and $\dot{x}_1(t)$ and $\dot{x}_2(t)$ denote the velocities. Without any loss of generality, we assume that $x_2(t) > x_1(t)$ and $\dot{x}_2(t) > \dot{x}_1(t)$. Using the free body diagrams shown in Fig. 6, one can verify that the equations of motion of the two masses can be written as

$$m_1\ddot{x}_1 = -k_1x_1 + k_2(x_2 - x_1) - c_1\dot{x}_1 + c_2(\dot{x}_2 - \dot{x}_1)$$

$$m_2\ddot{x}_2 = -k_2(x_2 - x_1) - c_2(\dot{x}_2 - \dot{x}_1)$$

which can be written as

$$m_1\ddot{x}_1 + (c_1 + c_2)\,\dot{x}_1 - c_2\dot{x}_2 + (k_1 + k_2)\,x_1 - k_2x_2 = 0$$

$$m_2\ddot{x}_2 + c_2\dot{x}_2 - c_2\dot{x}_1 + k_2x_2 - k_2x_1 = 0$$

These two equations can be written in matrix form as

$$\begin{bmatrix} m_1 & 0 \\ 0 & m_2 \end{bmatrix}\begin{bmatrix} \ddot{x}_1 \\ \ddot{x}_2 \end{bmatrix} + \begin{bmatrix} c_1 + c_2 & -c_2 \\ -c_2 & c_2 \end{bmatrix}\begin{bmatrix} \dot{x}_1 \\ \dot{x}_2 \end{bmatrix} + \begin{bmatrix} k_1 + k_2 & -k_2 \\ -k_2 & k_2 \end{bmatrix}\begin{bmatrix} x_1 \\ x_2 \end{bmatrix} = \begin{bmatrix} 0 \\ 0 \end{bmatrix}$$

or, equivalently,

$$\mathbf{M}\ddot{\mathbf{x}} + \mathbf{C}\dot{\mathbf{x}} + \mathbf{K}\mathbf{x} = \mathbf{0} \tag{6.54}$$

The matrix $\mathbf{M}$ is recognized as the mass matrix

$$\mathbf{M} = \begin{bmatrix} m_{11} & m_{12} \\ m_{21} & m_{22} \end{bmatrix} = \begin{bmatrix} m_1 & 0 \\ 0 & m_2 \end{bmatrix} \tag{6.55}$$

The matrix $\mathbf{C}$ is defined as the *damping matrix*

$$\mathbf{C} = \begin{bmatrix} c_{11} & c_{12} \\ c_{21} & c_{22} \end{bmatrix} = \begin{bmatrix} c_1 + c_2 & -c_2 \\ -c_2 & c_2 \end{bmatrix}, \tag{6.56}$$

and $\mathbf{K}$ is the stiffness matrix

$$\mathbf{K} = \begin{bmatrix} k_{11} & k_{12} \\ k_{21} & k_{22} \end{bmatrix} = \begin{bmatrix} k_1 + k_2 & -k_2 \\ -k_2 & k_2 \end{bmatrix} \tag{6.57}$$

The acceleration, velocity, and displacement vectors $\ddot{\mathbf{x}}$, $\dot{\mathbf{x}}$, and $\mathbf{x}$ are given by

$$\ddot{\mathbf{x}} = \begin{bmatrix} \ddot{x}_1 \\ \ddot{x}_2 \end{bmatrix}, \quad \dot{\mathbf{x}} = \begin{bmatrix} \dot{x}_1 \\ \dot{x}_2 \end{bmatrix}, \quad \mathbf{x} = \begin{bmatrix} x_1 \\ x_2 \end{bmatrix} \tag{6.58}$$

Equation 54 is the general form of the differential equations of motion that govern the linear free vibration of the damped two degree of freedom systems.

Solution Procedure Following a procedure similar to the one used with the single degree of freedom system, we assume a solution in the form

$$\mathbf{x}(t) = \mathbf{X}e^{st} = \begin{bmatrix} X_1 \\ X_2 \end{bmatrix} e^{st} \tag{6.59}$$

or

$$x_1(t) = X_1 e^{st}$$

$$x_2(t) = X_2 e^{st}$$

Differentiating Eq. 59 with respect to time yields the velocity and acceleration vectors

$$\dot{\mathbf{x}}(t) = s\mathbf{X}e^{st} \tag{6.60}$$

$$\ddot{\mathbf{x}}(t) = s^2\mathbf{X}e^{st} \tag{6.61}$$

Substituting Eqs. 59–61 into the differential equation of motion given by Eq. 54 yields

$$s^2\mathbf{M}\mathbf{X}e^{st} + s\mathbf{C}\mathbf{X}e^{st} + \mathbf{K}\mathbf{X}e^{st} = \mathbf{0}$$

which can be written as

$$[s^2\mathbf{M} + s\mathbf{C} + \mathbf{K}]\,\mathbf{X}e^{st} = \mathbf{0} \tag{6.62}$$

Since this equation must be satisfied all the time, one has

$$[s^2\mathbf{M} + s\mathbf{C} + \mathbf{K}]\,\mathbf{X} = \mathbf{0} \tag{6.63}$$

This is a homogeneous linear algebraic matrix equation in the two unknowns X_1 and X_2. This equation can be written in a more explicit form as

$$\begin{bmatrix} s^2 m_{11} + sc_{11} + k_{11} & s^2 m_{12} + sc_{12} + k_{12} \\ s^2 m_{21} + sc_{21} + k_{21} & s^2 m_{22} + sc_{22} + k_{22} \end{bmatrix} \begin{bmatrix} X_1 \\ X_2 \end{bmatrix} = \begin{bmatrix} 0 \\ 0 \end{bmatrix} \tag{6.64}$$

Equation 63 or, equivalently, Eq. 64 has a nontrivial solution if and only if the determinant of the coefficient matrix is equal to zero, that is,

$$|s^2\mathbf{M} + s\mathbf{C} + \mathbf{K}| = 0 \tag{6.65}$$

or

$$\begin{vmatrix} s^2 m_{11} + sc_{11} + k_{11} & s^2 m_{12} + sc_{12} + k_{12} \\ s^2 m_{21} + sc_{21} + k_{21} & s^2 m_{22} + sc_{22} + k_{22} \end{vmatrix} = 0 \tag{6.66}$$

This leads to the following *characteristic equation*

$$(s^2 m_{11} + sc_{11} + k_{11})(s^2 m_{22} + sc_{22} + k_{22}) - (s^2 m_{12} + sc_{12} + k_{12})(s^2 m_{21} + sc_{21} + k_{21}) = 0$$

which can be written as

$$\begin{aligned} &(m_{11}m_{22} - m_{12}m_{21})\,s^4 + (m_{22}c_{11} + m_{11}c_{22} - m_{12}c_{21} - m_{21}c_{12})\,s^3 \\ &\quad + (m_{11}k_{22} + m_{22}k_{11} + c_{11}c_{22} - m_{12}k_{21} - m_{21}k_{12} - c_{12}c_{21})\,s^2 \\ &\quad + (c_{11}k_{22} + c_{22}k_{11} - c_{12}k_{21} - c_{21}k_{12})\,s + k_{11}k_{22} - k_{12}k_{21} = 0 \end{aligned} \tag{6.67}$$

This equation is of fourth degree in s, and it can be shown that if the damping coefficients are zero, this equation reduces to the characteristic equation of the undamped system given by Eq. 44. Equation 67 has four roots denoted as

s_1, s_2, s_3, and s_4. By examining the coefficients in this equation, it can be shown that there are three possibilities regarding the roots of this equation. These possibilities are:

(1) All four roots are negative real numbers.
(2) All four roots are complex numbers. In this case, the roots will be two pairs of complex conjugates having negative real parts.
(3) Two roots may be real and negative, and the other two roots are complex and conjugate.

In the following we discuss these three cases in more detail.

Negative Real Roots If the four roots are real and negative, there is an independent solution associated with each root, and the complete solution is the sum of the four independent solutions. In a similar manner to the case of the undamped free vibration, the displacement vector can be written as

$$x_1(t) = \beta_1 X_{21} e^{s_1 t} + \beta_2 X_{22} e^{s_2 t} + \beta_3 X_{23} e^{s_3 t} + \beta_4 X_{24} e^{s_4 t}$$
$$x_2(t) = X_{21} e^{s_1 t} + X_{22} e^{s_2 t} + X_{23} e^{s_3 t} + X_{24} e^{s_4 t} \tag{6.68}$$

where β_1, β_2, β_3, and β_4 are the amplitude ratios, defined as

$$\beta_i = \left(\frac{X_1}{X_2}\right)_{s=s_i} = \frac{X_{1i}}{X_{2i}} \tag{6.69}$$

The amplitude ratios $\beta_i\,(i = 1, 2, 3, 4)$ are obtained from Eq. 64, by substituting $s = s_i\,(i = 1, 2, 3, 4)$.

Since, in this case, the roots s_1, s_2, s_3, and s_4 are all real and negative, the solution obtained is a decaying exponential function and, accordingly, the displacements exhibit no oscillations. This case represents the case of large damping, and the system returns to its equilibrium position without oscillation in a similar manner to the overdamped single degree of freedom system.

Complex Roots In this case, the roots, which occur as pairs of complex conjugates with negative real parts, can be written in the following form

$$\left.\begin{aligned} s_1 &= -p_1 + i\Omega_1 \\ s_2 &= -p_1 - i\Omega_1 \\ s_3 &= -p_2 + i\Omega_2 \\ s_4 &= -p_2 - i\Omega_2 \end{aligned}\right\} \tag{6.70}$$

where p_1, p_2, Ω_1, and Ω_2 are positive real numbers. The solution can then be written as

$$x_1(t) = X_{11} e^{s_1 t} + X_{12} e^{s_2 t} + X_{13} e^{s_3 t} + X_{14} e^{s_4 t} \tag{6.71}$$

$$x_2(t) = X_{21} e^{s_1 t} + X_{22} e^{s_2 t} + X_{23} e^{s_3 t} + X_{24} e^{s_4 t} \tag{6.72}$$

Substituting the value of s_1, s_2, s_3, and s_4 into Eq. 64, the amplitudes X_{1i} $(i = 1, 2, 3, 4)$ can be written in terms of the amplitudes X_{2i} $(i = 1, 2, 3, 4)$ using the amplitude ratios β_i defined as

$$\beta_i = \left(\frac{X_1}{X_2}\right)_{s=s_i} = \frac{X_{1i}}{X_{2i}} \tag{6.73}$$

Equations 71 and 72 can then be written as

$$x_1(t) = \beta_1 X_{21} e^{s_1 t} + \beta_2 X_{22} e^{s_2 t} + \beta_3 X_{23} e^{s_3 t} + \beta_4 X_{24} e^{s_4 t} \tag{6.74}$$

$$x_2(t) = X_{21} e^{s_1 t} + X_{22} e^{s_2 t} + X_{23} e^{s_3 t} + X_{24} e^{s_4 t} \tag{6.75}$$

These two equations have four arbitrary constants X_{21}, X_{22}, X_{23}, and X_{24} which can be determined from the initial conditions. It is, however, more convenient to express the solutions $x_1(t)$ and $x_2(t)$ in terms of the harmonic functions. To this end, we follow a procedure similar to the one used in the case of the underdamped free vibration of single degree of freedom systems.

Substituting Eq. 70 into Eqs. 74 and 75 yields

$$x_1(t) = \beta_1 X_{21} e^{(-p_1 + i\Omega_1)t} + \beta_2 X_{22} e^{(-p_1 - i\Omega_1)t} + \beta_3 X_{23} e^{(-p_2 + i\Omega_2)t} + \beta_4 X_{24} e^{(-p_2 - i\Omega_2)t}$$

$$x_2(t) = X_{21} e^{(-p_1 + i\Omega_1)t} + X_{22} e^{(-p_1 - i\Omega_1)t} + X_{23} e^{(-p_2 + i\Omega_2)t} + X_{24} e^{(-p_2 - i\Omega_2)t}$$

These two equations can be rewritten as

$$x_1(t) = e^{-p_1 t}(\beta_1 X_{21} e^{i\Omega_1 t} + \beta_2 X_{22} e^{-i\Omega_1 t}) + e^{-p_2 t}(\beta_3 X_{23} e^{i\Omega_2 t} + \beta_4 X_{24} e^{-i\Omega_2 t})$$

$$x_2(t) = e^{-p_1 t}(X_{21} e^{i\Omega_1 t} + X_{22} e^{-i\Omega_1 t}) + e^{-p_2 t}(X_{23} e^{i\Omega_2 t} + X_{24} e^{-i\Omega_2 t})$$

Euler's formula yields

$$e^{i\Omega_1 t} = \cos \Omega_1 t + i \sin \Omega_1 t$$

$$e^{-i\Omega_1 t} = \cos \Omega_1 t - i \sin \Omega_1 t$$

$$e^{i\Omega_2 t} = \cos \Omega_2 t + i \sin \Omega_2 t$$

$$e^{-i\Omega_2 t} = \cos \Omega_2 t - i \sin \Omega_2 t$$

Substituting these equations into the equations for x_1 and x_2 yields

$$x_1(t) = e^{-p_1 t}[(\beta_1 X_{21} + \beta_2 X_{22}) \cos \Omega_1 t + i(\beta_1 X_{21} - \beta_2 X_{22}) \sin \Omega_1 t] + e^{-p_2 t}[(\beta_3 X_{23} + \beta_4 X_{24}) \cos \Omega_2 t + i(\beta_3 X_{23} - \beta_4 X_{24}) \sin \Omega_2 t] \tag{6.76}$$

$$x_2(t) = e^{-p_1 t}[(X_{21} + X_{22}) \cos \Omega_1 t + i(X_{21} - X_{22}) \sin \Omega_1 t] + e^{-p_2 t}[(X_{23} + X_{24}) \cos \Omega_2 t + i(X_{23} - X_{24}) \sin \Omega_2 t] \tag{6.77}$$

Since the displacements $x_1(t)$ and $x_2(t)$ are real, one can show that the coefficients of the sine and cosine functions in the above two equations are real.

One can also show that the above two equations can be written as

$$\left.\begin{aligned} x_1(t) &= C_{11}e^{-p_1t}\sin(\Omega_1 t + \phi_{11}) + C_{12}e^{-p_2t}\sin(\Omega_2 t + \phi_{12}) \\ x_2(t) &= C_{21}e^{-p_1t}\sin(\Omega_1 t + \phi_{21}) + C_{22}e^{-p_2t}\sin(\Omega_2 t + \phi_{22}) \end{aligned}\right\} \tag{6.78}$$

where the coefficients $C_{11}, C_{12}, C_{21}, C_{22}, \phi_{11}, \phi_{12}, \phi_{21}$, and ϕ_{22} can be written in terms of the four arbitrary constants X_{21}, X_{22}, X_{23}, and X_{24}. The solution in this case is represented by exponentially decaying harmonic oscillations, which is a case similar to the case of the free vibration of underdamped single degree of freedom systems.

Real and Complex Roots This case occurs when two roots, say s_1 and s_2, are real and negative and the other two roots, s_3 and s_4, are complex conjugates. The roots s_3 and s_4 can be written, in this case, as

$$s_3 = -p + i\Omega$$

$$s_4 = -p - i\Omega$$

where p and Ω are positive numbers. The displacements $x_1(t)$ and $x_2(t)$ can then be written as

$$x_1(t) = \beta_1 X_{21}e^{s_1t} + \beta_2 X_{22}e^{s_2t} + \beta_3 X_{23}e^{(-p+i\Omega)t} + \beta_4 X_{24}e^{(-p-i\Omega)t}$$

$$x_2(t) = X_{21}e^{s_1t} + X_{22}e^{s_2t} + X_{23}e^{(-p+i\Omega)t} + X_{24}e^{(-p-i\Omega)t}$$

where $\beta_1, \beta_2, \beta_3$, and β_4 are the amplitude ratios and X_{21}, X_{22}, X_{23}, and X_{24} are constants to be determined from the initial conditions. The above two equations can be rewritten as

$$\begin{aligned} x_1(t) = \beta_1 X_{21}e^{s_1t} + \beta_2 X_{22}e^{s_2t} + e^{-pt}[(\beta_3 X_{23} + \beta_4 X_{24})\cos\Omega t \\ + i(\beta_3 X_{23} - \beta_4 X_{24})\sin\Omega t] \end{aligned} \tag{6.79}$$

$$x_2(t) = X_{21}e^{s_1t} + X_{22}e^{s_2t} + e^{-pt}[(X_{23} + X_{24})\cos\Omega t + i(X_{23} - X_{24})\sin\Omega t] \tag{6.80}$$

The displacements $x_1(t)$ and $x_2(t)$ can also be written in the following alternative form

$$x_1(t) = \beta_1 X_{21}e^{s_1t} + \beta_2 X_{22}e^{s_2t} + C_{11}e^{-pt}\sin(\Omega t + \phi_{11}) \tag{6.81}$$

$$x_2(t) = X_{21}e^{s_1t} + X_{22}e^{s_2t} + C_{22}e^{-pt}\sin(\Omega t + \phi_{22}) \tag{6.82}$$

where the constants C_{11}, C_{22}, ϕ_{11}, and ϕ_{22} can be expressed in terms of the two constants X_{23} and X_{24} and the amplitude ratios β_3 and β_4.

Example 6.6

Determine the matrix differential equations of motion of the damped two degree of freedom system shown in Fig. 7.

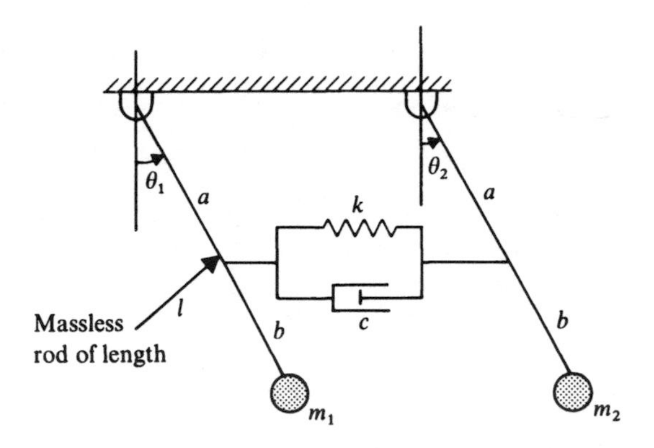

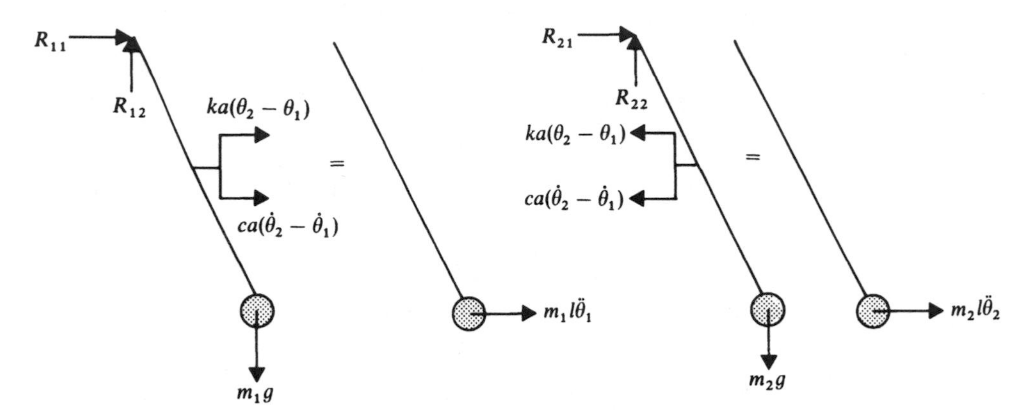

FIG. 6.7. Small oscillations of the two degree of freedom system.

Solution. Let θ_1 and θ_2 be the system degrees of freedom. Without loss of generality, we assume that $\theta_2 > \theta_1$ and $\dot{\theta}_2 > \dot{\theta}_1$. We also assume that the angular oscillation is small such that the motion of the masses in the vertical direction can be neglected. From the free body diagram shown in the figure and by taking the moments about the fixed point, one can show that the differential equation of motion of the first mass is given by

$$m_1 l^2 \ddot{\theta}_1 = ka^2(\theta_2 - \theta_1) + ca^2(\dot{\theta}_2 - \dot{\theta}_1) - m_1 g l \theta_1$$

Similarly, for the second mass, one has

$$m_2 l^2 \ddot{\theta}_2 = -ka^2(\theta_2 - \theta_1) - ca^2(\dot{\theta}_2 - \dot{\theta}_1) - m_2 g l \theta_2$$

These differential equations can be rewritten as

$$m_1 l^2 \ddot{\theta}_1 + ca^2 \dot{\theta}_1 - ca^2 \dot{\theta}_2 + (ka^2 + m_1 g l)\,\theta_1 - ka^2 \theta_2 = 0$$

$$m_2 l^2 \ddot{\theta}_2 + ca^2 \dot{\theta}_2 - ca^2 \dot{\theta}_1 + (ka^2 + m_2 g l)\,\theta_2 - ka^2 \theta_1 = 0$$

which can be written in matrix form as

$$\begin{bmatrix} m_1 l^2 & 0 \\ 0 & m_2 l^2 \end{bmatrix}\begin{bmatrix} \ddot{\theta}_1 \\ \ddot{\theta}_2 \end{bmatrix} + \begin{bmatrix} ca^2 & -ca^2 \\ -ca^2 & ca^2 \end{bmatrix}\begin{bmatrix} \dot{\theta}_1 \\ \dot{\theta}_2 \end{bmatrix} + \begin{bmatrix} (ka^2 + m_1 gl) & -ka^2 \\ -ka^2 & (ka^2 + m_2 gl) \end{bmatrix}\begin{bmatrix} \theta_1 \\ \theta_2 \end{bmatrix} = \begin{bmatrix} 0 \\ 0 \end{bmatrix}$$

This equation can be written in compact form as

$$\mathbf{M}\ddot{\boldsymbol{\theta}} + \mathbf{C}\dot{\boldsymbol{\theta}} + \mathbf{K}\boldsymbol{\theta} = \mathbf{0}$$

where $\ddot{\boldsymbol{\theta}}$, $\dot{\boldsymbol{\theta}}$, and $\boldsymbol{\theta}$ are the vectors

$$\ddot{\boldsymbol{\theta}} = \begin{bmatrix} \ddot{\theta}_1 \\ \ddot{\theta}_2 \end{bmatrix}, \quad \dot{\boldsymbol{\theta}} = \begin{bmatrix} \dot{\theta}_1 \\ \dot{\theta}_2 \end{bmatrix}, \quad \boldsymbol{\theta} = \begin{bmatrix} \theta_1 \\ \theta_2 \end{bmatrix},$$

and $\mathbf{M}$, $\mathbf{C}$, and $\mathbf{K}$ are, respectively, the mass, damping, and stiffness matrices defined as

$$\mathbf{M} = \begin{bmatrix} m_1 l^2 & 0 \\ 0 & m_2 l^2 \end{bmatrix}, \quad \mathbf{C} = \begin{bmatrix} ca^2 & -ca^2 \\ -ca^2 & ca^2 \end{bmatrix}, \quad \mathbf{K} = \begin{bmatrix} ka^2 + m_1 gl & -ka^2 \\ -ka^2 & ka^2 + m_2 gl \end{bmatrix}$$

Example 6.7

Determine the matrix differential equations of motion of the damped two degree of freedom system shown in Fig. 8 and identify the mass, damping, and stiffness matrices.

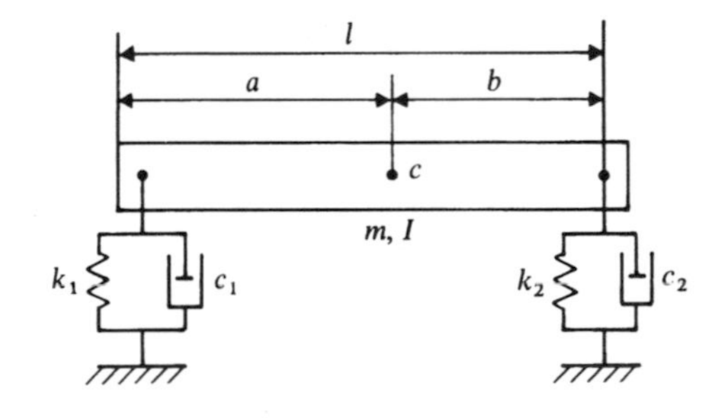

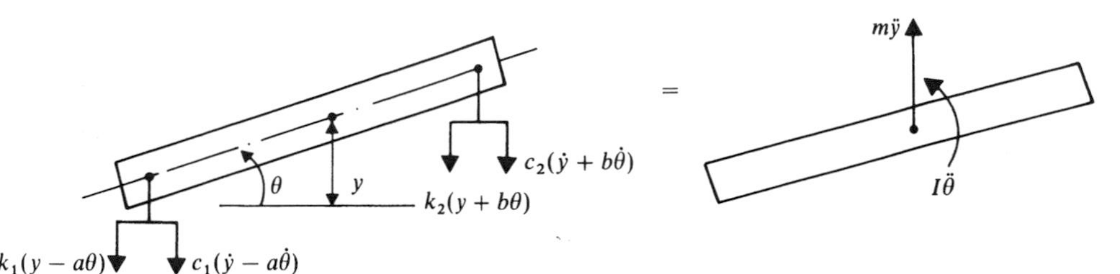

FIG. 6.8. Coupled linear and angular oscillations.

Solution. Assuming that the motion in the horizontal direction can be neglected, and applying D'Alembert's principle, the differential equations of motion can be written as

$$m\ddot{y} = -c_1(\dot{y} - a\dot{\theta}) - k_1(y - a\theta) - c_2(\dot{y} + b\dot{\theta}) - k_2(y + b\theta)$$

$$I\ddot{\theta} = c_1(\dot{y} - a\dot{\theta})a + k_1(y - a\theta)a - c_2(\dot{y} + b\dot{\theta})b - k_2(y + b\theta)b$$

In developing these equations, we assumed that the weight and the static deflections at the equilibrium position can be eliminated from the dynamic equations by using the static equations of equilibrium. The above differential equations can be rewritten as

$$m\ddot{y} + (c_1 + c_2)\dot{y} - (c_1 a - c_2 b)\dot{\theta} + (k_1 + k_2)y - (k_1 a - k_2 b)\theta = 0$$

$$I\ddot{\theta} + (c_1 a^2 + c_2 b^2)\dot{\theta} - (c_1 a - c_2 b)\dot{y} + (k_1 a^2 + k_2 b^2)\theta - (k_1 a - k_2 b)y = 0$$

which can be written in the following matrix form

$$\begin{bmatrix} m & 0 \\ 0 & I \end{bmatrix}\begin{bmatrix} \ddot{y} \\ \ddot{\theta} \end{bmatrix} + \begin{bmatrix} c_1 + c_2 & -(c_1 a - c_2 b) \\ -(c_1 a - c_2 b) & c_1 a^2 + c_2 b^2 \end{bmatrix}\begin{bmatrix} \dot{y} \\ \dot{\theta} \end{bmatrix}$$

$$+ \begin{bmatrix} k_1 + k_2 & k_2 b - k_1 a \\ k_2 b - k_1 a & k_1 a^2 + k_2 b^2 \end{bmatrix}\begin{bmatrix} y \\ \theta \end{bmatrix} = \begin{bmatrix} 0 \\ 0 \end{bmatrix}$$

where the mass matrix **M**, the damping matrix **C**, and the stiffness matrix **K** can be identified as

$$\mathbf{M} = \begin{bmatrix} m & 0 \\ 0 & I \end{bmatrix}, \quad \mathbf{C} = \begin{bmatrix} c_1 + c_2 & c_2 b - c_1 a \\ c_2 b - c_1 a & c_1 a^2 + c_2 b^2 \end{bmatrix},$$

$$\mathbf{K} = \begin{bmatrix} k_1 + k_2 & k_2 b - k_1 a \\ k_2 b - k_1 a & k_1 a^2 + k_2 b^2 \end{bmatrix}$$

Example 6.8

In the preceding example, let $m = 1000$ kg, $l = 4$ m, $a = b = l/2 = 2$ m, $I = 1300$ kg·m^2, $k_1 = 50 \times 10^3$ N/m, $k_2 = 70 \times 10^3$ N/m, and $c_1 = c_2 = 10$ N·s/m. Determine the system response as a function of time if the initial conditions are given by

$$y(t = 0) = y_0 = 0.03 \qquad \dot{y}(t = 0) = \dot{y}_0 = 0$$

$$\theta(t = 0) = \theta_0 = 0 \qquad \dot{\theta}(t = 0) = \dot{\theta}_0 = 0$$

Solution. The mass matrix **M**, the damping matrix **C**, and the stiffness matrix **K** are given by

$$\mathbf{M} = \begin{bmatrix} m & 0 \\ 0 & I \end{bmatrix} = \begin{bmatrix} 1000 & 0 \\ 0 & 1300 \end{bmatrix}$$

$$\mathbf{C} = \begin{bmatrix} c_1 + c_2 & c_2 b - c_1 a \\ c_2 b - c_1 a & c_1 a^2 + c_2 b^2 \end{bmatrix} = \begin{bmatrix} 20 & 0 \\ 0 & 80 \end{bmatrix}$$

$$\mathbf{K} = \begin{bmatrix} k_1 + k_2 & k_2 b - k_1 a \\ k_2 b - k_1 a & k_1 a^2 + k_2 b^2 \end{bmatrix} = \begin{bmatrix} 120 \times 10^3 & 40 \times 10^3 \\ 40 \times 10^3 & 480 \times 10^3 \end{bmatrix}$$

One can show that, in this example, the characteristic equation is given by

$$mIs^4 + [I(c_1 + c_2) + m(c_1a^2 + c_2b^2)]s^3 + [m(k_1a^2 + k_2b^2) + I(k_1 + k_2) + (c_1 + c_2)(c_1a^2 + c_2b^2)]s^2 + [(c_1 + c_2)(k_1a^2 + k_2b^2) + (c_1a^2 + c_2b^2)(k_1 + k_2)]s + (k_1 + k_2)(k_1a^2 + k_2b^2) - (k_2b - k_1a)^2 = 0$$

that is,

$$13 \times 10^5 s^4 + 1.06 \times 10^5 s^3 + 63.6 \times 10^7 s^2 + 192 \times 10^5 s + 560 \times 10^8 = 0$$

or

$$s^4 + 0.08154s^3 + 4.892 \times 10^2 s^2 + 14.769s + 4.3077 = 0$$

The roots of this polynomial are

$$s_1 = -0.0104 + 10.7i, \qquad s_2 = -0.0104 - 10.7i$$

$$s_3 = -0.0304 + 19.3i, \qquad s_4 = -0.0304 - 19.3i$$

The amplitude ratios β_i defined by Eq. 69 can be obtained using Eq. 64 as

$$\beta_i = \frac{X_{1i}}{X_{2i}} = -\frac{m_{12}s_i^2 + c_{12}s_i + k_{12}}{m_{11}s_i^2 + c_{11}s_i + k_{11}} = -\frac{m_{22}s_i^2 + c_{22}s_i + k_{22}}{m_{21}s_i^2 + c_{21}s_i + k_{21}}$$

which yields

$$\beta_1 = -8.279 - 0.01417i$$

$$\beta_2 = -8.279 + 0.01417i$$

$$\beta_3 = 0.106 - 4.645 \times 10^{-4}i$$

$$\beta_4 = 0.106 + 4.645 \times 10^{-4}i$$

The solution can then be written using Eqs. 76 and 77 as

$$y(t) = e^{-0.0104t}[(\beta_1X_{21} + \beta_2X_{22})\cos 10.7t + i(\beta_1X_{21} - \beta_2X_{22})\sin 10.7t] + e^{-0.0304t}[(\beta_3X_{23} + \beta_4X_{24})\cos 19.3t + i(\beta_3X_{23} - \beta_4X_{24})\sin 19.3t]$$

$$\theta(t) = e^{-0.0104t}[(X_{21} + X_{22})\cos 10.7t + i(X_{21} - X_{22})\sin 10.7t] + e^{-0.0304t}[(X_{23} + X_{24})\cos 19.3t + i(X_{23} - X_{24})\sin 19.3t]$$

Note that these two equations have only four unknown amplitudes X_{21}, X_{22}, X_{23}, and X_{24} which can be determined by using the initial conditions. To this end, we introduce the new real constants B_1, B_2, B_3, and B_4 defined as

$$B_1 = X_{21} + X_{22}, \qquad B_2 = i(X_{21} - X_{22})$$

$$B_3 = X_{23} + X_{24}, \qquad B_4 = i(X_{23} - X_{24})$$

Observe that since the amplitude ratios appear as complex conjugates in the form

$\beta_i = \alpha_i \pm i\gamma_i$, one has

$$\beta_1 X_{21} + \beta_2 X_{22} = \alpha_1 B_1 + \gamma_1 B_2$$
$$i(\beta_1 X_{21} - \beta_2 X_{22}) = -\gamma_1 B_1 + \alpha_1 B_2$$
$$\beta_3 X_{23} + \beta_4 X_{24} = \alpha_3 B_3 + \gamma_3 B_4$$
$$i(\beta_3 X_{23} - \beta_4 X_{24}) = -\gamma_3 B_3 + \alpha_3 B_4$$

where $\alpha_1 = \alpha_2 = -8.279$, $\gamma_1 = \gamma_2 = -0.01417$, $\alpha_3 = \alpha_4 = 0.1058$, and $\gamma_3 = \gamma_4 = -4.645 \times 10^{-4}$. Therefore, in terms of the new constants B_1, B_2, B_3, and B_4, the coordinates y and θ can be written as

$$y(t) = e^{-0.0104t}[(\alpha_1 B_1 + \gamma_1 B_2)\cos 10.7t + (-\gamma_1 B_1 + \alpha_1 B_2)\sin 10.7t]$$
$$+ e^{-0.0304t}[(\alpha_3 B_3 + \gamma_3 B_4)\cos 19.3t + (-\gamma_3 B_3 + \alpha_3 B_4)\sin 19.3t]$$
$$\theta(t) = e^{-0.0104t}[B_1 \cos 10.7t + B_2 \sin 10.7t]$$
$$+ e^{-0.0304t}[B_3 \cos 19.3t + B_4 \sin 19.3t]$$

Observe that these two equations can be expressed in the form given by Eq. 78. By using the initial conditions, one can verify that the constants B_1, B_2, B_3, and B_4 are

$$B_1 = -0.358 \times 10^{-2}, \qquad B_2 = -0.105 \times 10^{-4}$$
$$B_3 = 0.358 \times 10^{-2}, \qquad B_4 = 0.915 \times 10^{-5}$$

6.4 UNDAMPED FORCED VIBRATION

Thus far, we have only considered the undamped and damped free vibration of two degree of freedom systems. In this section, we consider the undamped forced vibration of such systems due to harmonic excitations. In Fig. 9, which shows an example of a two degree of freedom system, the harmonic forces $F_1(t)$ and $F_2(t)$ are applied to the masses m_1 and m_2, respectively. From the

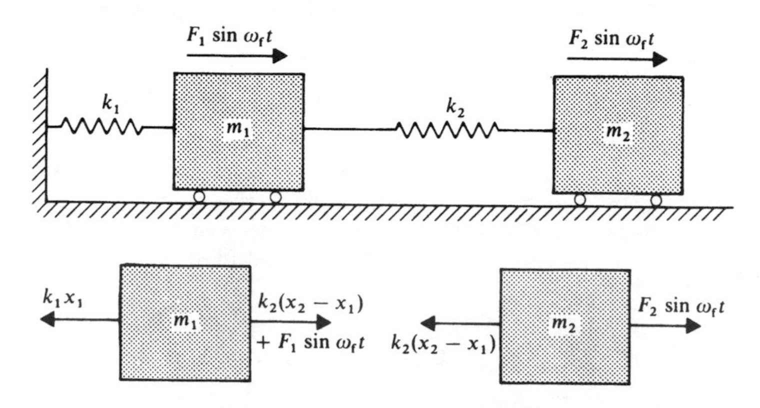

FIG. 6.9. Forced undamped vibration.

free body diagram shown in the figure, the two differential equations that govern the motion of the two masses are

$$m_1\ddot{x}_1 = -k_1x_1 + k_2(x_2 - x_1) + F_1 \sin \omega_f t$$

$$m_2\ddot{x}_2 = -k_2(x_2 - x_1) + F_2 \sin \omega_f t$$

These two equations can be written as

$$m_1\ddot{x}_1 + (k_1 + k_2)x_1 - k_2x_2 = F_1 \sin \omega_f t \quad (6.83)$$

$$m_2\ddot{x}_2 + k_2x_2 - k_2x_1 = F_2 \sin \omega_f t \quad (6.84)$$

These equations are linear second-order nonhomogeneous coupled differential equations which can be written in matrix form as

$$\begin{bmatrix} m_1 & 0 \\ 0 & m_2 \end{bmatrix}\begin{bmatrix} \ddot{x}_1 \\ \ddot{x}_2 \end{bmatrix} + \begin{bmatrix} k_1 + k_2 & -k_2 \\ -k_2 & k_2 \end{bmatrix}\begin{bmatrix} x_1 \\ x_2 \end{bmatrix} = \begin{bmatrix} F_1 \\ F_2 \end{bmatrix} \sin \omega_f t \quad (6.85)$$

This matrix equation is a special case of the general matrix equation which governs the motion of two degree of freedom systems and which can be written in compact form as

$$\mathbf{M}\ddot{\mathbf{x}} + \mathbf{K}\mathbf{x} = \mathbf{F} \sin \omega_f t \quad (6.86)$$

where the mass matrix $\mathbf{M}$ and the stiffness matrix $\mathbf{K}$ are given by

$$\mathbf{M} = \begin{bmatrix} m_{11} & m_{12} \\ m_{21} & m_{22} \end{bmatrix}, \qquad \mathbf{K} = \begin{bmatrix} k_{11} & k_{12} \\ k_{21} & k_{22} \end{bmatrix} \quad (6.87)$$

The forcing function $\mathbf{F}$ and the vectors $\ddot{\mathbf{x}}$ and $\mathbf{x}$ are given by

$$\mathbf{F} = \begin{bmatrix} F_1 \\ F_2 \end{bmatrix}, \qquad \ddot{\mathbf{x}} = \begin{bmatrix} \ddot{x}_1 \\ \ddot{x}_2 \end{bmatrix}, \qquad \mathbf{x} = \begin{bmatrix} x_1 \\ x_2 \end{bmatrix} \quad (6.88)$$

As in the case of the single degree of freedom system, we assume a solution in the form

$$\mathbf{x}(t) = \bar{\mathbf{X}} \sin \omega_f t \quad (6.89)$$

where $\bar{\mathbf{X}}$ is the vector of amplitudes given by

$$\bar{\mathbf{X}} = \begin{bmatrix} \bar{X}_1 \\ \bar{X}_2 \end{bmatrix} \quad (6.90)$$

that is,

$$x_1(t) = \bar{X}_1 \sin \omega_f t$$

$$x_2(t) = \bar{X}_2 \sin \omega_f t$$

Differentiating Eq. 89 twice, with respect to time, yields the acceleration vector

$$\ddot{\mathbf{x}}(t) = -\omega_f^2 \bar{\mathbf{X}} \sin \omega_f t \quad (6.91)$$

Substituting Eqs. 89 and 91 into Eq. 86 yields

$$-\omega_f^2 \mathbf{M}\bar{\mathbf{X}} \sin \omega_f t + \mathbf{K}\bar{\mathbf{X}} \sin \omega_f t = \mathbf{F} \sin \omega_f t$$

which yields the following matrix equation

$$[\mathbf{K} - \omega_f^2 \mathbf{M}]\bar{\mathbf{X}} = \mathbf{F} \tag{6.92}$$

This equation can be written in a more explicit form using the definition of **M** and **K** given by Eq. 87 as

$$\begin{bmatrix} k_{11} - \omega_f^2 m_{11} & k_{12} - \omega_f^2 m_{12} \\ k_{21} - \omega_f^2 m_{21} & k_{22} - \omega_f^2 m_{22} \end{bmatrix} \begin{bmatrix} \bar{X}_1 \\ \bar{X}_2 \end{bmatrix} = \begin{bmatrix} F_1 \\ F_2 \end{bmatrix} \tag{6.93}$$

Using *Cramer's rule*, it is an easy matter to verify that the amplitudes $\bar{X}_1$ and $\bar{X}_2$ are given by

$$\begin{aligned} \bar{X}_1 &= \frac{\begin{vmatrix} F_1 & k_{12} - \omega_f^2 m_{12} \\ F_2 & k_{22} - \omega_f^2 m_{22} \end{vmatrix}}{\begin{vmatrix} k_{11} - \omega_f^2 m_{11} & k_{12} - \omega_f^2 m_{12} \\ k_{21} - \omega_f^2 m_{21} & k_{22} - \omega_f^2 m_{22} \end{vmatrix}} \\ &= \frac{F_1(k_{22} - \omega_f^2 m_{22}) - F_2(k_{12} - \omega_f^2 m_{12})}{(k_{11} - \omega_f^2 m_{11})(k_{22} - \omega_f^2 m_{22}) - (k_{12} - \omega_f^2 m_{12})(k_{21} - \omega_f^2 m_{21})} \end{aligned} \tag{6.94}$$

$$\begin{aligned} \bar{X}_2 &= \frac{\begin{vmatrix} k_{11} - \omega_f^2 m_{11} & F_1 \\ k_{21} - \omega_f^2 m_{21} & F_2 \end{vmatrix}}{\begin{vmatrix} k_{11} - \omega_f^2 m_{11} & k_{12} - \omega_f^2 m_{12} \\ k_{21} - \omega_f^2 m_{21} & k_{22} - \omega_f^2 m_{22} \end{vmatrix}} \\ &= \frac{F_2(k_{11} - \omega_f^2 m_{11}) - F_1(k_{21} - \omega_f^2 m_{21})}{(k_{11} - \omega_f^2 m_{11})(k_{22} - \omega_f^2 m_{22}) - (k_{12} - \omega_f^2 m_{12})(k_{21} - \omega_f^2 m_{21})} \end{aligned} \tag{6.95}$$

provided that the determinant of the coefficient matrix in Eq. 93 is not equal to zero, that is,

$$(k_{11} - \omega_f^2 m_{11})(k_{22} - \omega_f^2 m_{22}) - (k_{12} - \omega_f^2 m_{12})(k_{21} - \omega_f^2 m_{21}) \neq 0 \tag{6.96}$$

Clearly, if the determinant of the coefficient matrix in Eq. 93 is equal to zero, one has the characteristic equation which was presented in its general form in Section 2. This characteristic equation can be solved for the natural frequencies ω_1 and ω_2. It is therefore clear that if $\omega_f = \omega_1$ or $\omega_f = \omega_2$, the denominators in Eqs. 94 and 95 are identically zero, and the system exhibits the *resonance phenomena* observed in the case of an undamped single degree of freedom system. This case, however, is different from the case of a single degree of freedom system, in the sense that there are two resonant frequencies which occur when $\omega_f = \omega_1$ or when $\omega_f = \omega_2$. This can also be illustrated by writing the denominators in Eqs. 94 and 95 in the following form

$$\begin{aligned} (m_{11}m_{22} - m_{12}m_{21})\omega_f^4 - (m_{11}k_{22} + m_{22}k_{11} - m_{12}k_{21} - m_{21}k_{12})\omega_f^2 \\ + k_{11}k_{22} - k_{12}k_{21} \end{aligned} \tag{6.97}$$

which can be written in compact form as

$$a\omega_f^4 + b\omega_f^2 + c \tag{6.98}$$

where

$$\left.\begin{aligned} a &= m_{11}m_{22} - m_{12}m_{21} \\ b &= -(m_{11}k_{22} + m_{22}k_{11} - m_{12}k_{21} - m_{21}k_{12}) \\ c &= k_{11}k_{22} - k_{12}k_{21} \end{aligned}\right\} \tag{6.99}$$

In terms of these constants, we have previously shown that the natural frequencies ω_1 and ω_2 can be written as

$$\omega_1^2 = \frac{-b + \sqrt{b^2 - 4ac}}{2a}, \qquad \omega_2^2 = \frac{-b - \sqrt{b^2 - 4ac}}{2a}$$

which show that

$$\omega_1^2 + \omega_2^2 = -\frac{b}{a}$$

and

$$\omega_1^2\omega_2^2 = \frac{1}{4a^2}[b^2 - b^2 + 4ac] = \frac{c}{a}$$

One can then write the denominators of Eqs. 94 and 95 which are given by Eq. 98 as

$$\begin{aligned} a\omega_f^4 + b\omega_f^2 + c &= \\ a\left(\omega_f^4 + \frac{b}{a}\omega_f^2 + \frac{c}{a}\right) &= a[\omega_f^4 - (\omega_1^2 + \omega_2^2)\omega_f^2 + \omega_1^2\omega_2^2] \\ &= a(\omega_f^2 - \omega_1^2)(\omega_f^2 - \omega_2^2) \end{aligned}$$

Therefore, Eqs. 94 and 95 can be written in a more simplified form as

$$\bar{X}_1 = \frac{1}{a}\frac{F_1(k_{22} - \omega_f^2 m_{22}) - F_2(k_{12} - \omega_f^2 m_{12})}{(\omega_f^2 - \omega_1^2)(\omega_f^2 - \omega_2^2)} \tag{6.100}$$

$$\bar{X}_2 = \frac{1}{a}\frac{F_2(k_{11} - \omega_f^2 m_{11}) - F_1(k_{21} - \omega_f^2 m_{21})}{(\omega_f^2 - \omega_1^2)(\omega_f^2 - \omega_2^2)} \tag{6.101}$$

where a is a constant defined by Eq. 99. It is clear that each mass will exhibit resonance, even in the special case in which a force acts only on one mass. For instance, if we consider the special case in which $F_2 = 0$, Eqs. 100 and 101 reduce to

$$\bar{X}_1 = \frac{1}{a}\frac{F_1(k_{22} - \omega_f^2 m_{22})}{(\omega_f^2 - \omega_1^2)(\omega_f^2 - \omega_2^2)}$$

$$\bar{X}_2 = -\frac{1}{a}\frac{F_1(k_{21} - \omega_f^2 m_{21})}{(\omega_f^2 - \omega_1^2)(\omega_f^2 - \omega_2^2)}$$

If the forced frequency ω_f of the force $F_1(t)$ is equal to either ω_1 or ω_2, the denominators in the above two equations are identically zero and $\bar{X}_1$ and $\bar{X}_2$ approach infinity.

Having determined $\bar{X}_1$ and $\bar{X}_2$ given by Eqs. 100 and 101, one can use Eq. 89 to write expressions for the forced response of the two degree of freedom systems.

Example 6.9

Determine the forced response of the undamped two degree of freedom system shown in Fig. 9, assuming that the force $F_2(t) = 0$.

Solution. In this case, we have

$$\mathbf{F} = \begin{bmatrix} F_1 \\ 0 \end{bmatrix}$$

$$\mathbf{M} = \begin{bmatrix} m_1 & 0 \\ 0 & m_2 \end{bmatrix}, \qquad \mathbf{K} = \begin{bmatrix} k_1 + k_2 & -k_2 \\ -k_2 & k_2 \end{bmatrix}$$

that is,

$$m_{11} = m_1, \qquad m_{22} = m_2, \qquad m_{12} = m_{21} = 0$$

$$k_{11} = k_1 + k_2, \qquad k_{22} = k_2, \qquad k_{12} = k_{21} = -k_2$$

The constant a of Eq. 99 is given by

$$a = m_{11}m_{22} - m_{12}m_{21} = m_1 m_2$$

Therefore, the amplitudes $\bar{X}_1$ and $\bar{X}_2$ of Eqs. 100 and 101 are, in this case, given by

$$\bar{X}_1 = \frac{1}{a}\frac{F_1(k_{22} - \omega_f^2 m_{22})}{(\omega_f^2 - \omega_1^2)(\omega_f^2 - \omega_2^2)} = \frac{1}{m_1 m_2}\frac{F_1(k_2 - \omega_f^2 m_2)}{(\omega_f^2 - \omega_1^2)(\omega_f^2 - \omega_2^2)}$$

$$\bar{X}_2 = -\frac{1}{a}\frac{F_1(k_{21} - \omega_f^2 m_{21})}{(\omega_f^2 - \omega_1^2)(\omega_f^2 - \omega_2^2)} = \frac{F_1 k_2}{m_1 m_2(\omega_f^2 - \omega_1^2)(\omega_f^2 - \omega_2^2)}$$

where the natural frequencies ω_1 and ω_2 are obtained using the constants of Eq. 99 as

$$\omega_1^2 = \frac{-b + \sqrt{b^2 - 4ac}}{2a}$$

$$= \frac{m_1 k_2 + m_2(k_1 + k_2) + \sqrt{[m_1 k_2 + m_2(k_1 + k_2)]^2 - 4m_1 m_2 k_1 k_2}}{2m_1 m_2}$$

$$\omega_2^2 = \frac{-b - \sqrt{b^2 - 4ac}}{2a}$$

$$= \frac{m_1 k_2 + m_2(k_1 + k_2) - \sqrt{[m_1 k_2 + m_2(k_1 + k_2)]^2 - 4m_1 m_2 k_1 k_2}}{2m_1 m_2}$$

The forced responses, $x_1(t)$ and $x_2(t)$, of the two masses m_1 and m_2 can then be

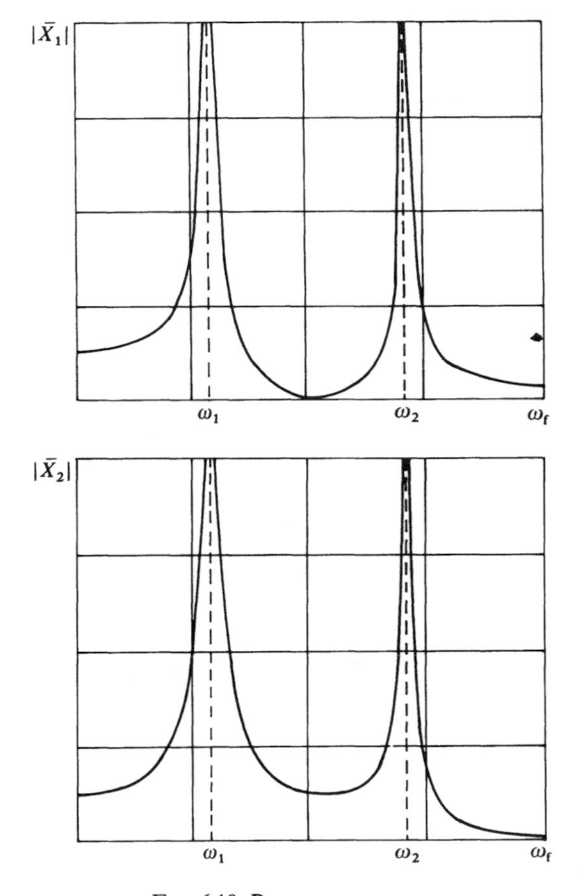

FIG. 6.10. Resonance curve.

written as

$$x_1(t) = \bar{X}_1 \sin \omega_f t$$

$$x_2(t) = \bar{X}_2 \sin \omega_f t$$

Plots of the steady state amplitudes $\bar{X}_1$ and $\bar{X}_2$ versus the frequency ω_f are shown in Fig. 10. In this figure, the amplitude $\bar{X}_1$ approaches zero when ω_f approaches $\sqrt{k_2/m_2}$.

Example 6.10

Determine the steady state response of the two degree of freedom system shown in Fig. 11, assuming small oscillations.

Solution. Assuming $\theta_2 > \theta_1$ and $\dot{\theta}_2 > \dot{\theta}_1$, and using the free body diagram shown in the figure, we obtain the following two differential equations

$$m_1 l^2 \ddot{\theta} = k l_1^2(\theta_2 - \theta_1) - m_1 g l \theta_1 + F l \sin \omega_f t$$

$$m_2 l^2 \ddot{\theta}_2 = -k l_1^2(\theta_2 - \theta_1) - m_2 g l \theta_2 + T \sin \omega_f t$$

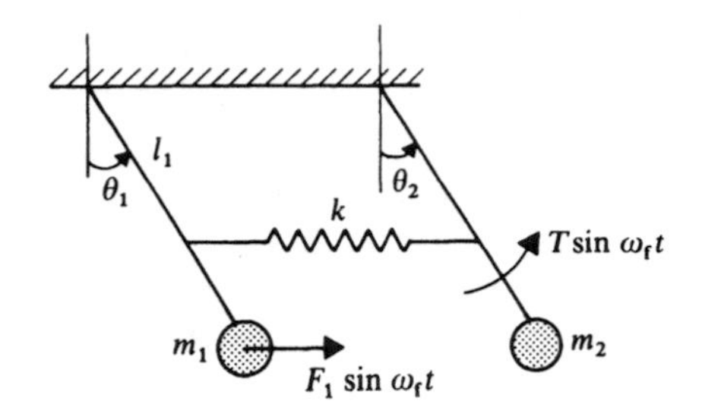

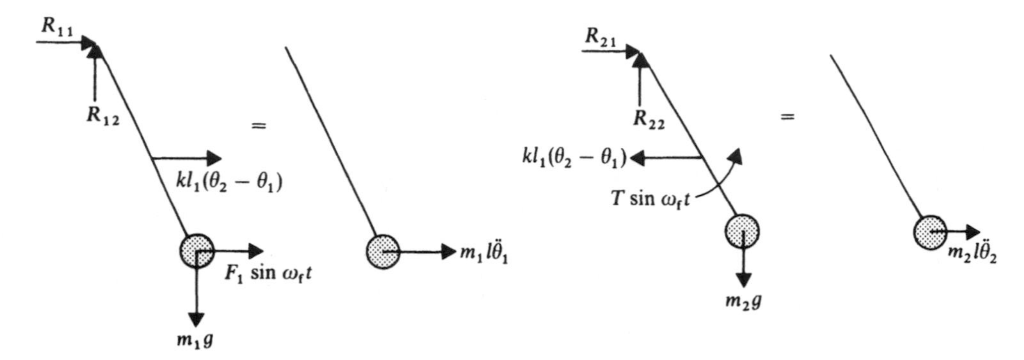

FIG. 6.11. Forced small oscillations of two degree of freedom system.

which can be written in matrix form as

$$\begin{bmatrix} m_1 l^2 & 0 \\ 0 & m_2 l^2 \end{bmatrix} \begin{bmatrix} \ddot{\theta}_1 \\ \ddot{\theta}_2 \end{bmatrix} + \begin{bmatrix} kl_1^2 + m_1 gl & -kl_1^2 \\ -kl_1^2 & kl_1^2 + m_2 gl \end{bmatrix} \begin{bmatrix} \theta_1 \\ \theta_2 \end{bmatrix} = \begin{bmatrix} Fl \\ T \end{bmatrix} \sin \omega_f t$$

in which we recognize

$$m_{11} = m_1 l^2, \qquad m_{22} = m_2 l^2, \qquad m_{12} = m_{21} = 0$$

$$k_{11} = kl_1^2 + m_1 gl, \qquad k_{22} = kl_1^2 + m_2 gl, \qquad k_{12} = k_{21} = -kl_1^2$$

$$F_1 = Fl, \qquad F_2 = T$$

The constants of Eq. 99 are then defined as

$$a = m_1 m_2 l^4$$

$$b = -[m_1 l^2(kl_1^2 + m_2 gl) + m_2 l^2(kl_1^2 + m_1 gl)]$$

$$= -2m_1 m_2 gl^3 - (m_1 + m_2)kl^2 l_1^2$$

$$c = (kl_1^2 + m_1 gl)(kl_1^2 + m_2 gl) - k^2 l_1^4$$

$$= m_1 m_2 (gl)^2 + kl_1^2 gl(m_1 + m_2)$$

In terms of these constants, the natural frequencies ω_1 and ω_2 are given by

$$\omega_1 = \frac{-b + \sqrt{b^2 - 4ac}}{2a}, \qquad \omega_2 = \frac{-b - \sqrt{b^2 - 4ac}}{2a}$$

In order to obtain the steady state response, assume a solution in the form

$$\boldsymbol{\theta} = \begin{bmatrix} \theta_1 \\ \theta_2 \end{bmatrix} = \boldsymbol{\Theta} \sin \omega_f t = \begin{bmatrix} \Theta_1 \\ \Theta_2 \end{bmatrix} \sin \omega_f t$$

Substituting this assumed solution into the differential equations, the amplitudes Θ_1 and Θ_2 can be determined using Eqs. 100 and 101 as

$$\Theta_1 = \frac{1}{m_1 m_2 l^4} \frac{F_1 l[kl_1^2 + m_2 gl - \omega_f^2 m_2 l^2] + Tkl_1^2}{(\omega_f^2 - \omega_1^2)(\omega_f^2 - \omega_2^2)}$$

$$\Theta_2 = \frac{1}{m_1 m_2 l^4} \frac{T[kl_1^2 + m_1 gl - \omega_f^2 m_1 l^2] + Flkl_1^2}{(\omega_f^2 - \omega_1^2)(\omega_f^2 - \omega_2^2)}$$

6.5 VIBRATION ABSORBER OF THE UNDAMPED SYSTEM

It was shown in the preceding chapters that a single degree of freedom system exhibits resonant conditions when the frequency of the forcing function is equal to the natural frequency of the system. In order to avoid undesirable resonance conditions in many applications, the system stiffness and inertia characteristics must be changed. Another approach, to alleviate the resonant conditions, is to convert the single degree of freedom system to a two degree of freedom system by adding an auxiliary spring and mass system. The parameters of the added system can be selected in such a manner that the vibration of the main mass is eliminated.

Consider the two degree of freedom system shown in Fig. 12. The frequency of the forcing function $F_1(t)$ is denoted as ω_f. The equations that govern the vibration of this system can be obtained as a special case from Eq. 85, where the force $F_2(t)$ is zero, that is, in this case, the matrix equation is given by

$$\begin{bmatrix} m_1 & 0 \\ 0 & m_2 \end{bmatrix} \begin{bmatrix} \ddot{x}_1 \\ \ddot{x}_2 \end{bmatrix} + \begin{bmatrix} k_1 + k_2 & -k_2 \\ -k_2 & k_2 \end{bmatrix} \begin{bmatrix} x_1 \\ x_2 \end{bmatrix} = \begin{bmatrix} F_1 \\ 0 \end{bmatrix} \sin \omega_f t \tag{6.102}$$

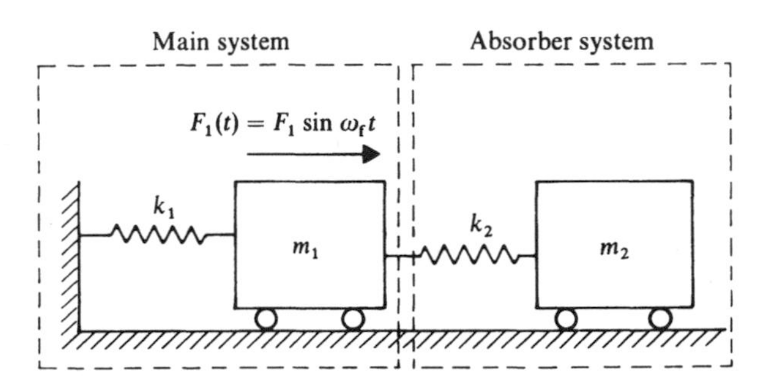

FIG. 6.12. Vibration absorber.

Following the procedure described in the preceding section, one can verify that the steady state amplitudes can be obtained from Eq. 93 as follows

$$\begin{bmatrix} k_1 + k_2 - \omega_f^2 m_1 & -k_2 \\ -k_2 & k_2 - \omega_f^2 m_2 \end{bmatrix} \begin{bmatrix} \bar{X}_1 \\ \bar{X}_2 \end{bmatrix} = \begin{bmatrix} F_1 \\ 0 \end{bmatrix} \tag{6.103}$$

This is a nonhomogeneous system of algebraic equations which can be solved for the steady state amplitudes $\bar{X}_1$ and $\bar{X}_2$ as follows

$$\begin{bmatrix} \bar{X}_1 \\ \bar{X}_2 \end{bmatrix} = \frac{1}{\Delta} \begin{bmatrix} k_2 - \omega_f^2 m_2 & k_2 \\ k_2 & k_1 + k_2 - \omega_f^2 m_1 \end{bmatrix} \begin{bmatrix} F_1 \\ 0 \end{bmatrix} \tag{6.104}$$

where Δ is the determinant of the coefficient matrix of Eq. 103 given by

$$\Delta = (k_1 + k_2 - \omega_f^2 m_1)(k_2 - \omega_f^2 m_2) - k_2^2 \tag{6.105}$$

Using this equation, the steady state amplitudes $\bar{X}_1$ and $\bar{X}_2$ can be written in a more explicit form as

$$\bar{X}_1 = \frac{(k_2 - m_2 \omega_f^2) F_1}{(k_1 + k_2 - \omega_f^2 m_1)(k_2 - \omega_f^2 m_2) - k_2^2} \tag{6.106}$$

$$\bar{X}_2 = \frac{k_2 F_1}{(k_1 + k_2 - \omega_f^2 m_1)(k_2 - \omega_f^2 m_2) - k_2^2} \tag{6.107}$$

It is clear from Eq. 106 that the steady state amplitude of the mass m_1 is zero if we select m_2 and k_2 such that

$$\frac{k_2}{m_2} = \omega_f^2 \tag{6.108}$$

If this condition is satisfied $\bar{X}_1$ is identically zero, and the determinant Δ of Eq. 105 or, equivalently, the denominator in Eq. 107 reduces to $-k_2^2$, that is, the steady state amplitude $\bar{X}_2$ of the second mass is given by

$$\bar{X}_2 = -\frac{F_1}{k_2} \tag{6.109}$$

and the steady state response of m_2 is given by

$$x_2(t) = \bar{X}_2 \sin \omega_f t = -\frac{F_1}{k_2} \sin \omega_f t \tag{6.110}$$

In this case, since $x_1(t) = 0$, the force exerted on the mass m_1 by the spring k_2 is given by

$$F_s = k_2(x_2 - x_1) = k_2 x_2 = k_2\left(-\frac{F_1}{k_2} \sin \omega_f t\right) = -F_1 \sin \omega_f t$$

which is a force equal in magnitude and opposite in direction to the applied force $F_1 \sin \omega_f t$.

It is therefore clear that, by a proper choice of the spring k_2 and the mass m_2, the motion of the mass m_1 can be brought to zero. Therefore, the vibration of the undamped single degree of freedom system can be alleviated by converting it to a two degree of freedom system, and selecting the added mass and spring in an appropriate manner to satisfy Eq. 108. The added system which consists of the mass m_2 and the spring k_2 is known as a *vibration absorber*. If the condition of Eq. 108 is not satisfied, the displacement of the main mass m_1 will not be equal to zero. It is also important to note that by adding the absorber system, the single degree of freedom system is converted to a two degree of freedom system which has two resonant frequencies instead of one. Therefore, the use of the vibration absorber is recommended when the frequency ω_f of the forcing function is known and constant. This is, in fact, the case in many engineering applications such as rotating machinery.

In order to better understand the relationships between the parameters of the main system and the parameters of the added system we define the following quantities

$$\left.\begin{aligned} \omega_m &= \sqrt{\frac{k_1}{m_1}} \\ \omega_a &= \sqrt{\frac{k_2}{m_2}} \\ r_a &= \frac{\omega_a}{\omega_m} \\ r_f &= \frac{\omega_f}{\omega_m} \\ r_{fa} &= \frac{\omega_f}{\omega_a} \\ \gamma &= \frac{m_2}{m_1} \\ \bar{X}_0 &= \frac{F_1}{k_1} \end{aligned}\right\} \tag{6.111}$$

where ω_m is the natural frequency of the main system alone, ω_a is the natural frequency of the absorber system alone, r_a and r_f are dimensionless frequency ratios, γ is a dimensionless parameter which represents the ratio of the absorber mass to the main mass, and $\bar{X}_0$ is the static deflection of the main system due to the force amplitude F_1. The steady state amplitudes $\bar{X}_1$ and $\bar{X}_2$ can be

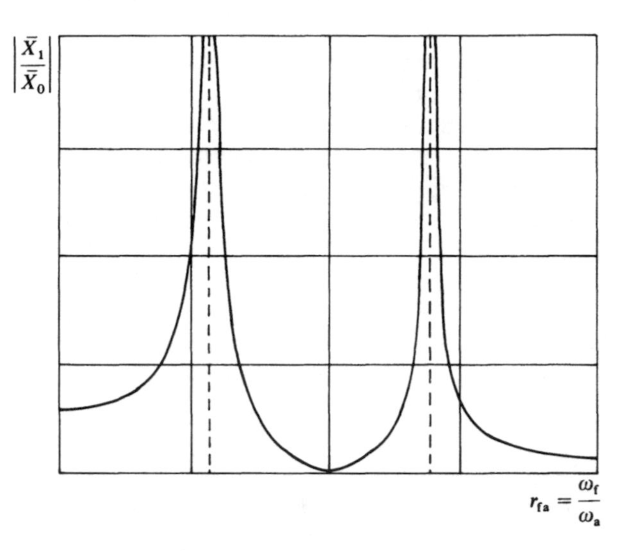

FIG. 6.13. Resonance curve.

written in terms of these parameters as

$$\bar{X}_1 = \frac{(1 - r_{fa}^2)\bar{X}_0}{(1 + \gamma r_a^2 - r_f^2)(1 - r_{fa}^2) - \gamma r_a^2} \tag{6.112}$$

$$\bar{X}_2 = \frac{\bar{X}_0}{(1 + \gamma r_a^2 - r_f^2)(1 - r_{fa}^2) - \gamma r_a^2} \tag{6.113}$$

Figure 13 shows the dimensionless amplitude ratio $(\bar{X}_1/\bar{X}_0)$ versus the frequency ratio $r_{fa} = \omega_f/\omega_a$ for a given mass ratio $\gamma = m_2/m_1$. If $r_{fa} = 1$, that is, $\omega_f = \omega_a$, the steady state amplitude of the main mass is identically zero. This is the case in which the parameters of the absorber system are selected to satisfy the condition of Eq. 108. It is also clear from the figure that the absorber is very effective over a small region in which $\omega_1 < \omega_f < \omega_2$, where ω_1 and ω_2 are natural frequencies of the two degree of freedom system. Therefore, the absorber will be useful in reducing the vibration in systems where there is no significant variation in the forced frequency ω_f.

6.6 FORCED VIBRATION OF DAMPED SYSTEMS

Figure 14 shows a two degree of freedom system which consists of the masses m_1 and m_2 connected by the springs k_1 and k_2 and the dampers c_1 and c_2. Let $F_1(t)$ and $F_2(t)$ be two harmonic forces that act on the masses m_1 and m_2,

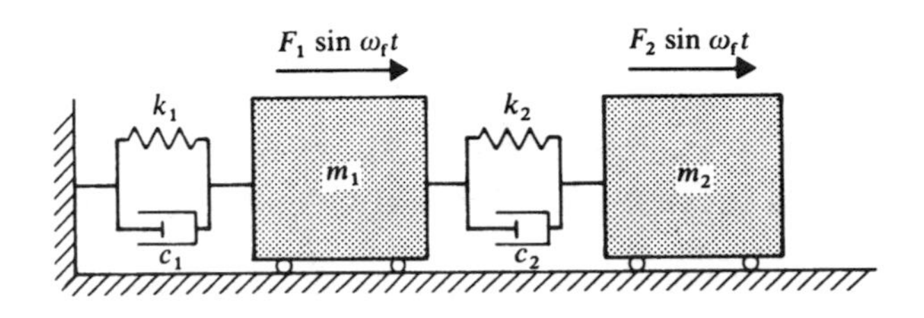

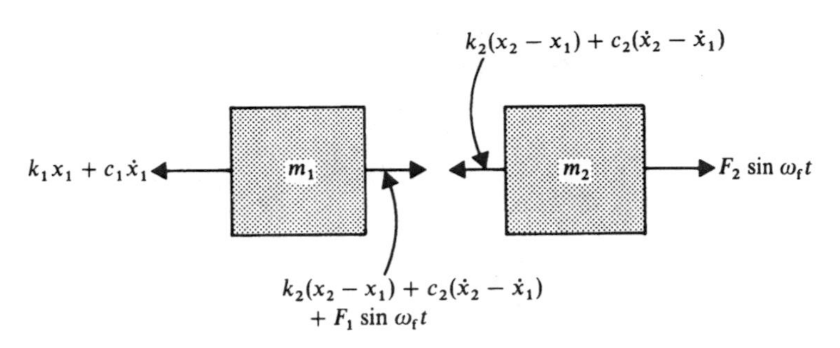

FIG. 6.14. Forced damped vibration of two degree of freedom system.

respectively. From the free body diagram shown in Fig. 14, and assuming that $x_2 > x_1$ and $\dot{x}_2 > \dot{x}_1$, it can be shown that the differential equations of motion of the two degree of freedom system shown in the figure can be written as

$$m_1\ddot{x}_1 = F_1 \sin \omega_f t + c_2(\dot{x}_2 - \dot{x}_1) - c_1\dot{x}_1 + k_2(x_2 - x_1) - k_1x_1$$
$$m_2\ddot{x}_2 = F_2 \sin \omega_f t - c_2(\dot{x}_2 - \dot{x}_1) - k_2(x_2 - x_1)$$

which can be written as

$$m_1\ddot{x}_1 + (c_1 + c_2)\dot{x}_1 - c_2\dot{x}_2 + (k_1 + k_2)x_1 - k_2x_2 = F_1 \sin \omega_f t$$
$$m_2\ddot{x}_2 + c_2\dot{x}_2 - c_2\dot{x}_1 + k_2x_2 - k_2x_1 = F_2 \sin \omega_f t$$

or in matrix form as

$$\begin{bmatrix} m_1 & 0 \\ 0 & m_2 \end{bmatrix}\begin{bmatrix} \ddot{x}_1 \\ \ddot{x}_2 \end{bmatrix} + \begin{bmatrix} c_1 + c_2 & -c_2 \\ -c_2 & c_2 \end{bmatrix}\begin{bmatrix} \dot{x}_1 \\ \dot{x}_2 \end{bmatrix} + \begin{bmatrix} k_1 + k_2 & -k_2 \\ -k_2 & k_2 \end{bmatrix}\begin{bmatrix} x_1 \\ x_2 \end{bmatrix}$$
$$= \begin{bmatrix} F_1 \\ F_2 \end{bmatrix} \sin \omega_f t$$

In general, given a two degree of freedom system which is subjected to a general harmonic excitation expressed in the following complex form

$$\mathbf{f}(t) = \mathbf{F}e^{i\omega_f t} = \mathbf{F}(\cos \omega_f t + i \sin \omega_f t)$$

where

$$\mathbf{F} = \begin{bmatrix} F_1 \\ F_2 \end{bmatrix},$$

the matrix differential equation that governs the vibration of this system can be written as

$$\mathbf{M}\ddot{\mathbf{x}} + \mathbf{C}\dot{\mathbf{x}} + \mathbf{K}\mathbf{x} = \mathbf{F}e^{i\omega_f t} \tag{6.114}$$

where **M**, **C**, and **K** are, respectively, the mass, the damping, and the stiffness matrices given in the following general form

$$\mathbf{M} = \begin{bmatrix} m_{11} & m_{12} \\ m_{21} & m_{22} \end{bmatrix}, \quad \mathbf{C} = \begin{bmatrix} c_{11} & c_{12} \\ c_{21} & c_{22} \end{bmatrix}, \quad \mathbf{K} = \begin{bmatrix} k_{11} & k_{12} \\ k_{21} & k_{22} \end{bmatrix} \tag{6.115}$$

and **x** and $\ddot{\mathbf{x}}$ are the vectors

$$\mathbf{x} = \begin{bmatrix} x_1 \\ x_2 \end{bmatrix}, \quad \ddot{\mathbf{x}} = \begin{bmatrix} \ddot{x}_1 \\ \ddot{x}_2 \end{bmatrix} \tag{6.116}$$

In order to obtain the steady state response of the two degree of freedom system due to harmonic excitation, we assume a solution in the following complex form

$$\mathbf{x} = \bar{\mathbf{X}}e^{i\omega_f t} \tag{6.117}$$

Substituting Eq. 117 into Eq. 114, we obtain

$$[-\omega_f^2\mathbf{M} + i\omega_f\mathbf{C} + \mathbf{K}]\bar{\mathbf{X}}e^{i\omega_f t} = \mathbf{F}e^{i\omega_f t}$$

which yields the matrix equation

$$[\mathbf{K} - \omega_f^2\mathbf{M} + i\omega_f\mathbf{C}]\bar{\mathbf{X}} = \mathbf{F} \tag{6.118}$$

Using the definition of the mass, the stiffness, and the damping matrices given by Eq. 115, Eq. 118 can be written in a more explicit form as

$$\begin{bmatrix} k_{11} - \omega_f^2 m_{11} + i\omega_f c_{11} & k_{12} - \omega_f^2 m_{12} + i\omega_f c_{12} \\ k_{21} - \omega_f^2 m_{21} + i\omega_f c_{21} & k_{22} - \omega_f^2 m_{22} + i\omega_f c_{22} \end{bmatrix} \begin{bmatrix} \bar{X}_1 \\ \bar{X}_2 \end{bmatrix} = \begin{bmatrix} F_1 \\ F_2 \end{bmatrix} \tag{6.119}$$

This equation can be used to determine the steady state response $\bar{X}_1$ and $\bar{X}_2$ as

$$\bar{\mathbf{X}} = \frac{1}{\Delta}\begin{bmatrix} k_{22} - \omega_f^2 m_{22} + i\omega_f c_{22} & \omega_f^2 m_{12} - k_{12} - i\omega_f c_{12} \\ \omega_f^2 m_{21} - k_{21} - i\omega_f c_{21} & k_{11} - \omega_f^2 m_{11} + i\omega_f c_{11} \end{bmatrix} \begin{bmatrix} F_1 \\ F_2 \end{bmatrix} \tag{6.120}$$

that is,

$$\bar{X}_1 = \frac{1}{\Delta}[F_1(k_{22} - \omega_f^2 m_{22} + i\omega_f c_{22}) + F_2(\omega_f^2 m_{12} - k_{12} - i\omega_f c_{12})] \tag{6.121}$$

$$\bar{X}_2 = \frac{1}{\Delta}[F_1(\omega_f^2 m_{21} - k_{21} - i\omega_f c_{21}) + F_2(k_{11} - \omega_f^2 m_{11} + i\omega_f c_{11})] \tag{6.122}$$

where Δ is the determinant of the coefficient matrix of Eq. 119 given by

$$\Delta = (k_{11} - \omega_f^2 m_{11} + i\omega_f c_{11})(k_{22} - \omega_f^2 m_{22} + i\omega_f c_{22})$$
$$-(k_{12} - \omega_f^2 m_{12} + i\omega_f c_{12})(k_{21} - \omega_f^2 m_{21} + i\omega_f c_{21}) \qquad (6.123)$$

Clearly, if the damping coefficients are all zeros, this determinant reduces to (Section 4)

$$\Delta = (k_{11} - \omega_f^2 m_{11})(k_{22} - \omega_f^2 m_{22}) - (k_{12} - \omega_f^2 m_{12})(k_{21} - \omega_f^2 m_{21})$$
$$= a(\omega_f^2 - \omega_1^2)(\omega_f^2 - \omega_2^2)$$

where $a = m_{11}m_{22} - m_{12}m_{21}$, and ω_1 and ω_2 are the natural frequencies of the system. In terms of these natural frequencies, Eq. 123 can be written as

$$\Delta = d_1 + id_2 \qquad (6.124)$$

where the constants d_1 and d_2 are defined in terms of the mass, the damping, and the stiffness coefficients as

$$d_1 = a(\omega_f^2 - \omega_1^2)(\omega_f^2 - \omega_2^2) - \omega_f^2(c_{11}c_{22} - c_{12}c_{21}) \qquad (6.125)$$

$$d_2 = \omega_f[c_{11}(k_{22} - \omega_f^2 m_{22}) + c_{22}(k_{11} - \omega_f^2 m_{11}) - c_{12}(k_{21} - \omega_f^2 m_{21})$$
$$- c_{21}(k_{12} - \omega_f^2 m_{12})] \qquad (6.126)$$

Therefore, the steady state amplitudes $\bar{X}_1$ and $\bar{X}_2$ of Eqs. 121 and 122 can be written as

$$\bar{X}_1 = \frac{b_1 + ib_2}{d_1 + id_2} = \frac{(b_1 d_1 + b_2 d_2) + i(d_1 b_2 - d_2 b_1)}{d_1^2 + d_2^2} \qquad (6.127)$$

$$\bar{X}_2 = \frac{\bar{c}_1 + i\bar{c}_2}{d_1 + id_2} = \frac{(\bar{c}_1 d_1 + \bar{c}_2 d_2) + i(d_1 \bar{c}_2 - d_2 \bar{c}_1)}{d_1^2 + d_2^2} \qquad (6.128)$$

where b_1, b_2, $\bar{c}_1$, and $\bar{c}_2$ are given by

$$b_1 = F_1(k_{22} - \omega_f^2 m_{22}) - F_2(k_{12} - \omega_f^2 m_{12})$$

$$b_2 = F_1 \omega_f c_{22} - F_2 \omega_f c_{12}$$

$$\bar{c}_1 = F_2(k_{11} - \omega_f^2 m_{11}) - F_1(k_{21} - \omega_f^2 m_{21})$$

$$\bar{c}_2 = F_2 \omega_f c_{11} - F_1 \omega_f c_{21}$$

The amplitudes $\bar{X}_1$ and $\bar{X}_2$ can be also expressed in exponential complex form as

$$\bar{X}_1 = A_1 e^{i\psi_1} \qquad (6.129)$$

$$\bar{X}_2 = A_2 e^{i\psi_2} \qquad (6.130)$$

where A_1, A_2, ψ_1, and ψ_2 are given by

$$A_1 = \frac{1}{d_1^2 + d_2^2}\sqrt{(b_1 d_1 + b_2 d_2)^2 + (d_1 b_2 - d_2 b_1)^2} \qquad (6.131)$$

$$A_2 = \frac{1}{d_1^2 + d_2^2}\sqrt{(\bar{c}_1 d_1 + \bar{c}_2 d_2)^2 + (d_1 \bar{c}_2 - d_2 \bar{c}_1)^2} \qquad (6.132)$$

or, equivalently,

$$A_1 = \frac{\sqrt{b_1^2 + b_2^2}}{\sqrt{d_1^2 + d_2^2}}, \qquad A_2 = \frac{\sqrt{\bar{c}_1^2 + \bar{c}_2^2}}{\sqrt{d_1^2 + d_2^2}}$$

$$\psi_1 = \tan^{-1}\left(\frac{d_1 b_2 - d_2 b_1}{b_1 d_1 + b_2 d_2}\right) \tag{6.133}$$

$$\psi_2 = \tan^{-1}\left(\frac{d_1 \bar{c}_2 - d_2 \bar{c}_1}{\bar{c}_1 d_1 + \bar{c}_2 d_2}\right) \tag{6.134}$$

By using Eqs. 117, 129, and 130, the steady state solution can then be written as

$$\mathbf{x}(t) = \begin{bmatrix} \bar{X}_1 \\ \bar{X}_2 \end{bmatrix} e^{i\omega_f t} = \begin{bmatrix} A_1 e^{i\psi_1} \\ A_2 e^{i\psi_2} \end{bmatrix} e^{i\omega_f t}$$

that is,

$$x_1(t) = A_1 e^{i(\omega_f t + \psi_1)}$$

$$x_2(t) = A_2 e^{i(\omega_f t + \psi_2)}$$

6.7 THE UNTUNED VISCOUS VIBRATION ABSORBER

In Section 5, we discussed a method for attenuating the vibration of single degree of freedom systems. In this method, the undamped single degree of freedom system is converted to a two degree of freedom system by adding the absorber system which consists of a mass and spring. The mass and spring stiffness of the absorber system were selected so as to eliminate the vibration at a certain known frequency, and as such the application of the undamped vibration absorber is limited only to the cases where the frequency of the forcing function is known. Therefore, the undamped vibration absorber is said to be *tuned* since it is effective only in a certain frequency range.

In this section, we consider a vibration absorber which can be used to reduce the vibration over a wider range of frequencies. The absorber system considered in this section is called the *untuned viscous vibration absorber*, and consists of a mass m_2 and a damper with damping coefficient c, as shown in Fig. 15. The main system consists of a mass m_1 and a spring with stiffness coefficient k. The

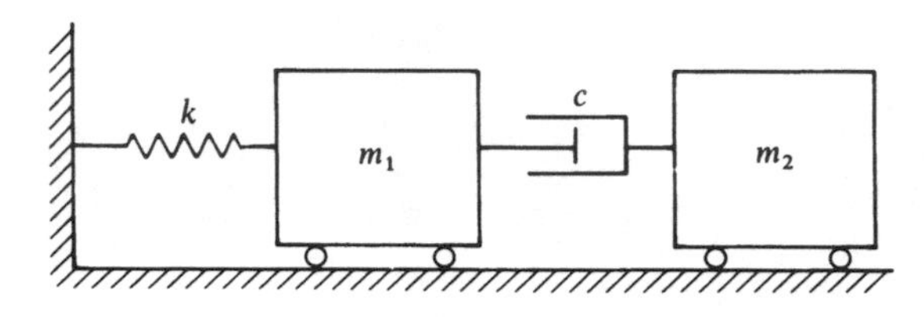

FIG. 6.15. Viscous vibration absorber.

differential equations of motion of the system shown in the figure are

$$m_1\ddot{x}_1 + c\dot{x}_1 - c\dot{x}_2 + kx_1 = F_0 e^{i\omega_f t}$$

$$m_2\ddot{x}_2 + c\dot{x}_2 - c\dot{x}_1 = 0$$

which can be written in matrix form as

$$\begin{bmatrix} m_1 & 0 \\ 0 & m_2 \end{bmatrix}\begin{bmatrix} \ddot{x}_1 \\ \ddot{x}_2 \end{bmatrix} + \begin{bmatrix} c & -c \\ -c & c \end{bmatrix}\begin{bmatrix} \dot{x}_1 \\ \dot{x}_2 \end{bmatrix} + \begin{bmatrix} k & 0 \\ 0 & 0 \end{bmatrix}\begin{bmatrix} x_1 \\ x_2 \end{bmatrix} = \begin{bmatrix} F_0 \\ 0 \end{bmatrix} e^{i\omega_f t} \tag{6.135}$$

One can verify that the natural frequencies of this system are

$$\omega_1 = \sqrt{\frac{k}{m_1}} \qquad \text{and} \qquad \omega_2 = 0$$

By using Eqs. 131 and 132, it is an easy matter to show that, in this case, the amplitudes of vibration A_1 and A_2 are given by

$$A_1 = \frac{\sqrt{b_1^2 + b_2^2}}{\sqrt{d_1^2 + d_2^2}}$$

$$A_2 = \frac{\sqrt{\bar{c}_1^2 + \bar{c}_2^2}}{\sqrt{d_1^2 + d_2^2}}$$

where

$$b_1 = -F_0\omega_f^2 m_2$$

$$b_2 = F_0\omega_f c$$

$$\bar{c}_1 = 0$$

$$\bar{c}_2 = F_0\omega_f c$$

$$d_1 = m_2\omega_f^2(m_1\omega_f^2 - k)$$

$$d_2 = c\omega_f[k - \omega_f^2(m_1 + m_2)]$$

that is,

$$A_1 = \frac{F_0\sqrt{(m_2\omega_f^2)^2 + (c\omega_f)^2}}{\sqrt{[m_2\omega_f^2(m_1\omega_f^2 - k)]^2 + (c\omega_f)^2[k - \omega_f^2(m_1 + m_2)]^2}}$$

$$A_2 = \frac{F_0 c\omega_f}{\sqrt{[m_2\omega_f^2(m_1\omega_f^2 - k)]^2 + (c\omega_f)^2[k - \omega_f^2(m_1 + m_2)]^2}}$$

which can be written as

$$A_1 = \frac{A_0\sqrt{\gamma^2 r^2 + 4\xi^2}}{\sqrt{\gamma^2 r^2(1 - r^2)^2 + 4\xi^2[\gamma r^2 - (1 - r^2)]^2}} \tag{6.136}$$

$$A_2 = \frac{A_0(2\xi)}{\sqrt{\gamma^2 r^2(1 - r^2)^2 + 4\xi^2[\gamma r^2 - (1 - r^2)]^2}} \tag{6.137}$$

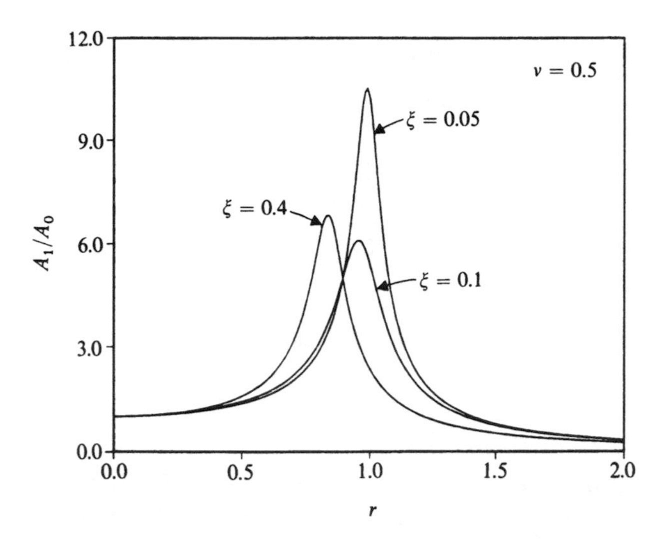

FIG. 6.16. Effect of the damping factor on the amplitude of vibration of the main mass.

where

$$A_0 = \frac{F_0}{k}, \qquad \xi = \frac{c}{2m_1\omega_1}, \qquad \gamma = \frac{m_2}{m_1}, \qquad r = \frac{\omega_f}{\omega_1} \tag{6.138}$$

Clearly, the amplitudes A_1 and A_2 depend on the three dimensionless parameters, the mass ratio γ, the frequency ratio r, and the damping factor ξ. It can be shown that for a given mass ratio γ, the curves (A_1/A_0), for different values for the damping factor ξ, intersect at one point, as shown in Fig. 16. To obtain the value of the frequency ratio at which these curves intersect, we equate (A_1/A_0) for two different values for the damping factor, that is,

$$\left(\frac{A_1}{A_0}\right)_{\xi=\xi_1} = \left(\frac{A_1}{A_0}\right)_{\xi=\xi_2}$$

This yields

$$\frac{\gamma^2 r^2 + 4\xi_1^2}{\gamma^2 r^2(1-r^2)^2 + 4\xi_1^2[\gamma r^2 - (1-r^2)]^2} = \frac{\gamma^2 r^2 + 4\xi_2^2}{\gamma^2 r^2(1-r^2)^2 + 4\xi_2^2[\gamma r^2 - (1-r^2)]^2}$$

which, after simplifying, yields the following equation

$$r^2(2+\gamma) - 2 = 0$$

which gives the value of r, at which the curves (A_1/A_0) intersect, as

$$r = \sqrt{\frac{2}{2+\gamma}} \tag{6.139}$$

As shown in Fig. 16, increasing the damping factor ξ does not necessarily result in reducing the maximum amplitude of vibration of the main mass. In fact, there is an optimum damping factor ξ_m for which the peak amplitude is minimum. It can be shown that the optimum damping factor ξ_m is given by

$$\xi_m = \frac{1}{\sqrt{2(1+\gamma)(2+\gamma)}} \tag{6.140}$$

and the minimum peak occurs at the value of r given by Eq. 139.

Houdaille Damper A similar concept can be used to reduce the vibration in rotating systems, such as in engine installations where the operation speed may vary over a wide range. A tuned viscous torsional damper referred to as the *Houdaille damper* or *viscous Lanchester damper* can be used to reduce the torsional oscillations of the crankshaft. As shown in Fig. 17, the damper consists of a disk with mass moment of inertia J_2. The disc is free to rotate inside a housing which is attached to the rotating shaft, and the housing and the rotating shaft are assumed to have equivalent mass moment of inertia J_1. The space between the housing and the disk is filled with viscous fluid. In most cases, the fluid is a silicon oil whose viscosity is of similar magnitude to oil. but which does not change significantly when the temperature changes. The damping effect is produced by the viscosity of the oil and is proportional to the relative angular velocity between the housing and the disk. Let θ_1 and θ_2 denote the rotations of the housing and disk, respectively, and let $M_0 e^{i\omega_f t}$ be the external harmonic torque which acts on the shaft whose torsional stiffness is equal to k. The damping torque resulting from the viscosity of the fluid is assumed to be proportional to the relative angular velocity $(\dot{\theta}_1 - \dot{\theta}_2)$ between the housing and the disk, and can be written as

$$M_d = c(\dot{\theta}_1 - \dot{\theta}_2) \tag{6.141}$$

where c is the viscous damping coefficient.

One can show that the differential equations of motion of the two degree of freedom system, shown in Fig. 16, can be written as

$$J_1\ddot{\theta}_1 + c(\dot{\theta}_1 - \dot{\theta}_2) + k\theta_1 = M_0 e^{i\omega_f t} \tag{6.142}$$

$$J_2\ddot{\theta}_2 - c(\dot{\theta}_1 - \dot{\theta}_2) = 0 \tag{6.143}$$

This equation can be written in matrix form as

$$\begin{bmatrix} J_1 & 0 \\ 0 & J_2 \end{bmatrix}\begin{bmatrix} \ddot{\theta}_1 \\ \ddot{\theta}_2 \end{bmatrix} + \begin{bmatrix} c & -c \\ -c & c \end{bmatrix}\begin{bmatrix} \dot{\theta}_1 \\ \dot{\theta}_2 \end{bmatrix} + \begin{bmatrix} k & 0 \\ 0 & 0 \end{bmatrix}\begin{bmatrix} \theta_1 \\ \theta_2 \end{bmatrix} = \begin{bmatrix} M_0 \\ 0 \end{bmatrix} e^{i\omega_f t} \tag{6.144}$$

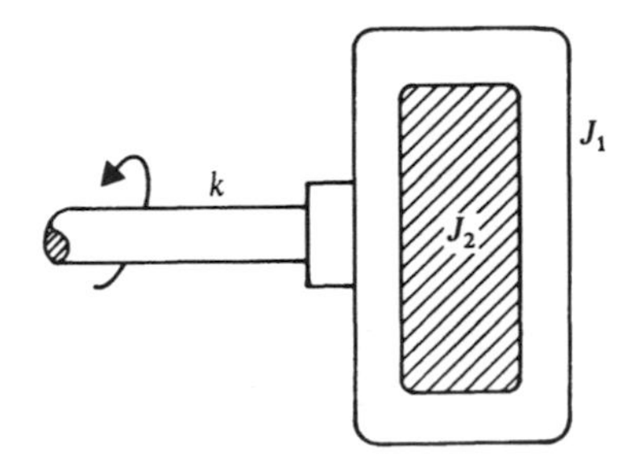

FIG. 6.17. Houdaille damper.

This matrix equation is similar to the matrix equation of Eq. 135. Therefore, similar comments to the ones previously made apply to the Houdaille damper. The solution of Eq. 144 is, therefore, left as an exercise.

6.8 MULTI-DEGREE OF FREEDOM SYSTEMS

As demonstrated in this chapter, the linear theory of vibration of the two degree of freedom systems can be developed to approximately the same level as was reached with the single degree of freedom systems. The methods, presented for the free and forced vibration analysis of undamped and damped two degree of freedom systems, can be considered as generalizations of the techniques presented in the preceding chapters for the vibration analysis of single degree of freedom systems. These methods can be generalized further to study the vibration of systems which have more than two degrees of freedom. In order to demonstrate this, we first present, in matrix form, some of the basic results and equations obtained in the preceding sections.

Free Vibration In the first two sections of this chapter, the free undamped vibration of the two degree of freedom systems is discussed, and the dynamic equations of motion that govern the free vibrations are developed and expressed in matrix form. It is shown that a system with two degrees of freedom has two natural frequencies that depend on the mass and stiffness coefficients of the system, and these natural frequencies can be obtained by solving the characteristic equations. It is also shown in the first two sections that the dynamic and elastic coupling terms that appear, respectively, in the mass and the stiffness matrices depend on the selection of the coordinates. A decoupled system of equations can be obtained by using the modal or principal coordinates. In this case, the coordinates of the system can be expressed in terms of the modal coordinates as

$$\mathbf{x} = \mathbf{\Phi q} \tag{6.145}$$

where $\mathbf{x}$ is the vector of system coordinates, $\mathbf{q}$ is the vector of modal (principal) coordinates, and $\mathbf{\Phi}$ is the modal transformation matrix defined in terms of the amplitude ratios β_1 and β_2 of Eqs. 15 and 18 as

$$\mathbf{\Phi} = \begin{bmatrix} \beta_1 & \beta_2 \\ 1 & 1 \end{bmatrix} \tag{6.146}$$

In the analysis presented in Section 2 of this chapter, it is shown that the use of the modal coordinates leads to decoupled equations because the modal transformation matrix $\mathbf{\Phi}$ is orthogonal to the mass and the stiffness matrices, that is

$$\mathbf{\Phi}^{\mathrm{T}}\mathbf{M}\mathbf{\Phi} = \mathbf{M}_{\mathrm{p}}$$

$$\mathbf{\Phi}^{\mathrm{T}}\mathbf{K}\mathbf{\Phi} = \mathbf{K}_{\mathrm{p}}$$

where $\mathbf{M}_{\mathrm{p}}$ and $\mathbf{K}_{\mathrm{p}}$ are two diagonal matrices. The diagonal matrices $\mathbf{M}_{\mathrm{p}}$ and

$\mathbf{K}_p$ are called, respectively, the *modal mass* and *stiffness matrices*. The diagonal modal mass and stiffness matrices can therefore be written as

$$\mathbf{M}_p = \begin{bmatrix} m_1 & 0 \\ 0 & m_2 \end{bmatrix}$$

$$\mathbf{K}_p = \begin{bmatrix} k_1 & 0 \\ 0 & k_2 \end{bmatrix}$$

where m_i and k_i ($i = 1, 2$) are called, respectively, the *modal mass* and *stiffness coefficients* and they are defined according to

$$m_i = \mathbf{A}_i^{\mathrm{T}}\mathbf{M}\mathbf{A}_i, \qquad i = 1, 2$$

$$k_i = \mathbf{A}_i^{\mathrm{T}}\mathbf{K}\mathbf{A}_i, \qquad i = 1, 2$$

in which $\mathbf{A}_i$ is the ith column in the modal matrix defined as

$$\mathbf{A}_i = \begin{bmatrix} \beta_i \\ 1 \end{bmatrix}$$

The vector $\mathbf{A}_i$ is called the ith eigenvector or mode shape. The mode shapes satisfy the following *orthogonality conditions*

$$\mathbf{A}_i^{\mathrm{T}}\mathbf{M}\mathbf{A}_i = \begin{cases} m_i & \text{if } i = j \\ 0 & \text{if } i \neq j \end{cases}$$

$$\mathbf{A}_i^{\mathrm{T}}\mathbf{K}\mathbf{A}_i = \begin{cases} k_i & \text{if } i = j \\ 0 & \text{if } i \neq j \end{cases}$$

Observe that in terms of the modal coordinates, the uncoupled equations of the free vibration of the two degree of freedom systems can be written as

$$m_i\ddot{q}_i + k_i q_i = 0, \qquad i = 1, 2 \tag{6.147}$$

These two equations are in a form similar to the equations that arise in the analysis of single degree of freedom systems. In this case, the two natural frequencies of the system can simply be evaluated as

$$\omega_i = \sqrt{\frac{k_i}{m_i}}, \qquad i = 1, 2 \tag{6.148}$$

Therefore, the free vibration of the two degree of freedom system can be considered as a combination of its principal modes of vibration. By solving Eq. 147 for the modal coordinates q_1 and q_2, the system displacements x_1 and x_2 can be determined by using the modal transformation of Eq. 145. More discussion on the use of the modal coordinates in the analysis of systems with more than one degree of freedom is presented in the second volume of this book (Shabana, 1991).

Forced Vibration The undamped and damped forced vibration of the two degree of freedom systems is discussed in the last four sections of this chapter, and it was demonstrated that, due to the fact that the two degree of freedom system has two natural frequencies, two resonant conditions are encountered. These resonant conditions occur when the frequency of the forcing function coincides with one of the natural frequencies of the system. As in the case of single degree of freedom systems, the damping has a significant effect on the forced response of the two degree of freedom systems. It is shown that the dynamic equations of forced vibration of the damped system can be written in a matrix form as

$$\mathbf{M}\ddot{\mathbf{x}} + \mathbf{C}\dot{\mathbf{x}} + \mathbf{K}\mathbf{x} = \mathbf{F} \tag{6.149}$$

where **M**, **C**, and **K** are, respectively, the mass, damping, and stiffness matrices of the system given by

$$\mathbf{M} = \begin{bmatrix} m_{11} & m_{12} \\ m_{21} & m_{22} \end{bmatrix}$$

$$\mathbf{C} = \begin{bmatrix} c_{11} & c_{12} \\ c_{21} & c_{22} \end{bmatrix}$$

$$\mathbf{K} = \begin{bmatrix} k_{11} & k_{12} \\ k_{21} & k_{22} \end{bmatrix}$$

and the vectors **x** and **F** are, respectively, the displacement and force vectors defined in the case of two degree of freedom systems as

$$\mathbf{x} = \begin{bmatrix} x_1 \\ x_2 \end{bmatrix}, \qquad \mathbf{F} = \begin{bmatrix} F_1 \\ F_2 \end{bmatrix}$$

Multi-Degree of Freedom Systems The equation of forced vibration of a single degree of freedom systems can be considered as a special case of Eq. 149 in which the matrices and vectors reduce to scalars. It also can be shown (Shabana, 1991) that an equation similar to Eq. 149 can be obtained for a system with an arbitrary finite number of degrees of freedom. For example, the equation that governs the forced vibration of a system with n degrees of freedom is in the same form as Eq. 149, with the matrices and vectors **M**, **C**, **K**, **x**, and **F** having dimension n, that is

$$\mathbf{M} = \begin{bmatrix} m_{11} & m_{12} & m_{13} & \cdots & m_{1n} \\ m_{21} & m_{22} & m_{23} & \cdots & m_{2n} \\ m_{31} & m_{32} & m_{33} & \cdots & m_{3n} \\ \vdots & \vdots & \vdots & \ddots & \vdots \\ m_{n1} & m_{n2} & m_{n3} & \cdots & m_{nn} \end{bmatrix}$$

$$\mathbf{C} = \begin{bmatrix} c_{11} & c_{12} & c_{13} & \cdots & c_{1n} \\ c_{21} & c_{22} & c_{23} & \cdots & c_{2n} \\ c_{31} & c_{32} & c_{33} & \cdots & c_{3n} \\ \vdots & \vdots & \vdots & \ddots & \vdots \\ c_{n1} & c_{n2} & c_{n3} & \cdots & c_{nn} \end{bmatrix}$$

$$\mathbf{K} = \begin{bmatrix} k_{11} & k_{12} & k_{13} & \dots & k_{1n} \\ k_{21} & k_{22} & k_{23} & \dots & k_{2n} \\ k_{31} & k_{32} & k_{33} & \dots & k_{3n} \\ \vdots & \vdots & \vdots & \ddots & \vdots \\ k_{n1} & k_{n2} & k_{n3} & \dots & k_{nn} \end{bmatrix}$$

$$\mathbf{x} = \begin{bmatrix} x_1 \\ x_2 \\ x_3 \\ \vdots \\ x_n \end{bmatrix}, \qquad \mathbf{F} = \begin{bmatrix} F_1 \\ F_2 \\ F_3 \\ \vdots \\ F_n \end{bmatrix}$$

where m_{ij}, c_{ij}, k_{ij} $(i, j = 1, 2, 3, \dots n)$ are, respectively, the mass, damping, and stiffness coefficients, and x_i and F_i are, respectively, the ith displacement and force components. With this definition of the matrices and vectors, a similar procedure as the one used in this chapter can be employed to study the vibration of multi degree of freedom systems. In this case, the modal matrix $\mathbf{\Phi}$ is an $n \times n$ matrix as demonstrated by the following example.

Example 6.11

The torsional system shown in Fig. 18 consists of three disks which have mass moment of inertia, $I_1 = 2.0 \times 10^3$ kg·m^2, $I_2 = 3.0 \times 10^3$ kg·m^2, and $I_3 = 4.0 \times 10^3$ kg·m^2. The stiffness coefficients of the shafts connecting these disks are $k_1 = 12 \times 10^5$ N·m, $k_2 = 24 \times 10^5$ N·m, and $k_3 = 36 \times 10^5$ N·m. The matrix equation of motion of the free vibration of this system is given by

$$\mathbf{M}\ddot{\boldsymbol{\theta}} + \mathbf{K}\boldsymbol{\theta} = \mathbf{0}$$

where, in this example, $\mathbf{M}$, $\mathbf{K}$, and $\boldsymbol{\theta}$ are given by

$$\mathbf{M} = \begin{bmatrix} I_1 & 0 & 0 \\ 0 & I_2 & 0 \\ 0 & 0 & I_3 \end{bmatrix} = 2 \times 10^3 \begin{bmatrix} 1.0 & 0 & 0 \\ 0 & 1.5 & 0 \\ 0 & 0 & 2 \end{bmatrix} \text{kg}\cdot\text{m}^2$$

$$\mathbf{K} = \begin{bmatrix} k_1 & -k_1 & 0 \\ -k_1 & k_1 + k_2 & -k_2 \\ 0 & -k_2 & k_2 + k_3 \end{bmatrix} = 12 \times 10^5 \begin{bmatrix} 1 & -1 & 0 \\ -1 & 3 & -2 \\ 0 & -2 & 5 \end{bmatrix} \text{N}\cdot\text{m}$$

$$\boldsymbol{\theta} = [\theta_1 \quad \theta_2 \quad \theta_3]^{\mathrm{T}}$$

in which θ_1, θ_2, and θ_3 are the torsional oscillations of the disks.

We assume a solution in the form

$$\boldsymbol{\theta} = \mathbf{A} \sin(\omega t + \phi)$$

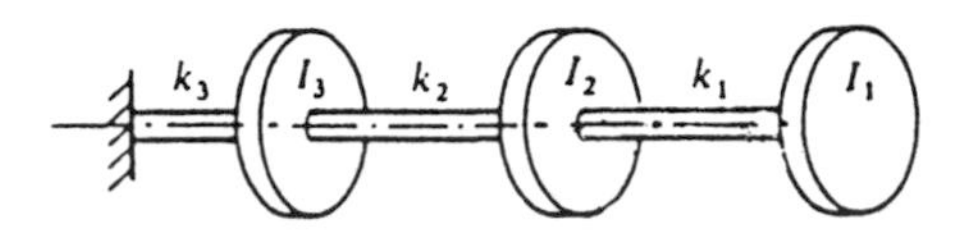

FIG. 6.18. Three degree of freedom torsional system.

By substituting this assumed solution into the differential equation, one obtains

$$[\mathbf{K} - \omega^2\mathbf{M}]\mathbf{A} = \mathbf{0}$$

Substituting for the mass and stiffness matrices, we get

$$\left[6 \times 10^5 \begin{bmatrix} 1 & -1 & 0 \\ -1 & 3 & -2 \\ 0 & -2 & 5 \end{bmatrix} - \omega^2 \times 10^3 \begin{bmatrix} 1 & 0 & 0 \\ 0 & 1.5 & 0 \\ 0 & 0 & 2 \end{bmatrix}\right]\mathbf{A} = \mathbf{0}$$

which can be rewritten as

$$\left[\begin{bmatrix} 1 & -1 & 0 \\ -1 & 3 & -2 \\ 0 & -2 & 5 \end{bmatrix} - \alpha \begin{bmatrix} 1 & 0 & 0 \\ 0 & 1.5 & 0 \\ 0 & 0 & 2 \end{bmatrix}\right]\mathbf{A} = \mathbf{0}$$

This system has a nontrivial solution if the determinant of the coefficient matrix is equal to zero. This leads to the following characteristic equation:

$$\alpha^3 - 5.5\alpha^2 + 7.5\alpha - 2 = 0$$

which has the following roots:

$$\alpha_1 = 0.3516, \qquad \alpha_2 = 1.606, \qquad \alpha_3 = 3.542$$

Since $\omega^2 = 600\alpha$, the natural frequencies associated with the roots α_1, α_2, and α_3 are given, respectively, by

$$\omega_1 = 14.52 \text{ rad/s}, \qquad \omega_2 = 31.05 \text{ rad/s}, \quad \text{and} \quad \omega_3 = 46.1 \text{ rad/s}$$

For a given root α_i, $i = 1, 2, 3$, the mode shapes can be determined using the equation

$$\begin{bmatrix} 1 - \alpha_i & -1 & 0 \\ -1 & 3 - 1.5\alpha_i & -2 \\ 0 & -2 & 5 - 2\alpha_i \end{bmatrix} \begin{bmatrix} A_{i1} \\ A_{i2} \\ A_{i3} \end{bmatrix} = \begin{bmatrix} 0 \\ 0 \\ 0 \end{bmatrix}$$

which, by partitioning the coefficient matrix, leads to

$$\begin{bmatrix} -1 \\ 0 \end{bmatrix} A_{i1} + \begin{bmatrix} (3 - 1.5\alpha_i) & -2 \\ -2 & 5 - 2\alpha_i \end{bmatrix} \begin{bmatrix} A_{i2} \\ A_{i3} \end{bmatrix} = \begin{bmatrix} 0 \\ 0 \end{bmatrix}$$

or

$$\begin{bmatrix} A_{i2} \\ A_{i3} \end{bmatrix} = \frac{1}{3\alpha_i^2 - 13.5\alpha_i + 11} \begin{bmatrix} 5 - 2\alpha_i \\ 2 \end{bmatrix} A_{i1}$$

Using the values obtained previously for α_i, $i = 1, 2, 3$, we have

$$\begin{bmatrix} A_{12} \\ A_{13} \end{bmatrix} = \begin{bmatrix} 0.649 \\ 0.302 \end{bmatrix} A_{11} \qquad \text{for} \quad \omega_1 = 14.52 \text{ rad/s}$$

$$\begin{bmatrix} A_{22} \\ A_{23} \end{bmatrix} = \begin{bmatrix} -0.607 \\ -0.679 \end{bmatrix} A_{21} \qquad \text{for} \quad \omega_2 = 31.05 \text{ rad/s}$$

$$\begin{bmatrix} A_{32} \\ A_{33} \end{bmatrix} = \begin{bmatrix} -2.54 \\ 2.438 \end{bmatrix} A_{31} \qquad \text{for} \quad \omega_3 = 46.1 \text{ rad/s}$$

Since the mode shapes are determined to within an arbitrary constant, we may assume $A_{i1} = 1$, for $i = 1, 2, 3$. This leads to the following mode shapes:

$$\mathbf{A}_1 = \begin{bmatrix} A_{11} \\ A_{12} \\ A_{13} \end{bmatrix} = \begin{bmatrix} 1 \\ 0.649 \\ 0.302 \end{bmatrix}$$

$$\mathbf{A}_2 = \begin{bmatrix} A_{21} \\ A_{22} \\ A_{23} \end{bmatrix} = \begin{bmatrix} 1 \\ -0.607 \\ -0.679 \end{bmatrix}$$

$$\mathbf{A}_3 = \begin{bmatrix} A_{31} \\ A_{32} \\ A_{33} \end{bmatrix} = \begin{bmatrix} 1 \\ -2.54 \\ 2.438 \end{bmatrix}$$

The modal matrix $\mathbf{\Phi}$ is then defined as

$$\mathbf{\Phi} = \begin{bmatrix} 1 & 1 & 1 \\ 0.649 & -0.607 & -2.54 \\ 0.302 & -0.679 & 2.438 \end{bmatrix}$$

Figure 19 shows the modes of vibration of the system.

The modal matrix is orthogonal with respect to the mass and stiffness matrices. One can show the following:

$$\mathbf{M}_p = \mathbf{\Phi}^{\mathrm{T}}\mathbf{M}\mathbf{\Phi} = 2 \times 10^3 \begin{bmatrix} 1.814 & 0 & 0 \\ 0 & 2.475 & 0 \\ 0 & 0 & 22.573 \end{bmatrix} = \begin{bmatrix} m_1 & 0 & 0 \\ 0 & m_2 & 0 \\ 0 & 0 & m_3 \end{bmatrix}$$

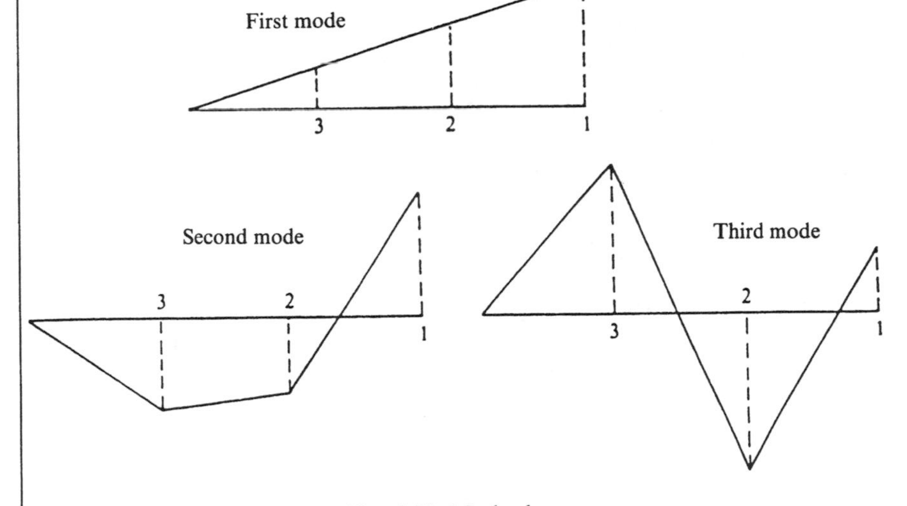

FIG. 6.19. Mode shapes.

$$\mathbf{K}_p = \mathbf{\Phi}^{\mathrm{T}}\mathbf{K}\mathbf{\Phi} = 2 \times 10^5 \begin{bmatrix} 3.828 & 0 & 0 \\ 0 & 23.86 & 0 \\ 0 & 0 & 497.7 \end{bmatrix}$$

$$= \begin{bmatrix} k_1 & 0 & 0 \\ 0 & k_2 & 0 \\ 0 & 0 & k_3 \end{bmatrix} = \begin{bmatrix} \omega_1^2 m_1 & 0 & 0 \\ 0 & \omega_2^2 m_2 & 0 \\ 0 & 0 & \omega_3^2 m_3 \end{bmatrix}$$

where m_i and k_i are, respectively, the modal mass and stiffness coefficients.

Using the modal mass and stiffness coefficients, an equation similar to Eq. 147 can be defined and solved for the uncoupled modal coordinates q_1, q_2, and q_3. The vector of torsional displacements $\boldsymbol{\theta}$ then can be determined using the modal transformation of Eq. 145 as $\boldsymbol{\theta} = \mathbf{\Phi}\mathbf{q}$.

For example, we consider the free vibration of the system as the result of the initial conditions

$$\boldsymbol{\theta}_0 = \begin{bmatrix} 0.1 \\ 0.05 \\ 0.01 \end{bmatrix} \text{rad} \quad \text{and} \quad \dot{\boldsymbol{\theta}}_0 = \begin{bmatrix} 10 \\ 15 \\ 20 \end{bmatrix} \text{rad/s}$$

In order to evaluate the initial modal coordinates and velocities, we first evaluate the inverse of the modal matrix $\mathbf{\Phi}^{-1}$ as

$$\mathbf{\Phi}^{-1} = \begin{bmatrix} 0.551 & 0.537 & 0.333 \\ 0.404 & -0.368 & -0.549 \\ 0.044 & -0.169 & 0.216 \end{bmatrix}$$

The initial modal coordinates are given by

$$\mathbf{q}_0 = \mathbf{\Phi}^{-1}\boldsymbol{\theta}_0 = \begin{bmatrix} 0.551 & 0.537 & 0.333 \\ 0.404 & -0.368 & -0.549 \\ 0.044 & -0.169 & 0.216 \end{bmatrix} \begin{bmatrix} 0.1 \\ 0.05 \\ 0.01 \end{bmatrix} = \begin{bmatrix} 0.085 \\ 0.017 \\ -0.002 \end{bmatrix} \text{rad}$$

and the initial modal velocities are

$$\dot{\mathbf{q}}_0 = \mathbf{\Phi}^{-1}\dot{\boldsymbol{\theta}}_0 = \begin{bmatrix} 0.551 & 0.537 & 0.333 \\ 0.404 & -0.368 & -0.549 \\ 0.044 & -0.169 & 0.216 \end{bmatrix} \begin{bmatrix} 10 \\ 15 \\ 20 \end{bmatrix} = \begin{bmatrix} 20.225 \\ -12.460 \\ 2.228 \end{bmatrix} \text{rad/s}$$

Using the initial modal coordinates and velocities, the modal coordinates can be defined as the solution of Eq. 147 as

$$q_i = q_{io} \cos \omega_i t + \frac{\dot{q}_{io}}{\omega_i} \sin \omega_i t, \qquad i = 1, 2, 3$$

which yields

$$\mathbf{q} = \begin{bmatrix} q_1 \\ q_2 \\ q_3 \end{bmatrix} = \begin{bmatrix} 0.085 \cos 14.52t + 1.393 \sin 14.52t \\ 0.017 \cos 31.05t - 0.401 \sin 31.05t \\ -0.002 \cos 46.1t + 0.048 \sin 46.1t \end{bmatrix}$$

The physical coordinates $\boldsymbol{\theta}$ can then be obtained using the relationship

$$\boldsymbol{\theta} = \mathbf{\Phi}\mathbf{q}$$

Problems

6.1. Determine the differential equation of motion of the two degree of freedom system shown in Fig. P1. Obtain the characteristic equation and determine the natural frequencies of the system in the special case of equal masses and spring constants.

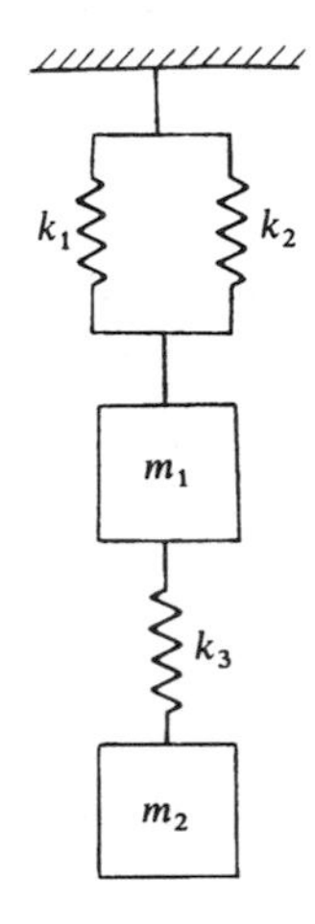

FIG. P6.1

6.2. Determine the differential equations of motion for the double pendulum shown in Fig. P2 in terms of the coordinates θ_1 and θ_2. Identify the system mass and stiffness matrices.

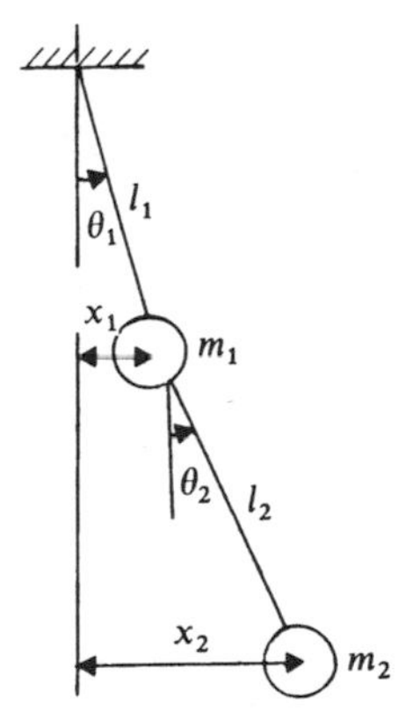

FIG. P6.2

6.3. Determine the differential equations of motion of the double pendulum shown in Fig. P2 in terms of the coordinates x_1 and x_2. Identify the system mass and stiffness matrices.

6.4. In the previous two problems, if $m_1 = m_2$ and $l_1 = l_2$, obtain the characteristic equation and determine the natural frequencies of the system and the amplitude ratios.

6.5. In Problem 1, if $m_1 = m_2 = 10$ kg and $k_1 = k_2 = k_3 = 1000$ N/m, determine the system response to the initial conditions

$$x_{10} = 0.02 \text{ m}, \quad x_{20} = 0, \quad \dot{x}_{10} = 0, \quad \dot{x}_{20} = 0$$

6.6. In Problem 2, if m_1 and $m_2 = 0.5$ kg and $l_1 = l_2 = 0.5$ m, determine θ_1 and θ_2 as function of time provided that the initial conditions are

$$\theta_{10} = 0, \quad \theta_{20} = 3^\circ, \quad \dot{\theta}_{10} = \dot{\theta}_{20} = 0$$

6.7. Write down the differential equations of motion of the two degree of freedom system shown in Fig. P3. Identify the system mass and stiffness matrices. Obtain the characteristic equation and determine the system natural frequencies and amplitude ratios in the following special case

$$m_1 = m_2, \quad a = b$$

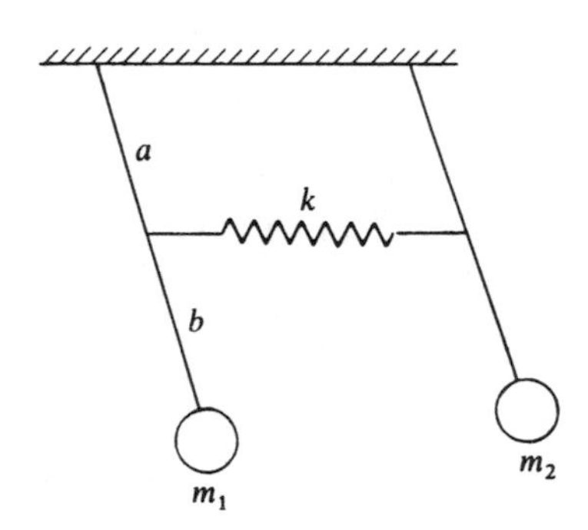

FIG. P6.3

6.8. In Problem 7, if $m_1 = m_2 = 0.5$ kg, $a = b = 0.25$ m, and $k = 1000$ N/m, determine the response of the system to the initial conditions

$$\theta_{10} = \theta_{20} = \dot{\theta}_{10} = 0, \quad \dot{\theta}_{20} = 3 \text{ rad/s}$$

6.9. Determine the differential equations of motion of the two degree of freedom system shown in Fig. P4. Identify the system mass and stiffness matrices. Obtain the characteristic equation and determine the system natural frequencies ω_1 and ω_2.

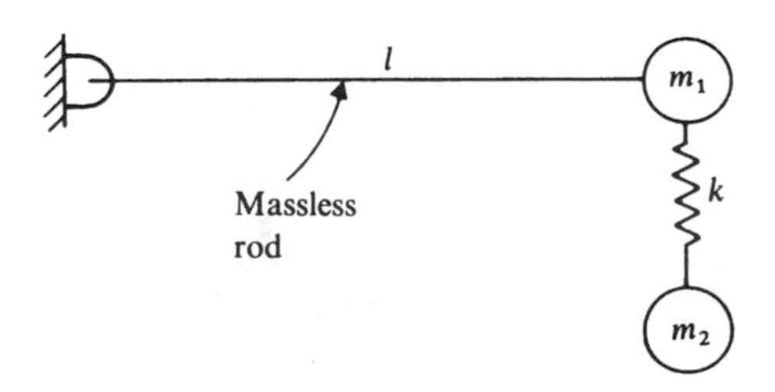

FIG. P6.4

6.10. In Problem 9, if $m_1 = m_2 = 0.5$ kg, $l = 0.5$ m, and $k = 1000$ N/m, determine the system response to an initial displacement of 0.02 m to the mass m_2.

6.11. Determine the differential equations of motion of the two degree of freedom system shown in Fig. P5. Identify the system mass and stiffness matrices. Obtain the characteristic equation and determine the natural frequencies of the system.

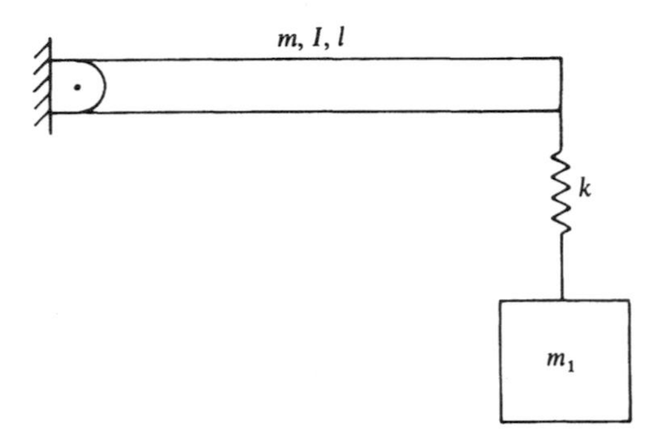

FIG. P6.5

6.12. Determine the differential equations of motion of the two degree of freedom system shown in Fig. P6. Identify the system mass and stiffness matrices. Obtain the characteristic equation and determine the natural frequencies ω_1 and ω_2. Assume small oscillations for the two rigid bodies.

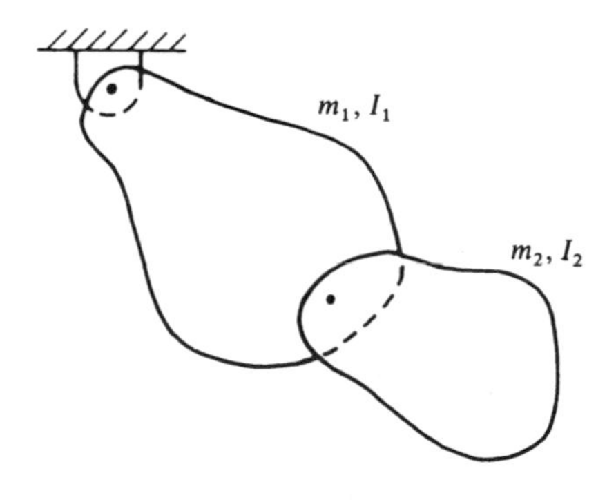

FIG. P6.6

6.13. Determine the differential equations of motion of the two degree of freedom system shown in Fig. P7. Identify the system mass and stiffness matrices.

6.14. Derive the differential equations of motion of the two degree of freedom system shown in Fig. P8. Assume small oscillations.

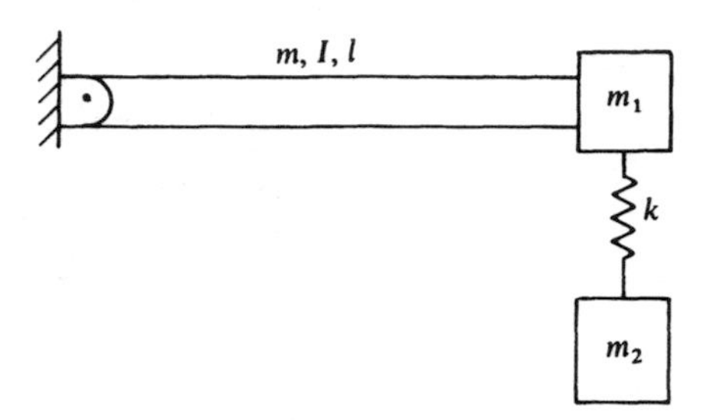

FIG. P6.7

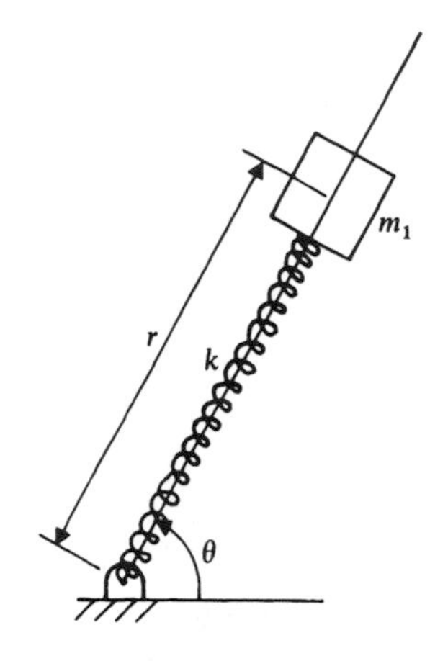

FIG. P6.8

6.15. Determine the differential equations of motion of the two degree of freedom system shown in Fig. P9. Identify the system mass, stiffness, and damping matrices.

6.16. In Problem 15, if $m_1 = m_2 = 5$ kg, $k_1 = k_2 = 1000$ N/m, and $c_1 = c_2 = c_3 = 10$ N·s/m, determine the system response as a function of time due to an initial displacement of m_2 equal to 0.01 m.

6.17. Determine the differential equations of motion of the system shown in Fig. P10. Determine the system natural frequencies.

6.18. Derive the differential equations of motion of the two degree of freedom system shown in Fig. P11. Identify the system mass, stiffness, and damping matrices.

6.19. Derive the differential equations of motion of the two degree of freedom system shown in Fig. P12.

6.20. Derive the system differential equations of motion of the two degree of freedom system shown in Fig. P13. Identify the system mass, stiffness, and damping matrices.

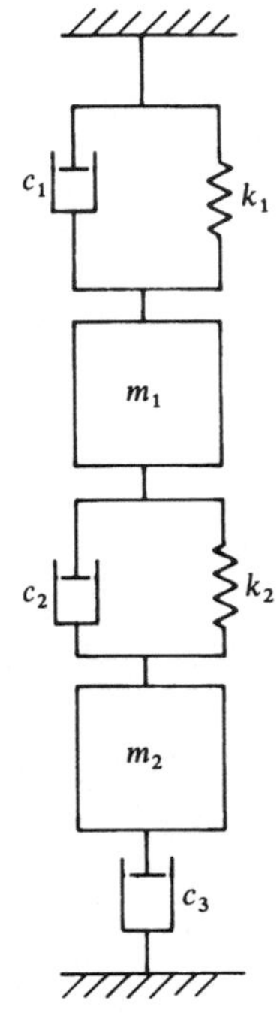

FIG. P6.9

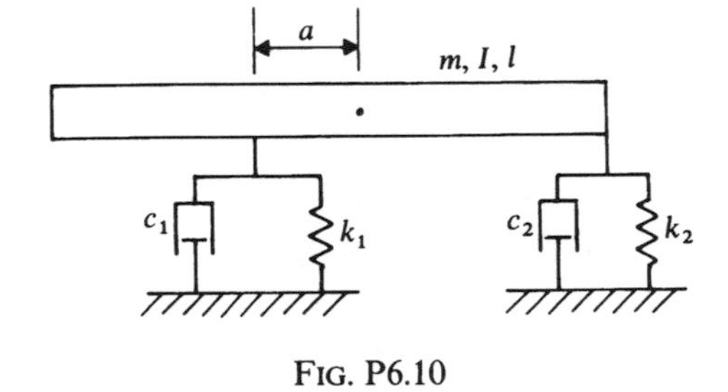

FIG. P6.10

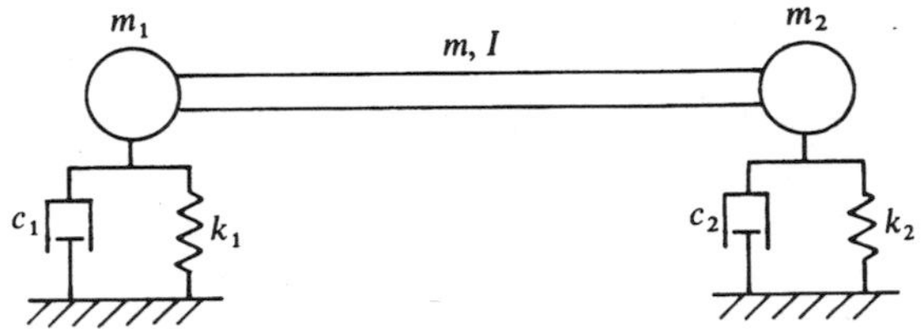

m_1 m, I m_2

c_1 k_1 c_2 k_2

FIG. P6.11

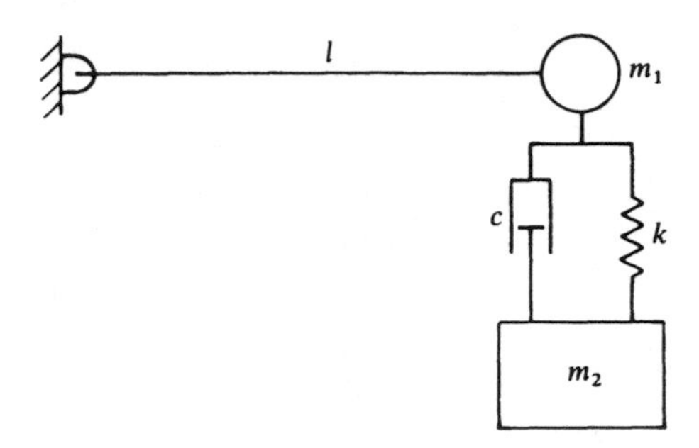

FIG. P6.12

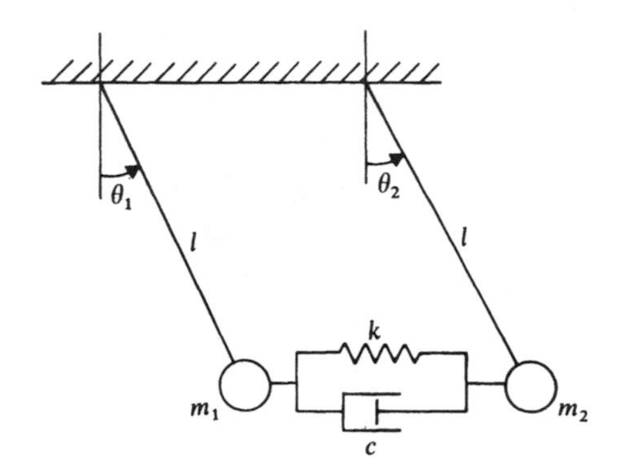

FIG. P6.13

6.21. In Problem 20, if $m_1 = m_2 = 0.5$ kg, $l = 0.5$ m, $k = 1000$ N/m, and $c = 10$ N · s/m, determine the system response as a function of time to an initial angular displacement $\theta_1 = 3°$.

6.22. Derive the dfferential equations of motion of the two degree of freedom system shown in Fig. P14.

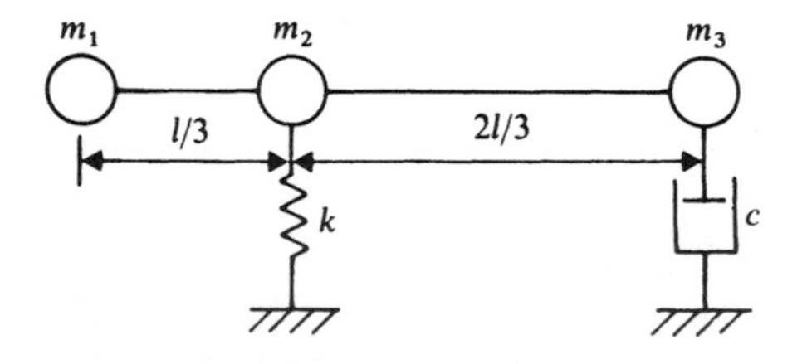

FIG. P6.14

6.23. In Fig. P12, let $m_1 = m_2 = 10$ kg, $l = 1$ m, $k = 1000$ N/m, and $c = 10$ N · s/m, determine the response of the system as a function of time due to an initial rotation of the rod equal to 2°.

6.24. Determine the forced response of the two degree of freedom system shown in Fig. P15 to the harmonic forcing function $F(t)$.

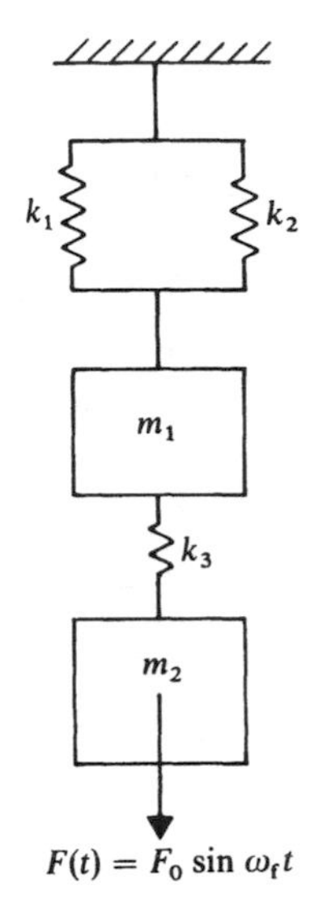

FIG. P6.15

6.25. Determine the forced response of the two degree of freedom system shown in Fig. P16 to the base excitation $y = Y_0 \sin \omega_f t$.

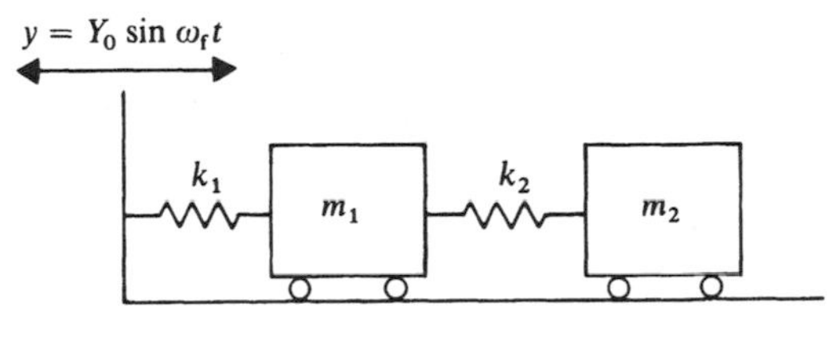

FIG. P6.16

6.26. Assuming small oscillations, derive the differential equations of motion of the two degree of freedom system shown in Fig. P17. Determine the response of the system, as a function of time, to the harmonic forcing function $T(t)$.

6.27. Assuming small oscillations, derive the differential equations of motion of the two degree of freedom system shown in Fig. P18. Determine the system response to the harmonic forcing function $F(t)$.

6.28. Derive the differential equations of motion of the system shown in Fig. P19, and determine the system response as a function of time.

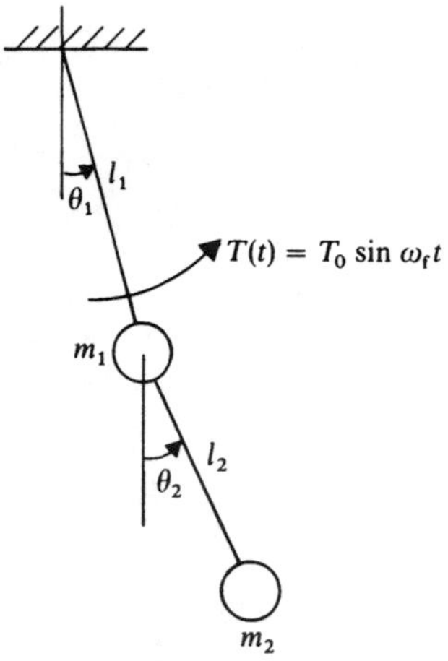

l_1
θ_1
$T(t) = T_0 \sin \omega_f t$
m_1
l_2
θ_2
m_2

FIG. P6.17

l_1
θ_1
m_1
l_2
θ_2
m_2
$F(t) = F_0 \sin \omega_f t$

FIG. P6.18

l
m_1
k
m_2
$F(t) =$
$F_0 \sin \omega_f t$

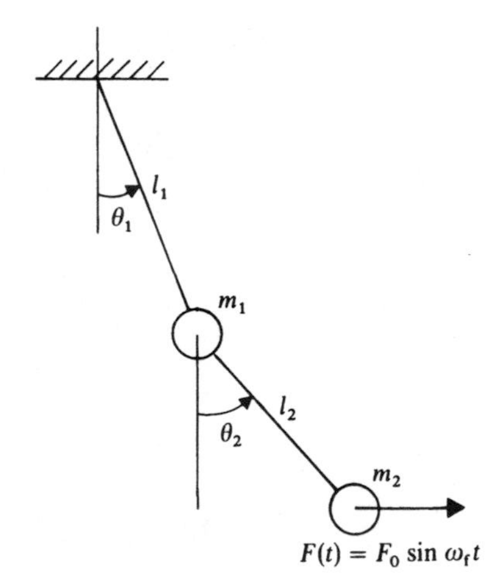

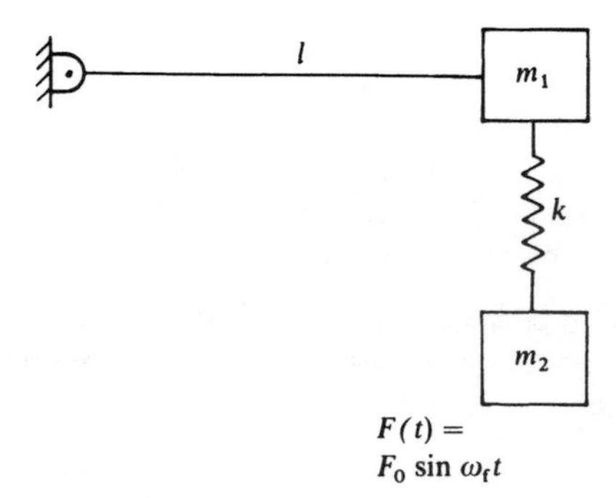

FIG. P6.19

6.29. Derive the differential equations of motion of the system shown in Fig. P20, and obtain the response of the system to the forcing function $T(t)$.

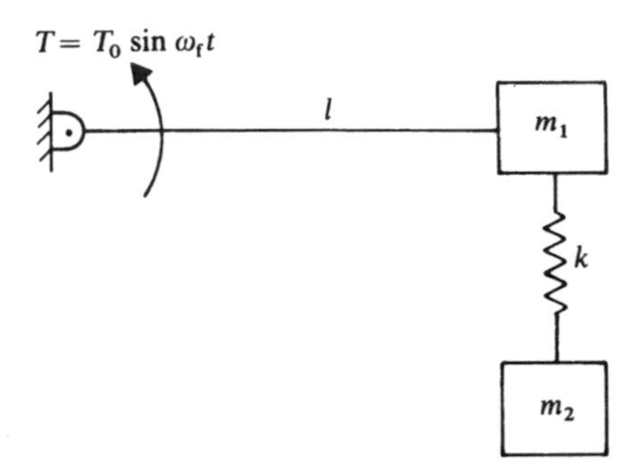

FIG. P6.20

6.30. Derive the differetial equations of motion of the two degree of freedom system shown in Fig. P21, and obtain the steady state solution as a function of time.

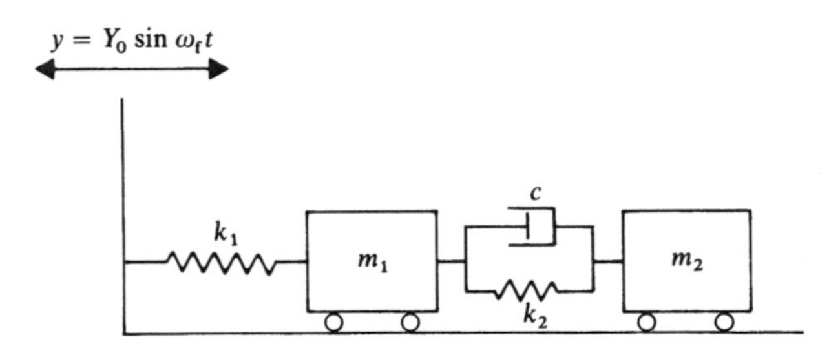

FIG. P6.21

6.31. Derive the differential equations of motion of the system shown in Fig. P22, and obtain the steady state solution as a function of time.

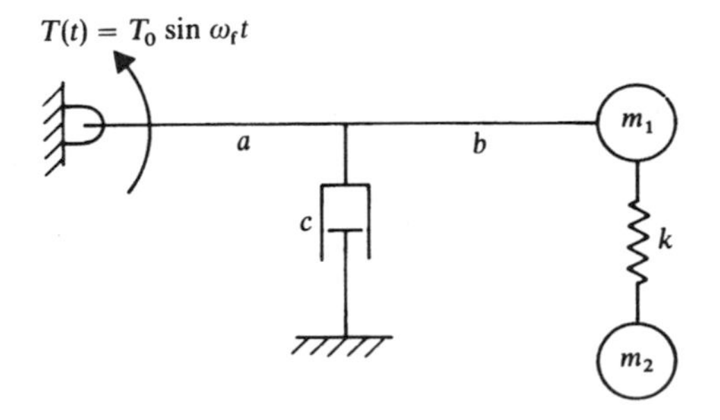

FIG. P6.22

6.32. Determine the steady state solution of the two degree of freedom system shown in Fig. P23.

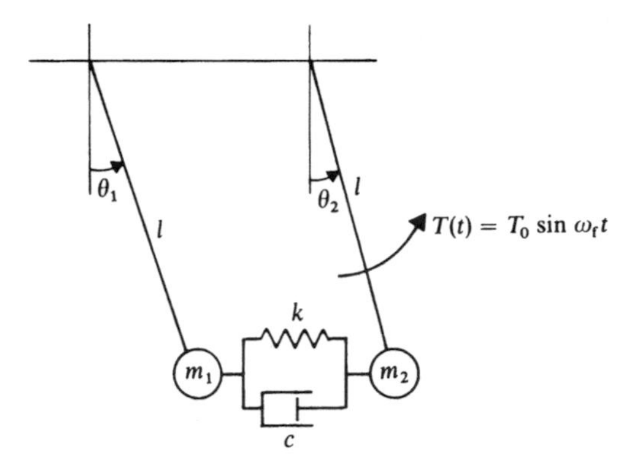

FIG. P6.23

6.33. Determine the steady state solution of the two degree of freedom system shown in Fig. P24.

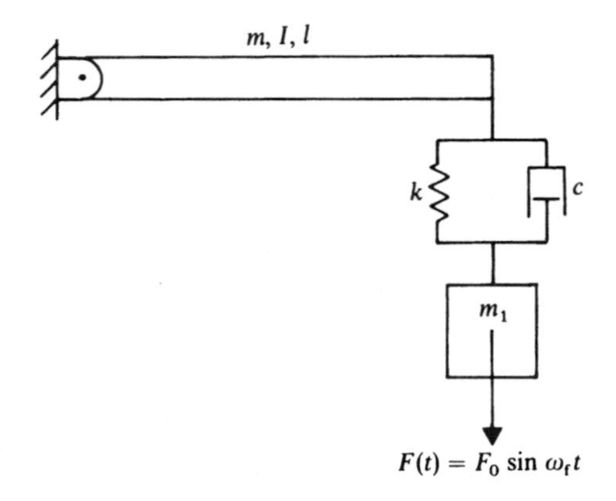

FIG. P6.24

6.34. Study the effect of the Houdaille (viscous Lanchester) damper on the vibration of a rotating system. Find the steady state solution of Eq. 144 and study the effect of the viscous damping coefficient c on the system response.

6.35. In Problem 34 determine the optimum damping factor for which the peak amplitude is minimum.

7
Continuous Systems

Mechanical systems in general consist of structural components which have distributed mass and elasticity. Examples of these structural components are rods, beams, plates, and shells. Our study of vibration thus far has been limited to discrete systems which have a finite number of degrees of freedom. As has been shown in the preceding chapters, the vibration of mechanical systems with lumped masses and discrete elastic elements is governed by a set of second-order ordinary differential equations. Rods, beams, and other structural components on the other hand are considered as continuous systems which have an infinite number of degrees of freedom, and as a consequence, the vibration of such systems is governed by partial differential equations which involve variables that depend on time as well as the spatial coordinates.

In this chapter, an introduction to the theory of vibration of continuous systems is presented. It is shown in the first two sections that the longitudinal and torsional vibration of rods can be described by second-order partial differential equations whose exact solutions are obtained using the method of separation of variables. In Section 3, the transverse vibrations of beams are examined and the fourth-order partial differential equation of motion that governs the transverse vibration is developed using the assumptions of the elementary beam theory. Solutions of the vibration equations are obtained for different boundary conditions. In Section 4, the orthogonality of the *eigenfunctions* (*mode shapes* or *principal modes*) is discussed and modal parameters such as mass and stiffness coefficients are introduced. The material covered in this section is used to study the forced vibration of continuous systems in Sections 5 and 6, where the solution of the vibration equations is expressed in terms of the principal modes of vibration.

7.1 FREE LONGITUDINAL VIBRATIONS

In this section, we study the longitudinal vibration of prismatic rods such as the one shown in Fig. 1. The rod, which has length l and cross-sectional area A, is assumed to be made of material which has a modulus of elasticity E

and mass density ρ and is subjected to a distributed external force $F(x, t)$ per unit length. Figure 1 also shows the forces that act on an infinitesimal volume of length δx. The geometric center of this infinitesimal volume is located at a distance $x + \delta x/2$ from the end of the rod. Let P be the axial force that results from the vibration of the rod. The application of Newton's second law leads to the following condition for the dynamic equilibrium of the infinitesimal volume:

$$\rho A \frac{\partial^2 u}{\partial t^2} \delta x = P + \frac{\partial P}{\partial x} \delta x - P + F(x, t)\delta x$$

which can be simplified to yield

$$\rho A \frac{\partial^2 u}{\partial t^2} \delta x = \frac{\partial P}{\partial x} \delta x + F(x, t)\delta x$$

Dividing this equation by δx leads to

$$\rho A \frac{\partial^2 u}{\partial t^2} = \frac{\partial P}{\partial x} + F(x, t) \tag{7.1}$$

The force P can be expressed in terms of the axial stress σ as

$$P = A\sigma \tag{7.2}$$

The stress σ can be written in terms of the axial strain ε using Hooke's law as

$$\sigma = E\varepsilon, \tag{7.3}$$

while the strain displacement relationship is

$$\varepsilon = \frac{\partial u}{\partial x} \tag{7.4}$$

Substituting Eqs. 3 and 4 into Eq. 2, the axial force can be expressed in terms of the longitudinal displacement as

$$P = EA \frac{\partial u}{\partial x} \tag{7.5}$$

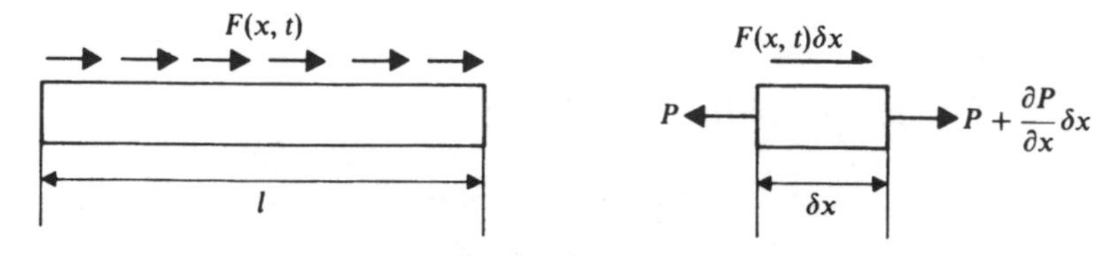

FIG. 7.1. Longitudinal vibration of rods.

Substituting Eq. 5 into Eq. 1 leads to

$$\rho A \frac{\partial^2 u}{\partial t^2} = \frac{\partial}{\partial x}\left(EA\frac{\partial u}{\partial x}\right) + F(x, t) \tag{7.6}$$

This is a partial differential equation that governs the forced longitudinal vibration of the rod.

Free Vibration The equation of free vibration can be obtained from Eq. 6 by letting $F(x, t) = 0$, that is,

$$\rho A \frac{\partial^2 u}{\partial t^2} = \frac{\partial}{\partial x}\left(EA\frac{\partial u}{\partial x}\right) \tag{7.7}$$

If the modulus of elasticity E and the cross-sectional area A are assumed to be constant, the partial differential equation of the longitudinal free vibration of the rod can be written as

$$\rho A \frac{\partial^2 u}{\partial t^2} = EA\frac{\partial^2 u}{\partial x^2} \tag{7.8}$$

or

$$\frac{\partial^2 u}{\partial t^2} = c^2 \frac{\partial^2 u}{\partial x^2} \tag{7.9}$$

where c is a constant defined by

$$c = \sqrt{\frac{E}{\rho}} \tag{7.10}$$

Separation of Variables The general solution of Eq. 9 can be obtained using the method of the *separation of variables*. In this case, we assume the solution in the form

$$u(x, t) = \phi(x)q(t) \tag{7.11}$$

where ϕ is a space-dependent function and q is a time-dependent function. The partial differentiation of Eq. 11 with respect to time and with respect to the spatial coordinate leads to

$$\frac{\partial^2 u}{\partial t^2} = \phi(x)\ddot{q}(t) \tag{7.12}$$

$$\frac{\partial^2 u}{\partial x^2} = \phi''(x)q(t) \tag{7.13}$$

where ($\dot{}$) denotes differentiation with respect to time and ($'$) denotes differentiation with respect to the spatial coordinate x, that is,

$$\ddot{q}(t) = \frac{d^2 q}{dt^2} \tag{7.14}$$

$$\phi''(x) = \frac{d^2\phi}{dx^2} \tag{7.15}$$

Substituting Eqs. 12 and 13 into Eq. 9 leads to

$$\phi\ddot{q} = c^2\phi''q \tag{7.16}$$

or

$$c^2\frac{\phi''}{\phi} = \frac{\ddot{q}}{q} \tag{7.17}$$

Since the left-hand side of this equation depends only on the spatial coordinate x and the right-hand side depends only on time, one concludes that Eq. 17 is satisfied only if both sides are equal to a constant, that is

$$c^2\frac{\phi''}{\phi} = \frac{\ddot{q}}{q} = -\omega^2 \tag{7.18}$$

where ω is a constant. Note that a negative constant $-\omega^2$ was selected, since this choice leads to oscillatory motion. The choice of zero or a positive constant does not lead to vibratory motion and, therefore, it must be rejected. For example, one can show that if the constant is selected to be zero the solution increases linearly with time, while if a positive constant is selected, the solution contains two terms; one exponentially increasing function and the second is an exponentially decreasing function. This leads to an unstable solution which does not represent an oscillatory motion. Therefore, in order to obtain an acceptable solution that describes the undamped vibration of the system, both sides of Eq. 17 must be equal to a negative constant.

Equation 18 leads to the following two equations:

$$\phi'' + \left(\frac{\omega}{c}\right)^2\phi = 0 \tag{7.19}$$

$$\ddot{q} + \omega^2 q = 0 \tag{7.20}$$

The solution of these two equations is given by

$$\phi(x) = A_1 \sin\frac{\omega}{c}x + A_2 \cos\frac{\omega}{c}x \tag{7.21}$$

$$q(t) = B_1 \sin\omega t + B_2 \cos\omega t \tag{7.22}$$

By using Eq. 11, the longitudinal displacement $u(x, t)$ can then be written as

$$u(x, t) = \phi(x)q(t) = \left(A_1 \sin\frac{\omega}{c}x + A_2 \cos\frac{\omega}{c}x\right)(B_1 \sin\omega t + B_2 \cos\omega t) \tag{7.23}$$

where A_1, A_2, B_1, B_2, and ω are arbitrary constants to be determined by using the boundary and initial conditions.

Boundary Conditions and the Orthogonality of the Eigenfunctions In order to demonstrate the procedure for determining the constants in Eq. 23, we consider the example shown in Fig. 2 where the rod is fixed at one end and is free at the other end. The boundary condition at the fixed end is given by

$$u(0, t) = 0 \tag{7.24}$$

while at the free end, the stress must be equal to zero, that is,

$$\sigma(l, t) = E\varepsilon(l, t) = E\frac{\partial u(l, t)}{\partial x} = 0$$

which defines the boundary condition at the free end as

$$\frac{\partial u(l, t)}{\partial x} = u'(l, t) = 0 \tag{7.25}$$

The type of boundary conditions of Eq. 24 that describes the state of displacement at the fixed end is called *geometric boundary condition.* On the other hand, Eq. 25 describes the state of force or stress at the free end of the rod, and this type of conditions is often refered to as *natural boundary condition.* That is, the geometric boundary conditions describe the specified displacements and slopes, while the natural boundary conditions describe the specified forces and moments. Substituting Eqs. 24 and 25 into Eq. 23 results in

$$u(0, t) = \phi(0)q(t) = A_2 q(t) = 0$$

$$u'(l, t) = \phi'(l)q(t) = \frac{\omega}{c}\left(A_1 \cos\frac{\omega}{c}l - A_2 \sin\frac{\omega}{c}l\right)q(t) = 0$$

which lead to the following two conditions:

$$A_2 = 0 \tag{7.26}$$

$$A_1 \cos\frac{\omega l}{c} = 0 \tag{7.27}$$

For a nontrivial solution, Eqs. 26 and 27 lead, respectively, to

$$\phi(x) = A_1 \sin\frac{\omega}{c}x \tag{7.28}$$

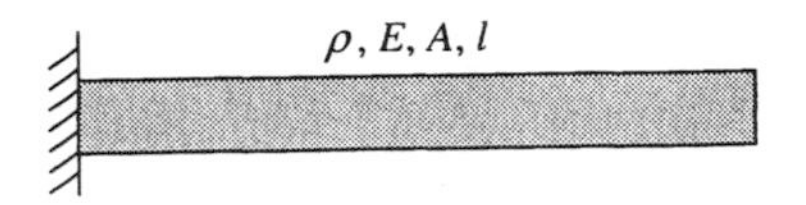

FIG. 7.2. Fixed-end conditions.

and

$$\cos\frac{\omega l}{c} = 0 \tag{7.29}$$

Equation 29, which is called the *frequency* or the *characteristic equation*, has roots which are given by

$$\frac{\omega l}{c} = \frac{\pi}{2}, \frac{3\pi}{2}, \frac{5\pi}{2}, \ldots, \frac{(2n-1)\pi}{2}, \ldots \tag{7.30}$$

This equation defines the *natural frequencies* of the rod as

$$\omega_j = \frac{(2j-1)\pi c}{2l}, \qquad j = 1, 2, 3, \ldots \tag{7.31}$$

Using the definition of the wave velocity c given by Eq. 10, the jth natural frequency ω_j can be defined as

$$\omega_j = \frac{(2j-1)\pi}{2l}\sqrt{\frac{E}{\rho}}, \qquad j = 1, 2, 3, \ldots \tag{7.32}$$

Thus, the continuous rod has an infinite number of natural frequencies. Corresponding to each of these natural frequencies, there is a *mode shape* or an *eigenfunction* ϕ_j defined by Eq. 28 as

$$\phi_j = A_{1j}\sin\frac{\omega_j}{c}x, \qquad j = 1, 2, 3, \ldots \tag{7.33}$$

where $A_{1j}, j = 1, 2, 3, \ldots$, are arbitrary constants. The eigenfunctions satisfy the following *orthogonality condition*:

$$\int_0^l \phi_i\phi_j\,dx = \begin{cases} 0 & \text{if } i \neq j \\ h_i & \text{if } i = j \end{cases} \tag{7.34}$$

where h_i is a positive constant. The longitudinal displacement $u(x, t)$ of the rod can then be expressed as

$$\begin{aligned} u(x,t) &= \sum_{j=1}^{\infty} \phi_j(x)q_j(t) \\ &= \sum_{j=1}^{\infty} (C_j\sin\omega_j t + D_j\cos\omega_j t)\sin\frac{\omega_j}{c}x \end{aligned} \tag{7.35}$$

where C_j and D_j are constants to be determined by using the initial conditions.

Initial Conditions Let us assume that the rod is subjected to the following initial conditions:

$$u(x, 0) = f(x) \tag{7.36}$$

$$\dot{u}(x, 0) = g(x) \tag{7.37}$$

Substituting these initial conditions into Eq. 35 leads to

$$u(x,0) = f(x) = \sum_{j=1}^{\infty} D_j\sin\frac{\omega_j}{c}x \tag{7.38}$$

$$\dot{u}(x, 0) = g(x) = \sum_{j=1}^{\infty} \omega_j C_j \sin \frac{\omega_j}{c} x \tag{7.39}$$

In order to determine the constants D_j in Eq. 38, we multiply this equation by $\sin(\omega_j/c)x$, integrate over the length of the rod, and use the orthogonality condition of Eq. 34, to obtain

$$D_j = \frac{2}{l} \int_0^l f(x) \sin \frac{\omega_j}{c} x \, dx, \qquad j = 1, 2, 3, \ldots \tag{7.40}$$

Similarly, in order to determine the constants C_j, we multiply Eq. 39 by $\sin(\omega_j x/c)$, integrate over the length of the rod, and use the orthogonality condition of Eq. 34. This leads to

$$C_j = \frac{2}{l\omega_j} \int_0^l g(x) \sin \frac{\omega_j}{c} x \, dx, \qquad j = 1, 2, 3, \ldots \tag{7.41}$$

Other Boundary Conditions From the analysis presented in this section, it is clear that the mode shapes and the natural frequencies of the rod depend on the boundary conditions. Even though we have considered only the case in which one end of the rod is fixed while the other end is free, by following a similar procedure to the one described in this section, the natural frequencies and mode shapes can be determined for other boundary conditions.

If the rod is assumed to be *free at both ends*, the boundary conditions are given by

$$\frac{\partial u(0, t)}{\partial x} = 0, \qquad \frac{\partial u(l, t)}{\partial x} = 0 \tag{7.42}$$

Using these boundary conditions and the separation of variables technique, one can show that the frequency equation is given by

$$\sin \frac{\omega l}{c} = 0 \tag{7.43}$$

which yields the natural frequencies of the longitudinal vibration of the rod with free ends as

$$\omega_j = \frac{j\pi c}{l} = \frac{j\pi}{l} \sqrt{\frac{E}{\rho}}, \qquad j = 1, 2, 3, \ldots \tag{7.44}$$

The associated eigenfunctions or mode shapes are given by

$$\phi_j = A_{2j} \cos \frac{\omega_j x}{c} \tag{7.45}$$

Another example is a rod with *both ends fixed*. The boundary conditions in this case are given by

$$u(0, t) = 0, \qquad u(l, t) = 0 \tag{7.46}$$

Using these boundary conditions and the separation of variables method, one can show that the frequency equation is given by

$$\sin\frac{\omega l}{c} = 0 \tag{7.47}$$

which yields the natural frequencies

$$\omega_j = \frac{j\pi c}{l} = \frac{j\pi}{l}\sqrt{\frac{E}{\rho}}, \qquad j = 1, 2, 3, \ldots \tag{7.48}$$

The mode shapes associated with these frequencies are

$$\phi_j = A_{1j}\sin\frac{j\pi x}{l}, \qquad j = 1, 2, 3, \ldots \tag{7.49}$$

Example 7.1

The system shown in Fig. 3 consists of a rigid mass m attached to a rod which has mass density ρ, length l, modulus of elasticity E, and cross-sectional area A. Determine the frequency equation and the mode shapes of the longitudinal vibration.

Solution. The end condition at the fixed end of the rod is given by

$$u(0, t) = 0 \tag{7.50}$$

The other end of the rod is attached to the mass m which exerts a force on the rod because of the inertia effect. From the free-body digram shown in the figure, the second boundary condition is given by

$$m\frac{\partial^2 u(l,t)}{\partial t^2} = -P(l, t) = -\sigma(l, t)A = -EA\frac{\partial u(l,t)}{\partial x}$$

which yields

$$m\frac{\partial^2 u(l,t)}{\partial t^2} = -EA\frac{\partial u(l,t)}{\partial x} \tag{7.51}$$

The longitudinal vibration of the rod is governed by the equation

$$u(x,t) = \phi(x)q(t)$$

$$= \left(A_1\sin\frac{\omega}{c}x + A_2\cos\frac{\omega}{c}x\right)(B_1\sin\omega t + B_2\cos\omega t) \tag{7.52}$$

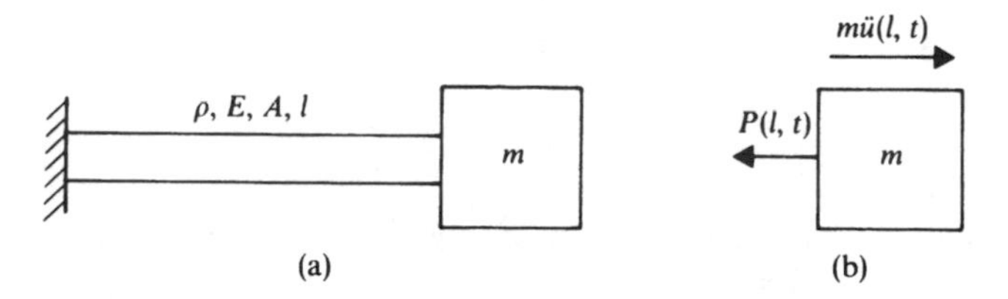

FIG. 7.3. Longitudinal vibration of a rod with a mass attached to its end.

Substituting Eq. 50 into Eq. 52 yields

$$A_2 = 0$$

It follows that

$$u(x, t) = \phi(x)q(t) = A_1 \sin\frac{\omega x}{c}(B_1 \sin \omega t + B_2 \cos \omega t)$$

which yields

$$\frac{\partial^2 u}{\partial t^2} = -\omega^2 A_1 \sin\frac{\omega x}{c}(B_1 \sin \omega t + B_2 \cos \omega t)$$

$$\frac{\partial u}{\partial x} = \frac{\omega}{c} A_1 \cos\frac{\omega x}{c}(B_1 \sin \omega t + B_2 \cos \omega t)$$

Therefore, the second boundary condition of Eq. 51 yields

$$-m\omega^2 A_1 \sin\frac{\omega l}{c}(B_1 \sin \omega t + B_2 \cos \omega t) = -EA\frac{\omega}{c} A_1 \cos\frac{\omega l}{c}(B_1 \sin \omega t + B_2 \cos \omega t)$$

This equation implies

$$\frac{m\omega c}{EA}\sin\frac{\omega l}{c} = \cos\frac{\omega l}{c}$$

That is,

$$\tan\frac{\omega l}{c} = \frac{EA}{m\omega c} \tag{7.53}$$

TABLE 7.1. Roots of Eq. 54b

Mode Number	γ				
	$1/\mu = 0$	$1/\mu = 0.01$	$1/\mu = 0.1$	$1/\mu = 1$	$1/\mu = 10$
1	1.5708	1.5552	1.4289	0.8603	0.3111
2	4.7124	4.6658	4.3058	3.4256	3.1731
3	7.8540	7.7764	7.2281	6.4372	6.2991
4	10.9956	10.8871	10.2003	9.5293	9.4354
5	14.1372	13.9981	13.2142	12.6453	12.5743
6	17.2788	17.1093	16.2594	15.7713	15.7143
7	20.4204	20.2208	19.3270	18.9024	18.8549
8	23.5619	23.3327	22.4108	22.0365	21.9957
9	26.7035	26.4451	25.5064	25.1725	25.1367
10	29.8451	29.5577	28.6106	28.3096	28.2779
11	32.9867	32.6710	31.7213	31.4477	31.4191
12	36.1283	35.7847	34.8371	34.5864	34.5604
13	39.2699	38.8989	37.9567	37.7256	37.7018
14	42.4115	42.0138	31.0795	40.8652	40.4832
15	45.5531	45.1292	44.2048	44.0050	43.9846
16	48.6947	48.2452	47.3321	47.1451	47.1260
17	51.8363	51.3618	50.4611	50.2854	50.2675
18	54.9779	54.4791	53.5916	53.4258	53.4089
19	58.1195	57.5969	56.7232	56.5663	56.5504
20	61.2611	60.7155	56.7232	59.7070	59.6919

This is the frequency equation which upon multiplying both of its sides by $\omega l/c$, one obtains

$$\frac{\omega l}{c}\tan\frac{\omega l}{c} = \frac{\omega l EA}{m\omega c^2} = \frac{lA\rho}{m} = \frac{M}{m}$$

where M is the mass of the rod. The preceding equation can be written as

$$\gamma \tan \gamma = \mu \tag{7.54a}$$

where

$$\gamma = \frac{\omega l}{c}, \qquad \mu = \frac{M}{m}$$

Note that the frequency equation is a transcendental equation which has an infinite number of roots, and therefore, its solution defines an infinite number of natural frequencies. This equation can be expressed for each root as

$$\gamma_j \tan \gamma_j = \mu, \qquad j = 1, 2, 3, \ldots \tag{7.54b}$$

where

$$\gamma_j = \frac{\omega_j l}{c}$$

and the eigenfunction associated with the natural frequency ω_j is

$$\phi_j = A_{1j} \sin\frac{\omega_j x}{c}$$

Table 1 shows the first twenty roots of Eq. 54b for different values of the mass ratio μ.

Example 7.2

The system shown in Fig. 4 consists of a prismatic rod which has one end fixed and the other end attached to a spring with stiffness k as shown in the figure. The rod has length l, cross-sectional area A, mass density ρ, and modulus of elasticity E. Obtain the frequency equation and the eigenfunctions of this system.

Solution. As in the preceding example, the boundary condition at the fixed end is given by

$$u(0, t) = 0$$

At the other end, the axial force of the rod is equal in magnitude and opposite in direction to the spring force, that is,

$$P(l, t) = -ku(l, t)$$

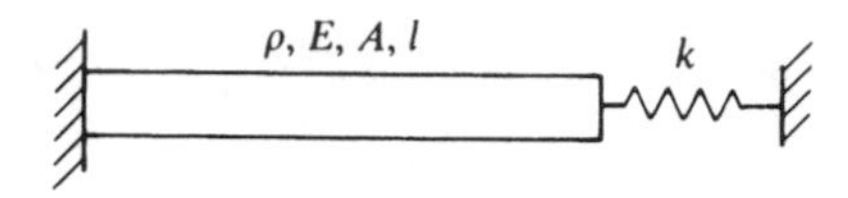

FIG. 7.4. Longitudinal vibration of a rod with one end attached to a spring.

Since $P = EAu'$, the preceding boundary condition is

$$EAu'(l, t) = -ku(l, t)$$

Using the technique of the separation of variables, the solution of the partial differential equation of the rod can be expressed as

$$\begin{aligned} u(x, t) &= \phi(x)q(t) \\ &= \left(A_1 \sin\frac{\omega}{c}x + A_2 \cos\frac{\omega}{c}x\right)(B_1 \sin \omega t + B_2 \cos \omega t) \end{aligned}$$

As in the preceding example, the boundary condition at the fixed end leads to

$$A_2 = 0$$

and the expression for the longitudinal displacement u may be simplified and written as

$$u = \phi(x)q(t) = A_1 \sin\frac{\omega}{c}x(B_1 \sin \omega t + B_2 \cos \omega t)$$

It follows that

$$u' = \frac{\omega}{c}A_1 \cos\frac{\omega}{c}x(B_1 \sin \omega t + B_2 \cos \omega t)$$

By using the second boundary condition, the following equation is obtained:

$$EA\frac{\omega}{c}\cos\frac{\omega}{c}l = -k \sin\frac{\omega}{c}l$$

or

$$\tan\frac{\omega}{c}l = -\frac{EA\omega}{kc}$$

Multiplying both sides of this equation by $\omega l/c$ and using the definition of $c = \sqrt{E/\rho}$, one obtains

$$\frac{\omega l}{c}\tan\frac{\omega l}{c} = -\frac{EA\omega^2 l}{kc^2} = -\frac{EA\omega^2 l\rho}{kE} = -\frac{\omega^2 M}{k}$$

where M is the mass of the rod. The above equation is the frequency equation which can be written as

$$\gamma \tan \gamma = -\frac{\omega^2 M}{k}$$

where

$$\gamma = \frac{\omega l}{c}$$

The roots of the frequency equation can be determined numerically and used to define the natural frequencies, ω_j, $j = 1, 2, 3, \ldots$. It is clear in this case that the eigenfunction associated with the jth natural frequency is

$$\phi_j = A_{1j} \sin\frac{\omega_j x}{c}$$

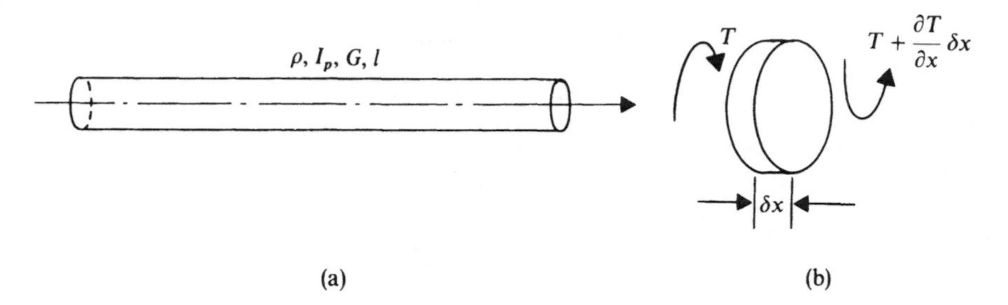

FIG. 7.5. Torsional vibration.

7.2 FREE TORSIONAL VIBRATIONS

In this section, we consider the torsional vibration of straight circular shafts as the one shown in Fig. 5(a). The shaft is assumed to have a length l and a cross section which has a polar moment of inertia I_p. The shaft is assumed to be made of material which has modulus of rigidity G and mass density ρ and is subjected to a distributed external torque defined per unit length by the function $T_e(x, t)$. In the analysis presented in this section, the cross sections of the shaft are assumed to remain in their planes and rotate as rigid surfaces about their centers. The equation of dynamic equilibrium of an infinitesimal volume of the shaft in torsional vibration is given by

$$\rho I_p \frac{\partial^2 \theta}{\partial t^2} \delta x = T + \frac{\partial T}{\partial x} \delta x - T + T_e \delta x \tag{7.55}$$

where $\theta = \theta(x, t)$ is the angle of torsional oscillation of the infinitesimal volume about the axis of the shaft, T is the internal torque acting on the cross section at a distance x from the end of the shaft, and δx is the length of the infinitesimal volume. Equation 55 can be simplified and written as

$$\rho I_p \frac{\partial^2 \theta}{\partial t^2} = \frac{\partial T}{\partial x} + T_e \tag{7.56}$$

From elementary strength of material theory, the internal torque T is proportional to the spatial derivative of the torsional oscillation θ. The relationship between the internal torque and the torsional oscillation is given in terms of the modulus of rigidity G as

$$T = G I_p \frac{\partial \theta}{\partial x} \tag{7.57}$$

Substituting Eq. 57 into Eq. 56, one obtains

$$\rho I_p \frac{\partial^2 \theta}{\partial t^2} = \frac{\partial}{\partial x}\left(G I_p \frac{\partial \theta}{\partial x}\right) + T_e \tag{7.58}$$

If the shaft is assumed to have a constant cross-sectional area and a constant modulus of rigidity, Eq. 58 can be expressed as

$$\rho I_{\mathrm{p}} \frac{\partial^2 \theta}{\partial t^2} = G I_{\mathrm{p}} \frac{\partial^2 \theta}{\partial x^2} + T_{\mathrm{e}} \tag{7.59}$$

Free Vibration The partial differential equation that governs the free torsional vibration of the shaft can be obtained from Eq. 59 if we let $T_{\mathrm{e}} = 0$, that is,

$$\rho I_{\mathrm{p}} \frac{\partial^2 \theta}{\partial t^2} = G I_{\mathrm{p}} \frac{\partial^2 \theta}{\partial x^2} \tag{7.60}$$

which can be rewritten as

$$\frac{\partial^2 \theta}{\partial t^2} = c^2 \frac{\partial^2 \theta}{\partial x^2} \tag{7.61}$$

where c is a constant that depends on the inertia properties and the material of the shaft and is given by

$$c = \sqrt{\frac{G}{\rho}} \tag{7.62}$$

Equation 61 is in the same form as Eq. 9 that governs the longitudinal vibration of prismatic rods, and therefore, the same solution procedure can be used to solve Eq. 61. Using the separation of variables technique, the torsional oscillation θ can be expressed as

$$\theta = \phi(x) q(t) \tag{7.63}$$

where $\phi(x)$ is a space-dependent function and $q(t)$ depends on time. Following the procedure described in the preceding section, one can show that

$$\phi(x) = A_1 \sin \frac{\omega x}{c} + A_2 \cos \frac{\omega x}{c} \tag{7.64}$$

$$q(t) = B_1 \sin \omega t + B_2 \cos \omega t \tag{7.65}$$

where A_1, A_2, B_1, B_2, and ω are arbitrary constants to be determined using the boundary and initial conditions. Substituting Eqs. 64 and 65 into Eq. 63 yields

$$\theta(x, t) = \left(A_1 \sin \frac{\omega x}{c} + A_2 \cos \frac{\omega x}{c} \right) (B_1 \sin \omega t + B_2 \cos \omega t) \tag{7.66}$$

Boundary Conditions Equation 66 is in the same form as Eq. 23 which describes the longitudinal vibration of prismatic rods. Therefore, one expects that the natural frequencies and mode shapes will also be in the same form. For example, in the case of a shaft with one end fixed and the other end free, the natural frequencies and eigenfunctions of torsional vibration are given by

$$\omega_j = \frac{(2j-1)\pi c}{2l} = \frac{(2j-1)\pi}{2l} \sqrt{\frac{G}{\rho}}, \qquad j = 1, 2, 3, \ldots \tag{7.67}$$

$$\phi_j = A_{1j} \sin \frac{\omega_j x}{c}, \qquad j = 1, 2, 3, \ldots \tag{7.68}$$

and the solution for the free torsional vibration is

$$\theta = \sum_{j=1}^{\infty} \sin \frac{\omega_j x}{c} (C_j \sin \omega_j t + D_j \cos \omega_j t) \tag{7.69}$$

where the constants C_j and D_j, $j = 1, 2, 3, \ldots$, can be determined by using the initial conditions as described in the preceding section.

In the case of a shaft with *free ends*, the natural frequencies and eigenfunctions are given by

$$\omega_j = \frac{j\pi c}{l} = \frac{j\pi}{l}\sqrt{\frac{G}{\rho}}, \qquad j = 1, 2, 3, \ldots \tag{7.70}$$

$$\phi_j = A_{2j} \cos \frac{\omega_j x}{c}, \qquad j = 1, 2, 3, \ldots \tag{7.71}$$

and the solution for the free torsional vibration of the shaft with free ends has the form

$$\theta = \sum_{j=1}^{\infty} \cos \frac{\omega_j x}{c} (C_j \sin \omega_j t + D_j \cos \omega_j t) \tag{7.72}$$

where C_j and D_j, $j = 1, 2, 3, \ldots$, are constants to be determined from the initial conditions.

Similarly one can show that the natural frequencies and mode shapes of the torsional vibration in the case of a shaft with *both ends fixed* are

$$\omega_j = \frac{j\pi c}{l} = \frac{j\pi}{l}\sqrt{\frac{G}{\rho}}, \qquad j = 1, 2, 3, \ldots \tag{7.73}$$

$$\phi_j = A_{1j} \sin \frac{\omega_j x}{c}, \qquad j = 1, 2, 3, \ldots \tag{7.74}$$

and the solution for the free torsional vibration is given by

$$\theta = \sum_{j=1}^{\infty} \sin \frac{\omega_j x}{c} (C_j \sin \omega_j t + D_j \cos \omega_j t) \tag{7.75}$$

As in the case of longitudinal vibrations discussed in the preceding section, the orthogonality of the mode shapes can be utilized in determining the arbitrary constants C_j and D_j using the initial conditions.

Example 7.3

The system shown in Fig. 6 consists of a disk with a mass moment of inertia I_d attached to a circular shaft which has mass density ρ, length l, cross-sectional polar moment of inertia I_p, and modulus of rigidity G. Obtain the frequency equation and the mode shapes of the torsional vibration.

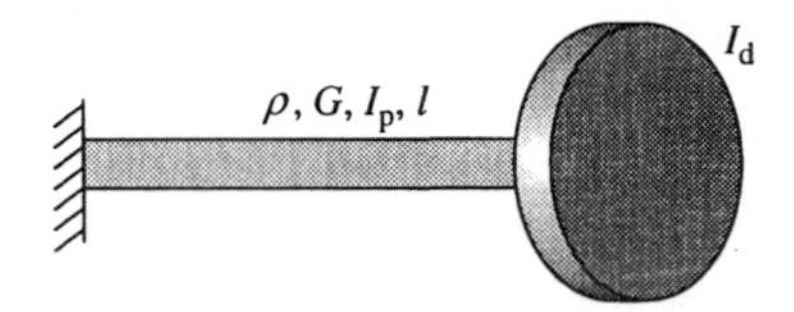

FIG. 7.6. Torsional vibration of a shaft with one end attached to a disk.

Solution. The two boundary conditions in this case are given by

$$\theta(0, t) = 0$$

$$T(l, t) = GI_p\theta'(l, t) = -I_d\ddot{\theta}(l, t)$$

The solution for the free torsional vibration can be assumed in the form

$$\theta(x, t) = \phi(x)q(t)$$

$$= \left(A_1 \sin\frac{\omega x}{c} + A_2 \cos\frac{\omega x}{c}\right)(B_1 \sin \omega t + B_2 \cos \omega t)$$

Substituting the two boundary conditions into this solution and following the procedure described in Example 1, it can be shown that the frequency equation is given by

$$\gamma \tan \gamma = \mu$$

where

$$\gamma = \frac{\omega l}{c}, \qquad \mu = \frac{\rho I_p l}{I_d}, \qquad c = \sqrt{\frac{G}{\rho}}$$

This equation has an infinite number of roots γ_j, $j = 1, 2, 3, \ldots$, which can be used to define the natural frequencies ω_j as

$$\omega_j = \frac{\gamma_j c}{l}, \qquad j = 1, 2, 3, \ldots$$

Example 7.4

The system shown in Fig. 7 consists of a straight cylindrical shaft which has one end fixed and the other end attached to a torsional spring with stiffness k_t as shown in the figure. The shaft has length l, mass density ρ, cross-sectional polar moment of inertia I_p, and modulus of rigidity G. Obtain the frequency equation and the eigenfunctions of this system.

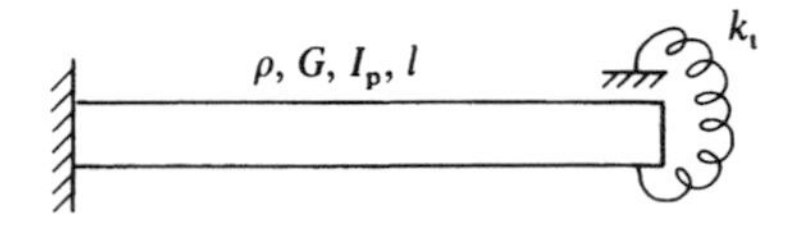

FIG. 7.7. Torsional vibration of a shaft with one end attached to a torsional spring.

Solution. The boundary conditions in this case are given by

$$\theta(0, t) = 0$$

$$T(l, t) = GI_p\theta'(l, t) = -k_t\theta(l, t)$$

The solution for the free torsional vibration is given by

$$\theta(x, t) = \phi(x)q(t)$$

$$= \left(A_1 \sin\frac{\omega x}{c} + A_2 \cos\frac{\omega x}{c}\right)(B_1 \sin \omega t + B_2 \cos \omega t)$$

Substituting the boundary conditions into this solution and following the procedure described in Example 2, it can be shown that the frequency equation is given by

$$\gamma \tan \gamma = -\mu$$

where

$$\gamma = \frac{\omega l}{c}, \qquad \mu = \frac{\omega^2 \rho l I_p}{k_t}, \qquad c = \sqrt{\frac{G}{\rho}}$$

The roots $\gamma_j, j = 1, 2, 3, \ldots$, of the frequency equation can be obtained numerically. These roots define the natural frequencies $\omega_j, j = 1, 2, 3, \ldots$, as

$$\omega_j = \frac{\gamma_j c}{l}, \qquad j = 1, 2, 3, \ldots$$

The associated mode shapes are

$$\phi_j = A_{1j} \sin\frac{\omega_j x}{c}$$

7.3 FREE TRANSVERSE VIBRATIONS

It was shown in the preceding two sections that the differential equations that govern the longitudinal and torsional vibration of rods have the same form. The theory of transverse vibration of beams is more difficult than that of the two types of vibration already considered in the preceding sections. There are several important applications of the theory of transverse vibration of beams; among these applications are the study of vibrations of rotating shafts and rotors and the transverse vibration of suspended cables. In this section, the equations that govern the transverse vibration of beams are developed and methods for obtaining their solutions are discussed.

Elementary Beam Theory In the elementary beam theory, all stresses are assumed to be equal to zero except the normal stress σ which is assumed to vary linearly over the cross section with the y-coordinate of the beam as shown in Fig. 8. The normal stress σ can be written as

$$\sigma = ky \tag{7.76}$$

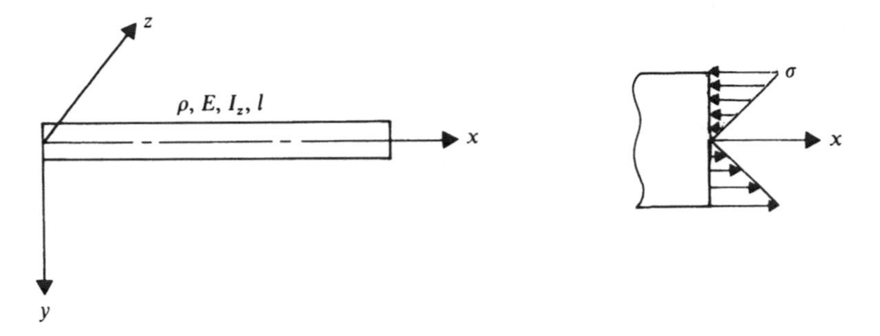

FIG. 7.8. Elementary beam theory.

where k is constant, and $y = 0$ contains the neutral surface along which the normal stress σ is equal to zero. The assumption that all other stresses are equal to zero requires that the resultant of the internal forces be zero and that the moments of the internal forces about the neutral axis equal the bending moment M. That is,

$$\int_A \sigma \, dA = 0, \qquad \int_A y\sigma \, dA = M \tag{7.77}$$

where A is the cross-sectional area of the beam. Substituting Eq. 76 into Eq. 77 yields

$$k \int_A y \, dA = 0 \tag{7.78}$$

$$k \int_A y^2 \, dA = M \tag{7.79}$$

Since k is a nonzero constant, the first equation implies that the neutral and centroidal axes of the cross section coincide. The second equation can be used to define k as

$$k = \frac{M}{I_z} \tag{7.80}$$

where I_z is the moment of inertia of the cross section about the z-axis of the beam cross section, that is,

$$I_z = \int_A y^2 \, dA$$

Substituting Eq. 80 into Eq. 76 yields

$$\sigma = \frac{My}{I_z} \tag{7.81}$$

Using Hooke's law, the strain ε is given by

$$\varepsilon = \frac{\sigma}{E} = \frac{My}{EI_z} \tag{7.82}$$

Let v denotes the transverse displacement of the beam. For small deformations, $dv/dx \ll 1$, and it can be shown that

$$\frac{1}{r} \approx \frac{d^2v}{dx^2} = -\frac{\varepsilon}{y} = -\frac{M}{EI_z} \tag{7.83}$$

where r is the radius of curvature of the beam. Equation 83 implies that

$$M = -EI_z v'' \tag{7.84}$$

This equation is known as the *Euler–Bernoulli law* of the elementary beam theory.

Partial Differential Equation In order to determine the differential equation for the transverse vibration of beams, we consider an infinitesimal volume at a distance x from the end of the beam as shown in Fig. 9. The length of this infinitesimal volume is assumed to be δx. Let V and M be, respectively, the shear force and bending moment, and let $F(x, t)$ be the loading per unit length of the beam. Neglecting the rotary inertia, the sum of the moments about the left end of the section yields

$$M + \frac{\partial M}{\partial x}\delta x - M - V\delta x - \frac{\partial V}{\partial x}(\delta x)^2 - F(x, t)\frac{(\delta x)^2}{2} - \rho A\frac{\partial^2 v}{\partial t^2}\frac{(\delta x)^2}{2} = 0 \tag{7.85a}$$

Taking the limit as δx approaches zero, the preceding equation leads to

$$V = \frac{\partial M}{\partial x} \tag{7.85b}$$

The dynamic equilibrium condition for the transverse vibration of the beam

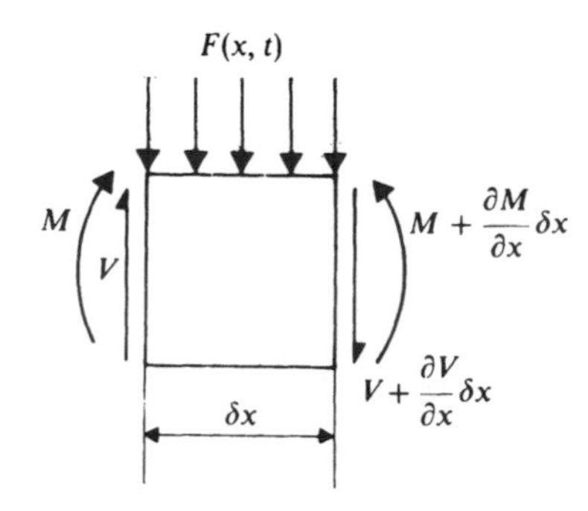

FIG. 7.9. Moments and shear forces.

is obtained by applying Newton's second law as

$$\rho A \delta x \frac{\partial^2 v}{\partial t^2} = V + \frac{\partial V}{\partial x}\delta x - V + F(x, t)\delta x \tag{7.86}$$

where ρ is the mass density and A is the cross-sectional area. Equation 86 can be rewritten after simplification as

$$\rho A \frac{\partial^2 v}{\partial t^2} = \frac{\partial V}{\partial x} + F(x, t) \tag{7.87}$$

Substituting Eq. 85b into Eq. 87 yields

$$\rho A \frac{\partial^2 v}{\partial t^2} = \frac{\partial^2 M}{\partial x^2} + F(x, t) \tag{7.88}$$

The moment M can be eliminated from this equation by using the moment displacement relationship. To this end, Eq. 84 is substituted into Eq. 88. This leads to

$$\rho A \frac{\partial^2 v}{\partial t^2} = -\frac{\partial^2}{\partial x^2}(EI_z v'') + F(x, t) \tag{7.89}$$

If E and I_z are assumed to be constant, Eq. 89 becomes

$$\rho A \frac{\partial^2 v}{\partial t^2} = -EI_z \frac{\partial^4 v}{\partial x^4} + F(x, t) \tag{7.90}$$

In the case of free vibration, $F(x, t) = 0$ and accordingly

$$\frac{\partial^2 v}{\partial t^2} = -c^2 \frac{\partial^4 v}{\partial x^4} \tag{7.91}$$

where c is a constant defined as

$$c = \sqrt{\frac{EI_z}{\rho A}} \tag{7.92}$$

Separation of Variables Equation 91 is a fourth-order partial differential equation that governs the free transverse vibration of the beam. The solution of this equation can be obtained by using the technique of the separation of variables. In this case, we assume a solution in the form

$$v = \phi(x)q(t) \tag{7.93}$$

where $\phi(x)$ is a space-dependent function, and $q(t)$ is a function that depends only on time. Equation 93 leads to

$$\frac{\partial^2 v}{\partial t^2} = \phi(x)\frac{d^2 q(t)}{dt^2} = \phi(x)\ddot{q}(t) \tag{7.94}$$

$$\frac{\partial^4 v}{\partial x^4} = \frac{d^4\phi(x)}{dx^4}q(t) = \phi^{iv}(x)q(t) \tag{7.95}$$

Substituting these equations into Eq. 91, one obtains

$$\phi(x)\ddot{q}(t) = -c^2\phi^{iv}(x)q(t)$$

which implies that

$$\frac{\ddot{q}(t)}{q(t)} = -c^2\frac{\phi^{iv}(x)}{\phi(x)} = -\omega^2 \tag{7.96}$$

where ω is a constant to be determined. Equation 96 leads to the following two equations:

$$\ddot{q} + \omega^2 q = 0 \tag{7.97}$$

$$\phi^{iv} - \left(\frac{\omega}{c}\right)^2\phi = 0 \tag{7.98}$$

The solution of Eq. 97 is given by

$$q = B_1 \sin \omega t + B_2 \cos \omega t \tag{7.99}$$

For Eq. 98, we assume a solution in the form

$$\phi = Ae^{\lambda x}$$

Substituting this assumed solution into Eq. 98 yields

$$\left[\lambda^4 - \left(\frac{\omega}{c}\right)^2\right]Ae^{\lambda x} = 0$$

or

$$\lambda^4 - \left(\frac{\omega}{c}\right)^2 = 0$$

which can be written as

$$\lambda^4 - \eta^4 = 0 \tag{7.100}$$

where

$$\eta = \sqrt{\frac{\omega}{c}} \tag{7.101}$$

The roots of Eq. 100 are

$$\lambda_1 = \eta, \qquad \lambda_2 = -\eta, \qquad \lambda_3 = i\eta, \qquad \lambda_4 = -i\eta$$

where $i = \sqrt{-1}$. Therefore, the general solution of Eq. 98 can be written as

$$\phi(x) = A_1 e^{\eta x} + A_2 e^{-\eta x} + A_3 e^{i\eta x} + A_4 e^{-i\eta x} \tag{7.102}$$

which can be rewritten as

$$\phi(x) = A_5 \frac{e^{\eta x} - e^{-\eta x}}{2} + A_6 \frac{e^{\eta x} + e^{-\eta x}}{2} + A_7(-i)\frac{e^{i\eta x} - e^{-i\eta x}}{2} + A_8 \frac{e^{i\eta x} + e^{-i\eta x}}{2} \tag{7.103}$$

where

$$A_1 = \frac{A_5 + A_6}{2}, \qquad A_2 = \frac{A_6 - A_5}{2}$$

$$A_3 = \frac{A_8 - iA_7}{2}, \qquad A_4 = \frac{A_8 + iA_7}{2}$$

Equation 103 can then be rewritten, using Euler's formula of the complex variables, as

$$\phi(x) = A_5 \sinh \eta x + A_6 \cosh \eta x + A_7 \sin \eta x + A_8 \cos \eta x \tag{7.104}$$

Substituting Eqs. 99 and 104 into Eq. 93 yields

$$v(x, t) = (A_5 \sinh \eta x + A_6 \cosh \eta x + A_7 \sin \eta x + A_8 \cos \eta x) \cdot (B_1 \sin \omega t + B_2 \cos \omega t) \tag{7.105}$$

where ω is defined by Eq. 101 as

$$\omega = c\eta^2 \tag{7.106}$$

Boundary Conditions The natural frequencies of the beam, as well as the constants that appear in Eq. 105, depend on the boundary and initial conditions. For example, if the beam is *simply supported at both ends* the boundary conditions are

$$v(0, t) = 0, \qquad v''(0, t) = 0$$

$$v(l, t) = 0, \qquad v''(l, t) = 0$$

which imply that

$$\left.\begin{aligned} \phi(0) = 0, \qquad \phi''(0) = 0 \\ \phi(l) = 0, \qquad \phi''(l) = 0 \end{aligned}\right. \tag{7.107}$$

It is clear that in this case, there are two geometric boundary conditions that specify the displacements at the two ends of the beam and there are two natural boundary conditions that specify the moments at the ends of the beam. Substituting these conditions into Eq. 104 yields

$$\left.\begin{aligned} A_6 + A_8 = 0 \\ A_6 - A_8 = 0 \\ A_5 \sinh \eta l + A_6 \cosh \eta l + A_7 \sin \eta l + A_8 \cos \eta l = 0 \\ A_5 \sinh \eta l + A_6 \cosh \eta l - A_7 \sin \eta l - A_8 \cos \eta l = 0 \end{aligned}\right\} \tag{7.108}$$

These equations are satisfied if $A_5 = A_6 = A_8 = 0$ and

$$A_7 \sin \eta l = 0 \tag{7.109}$$

The roots of Eq. 109 are

$$\eta l = j\pi, \qquad j = 1, 2, 3, \ldots \tag{7.110}$$

Therefore, the natural frequencies are given by

$$\omega_j = \frac{j^2\pi^2}{l^2} c = \frac{j^2\pi^2}{l^2} \sqrt{\frac{EI_z}{\rho A}}, \qquad j = 1, 2, 3, \ldots \tag{7.111}$$

and the corresponding modes of vibration are

$$\phi_j = A_{7j} \sin \eta_j x, \qquad j = 1, 2, 3, \ldots \tag{7.112}$$

The solution for the free vibration of the simply supported beam can then be written as

$$v(x, t) = \sum_{j=1}^{\infty} \phi_j q_j = \sum_{j=1}^{\infty} (C_j \sin \omega_j t + D_j \cos \omega_j t) \sin \eta_j x \tag{7.113}$$

The arbitrary constants C_j and D_j, $j = 1, 2, 3, \ldots$, can be determined using the initial conditions by the method described in Section 1 of this chapter.

In the case of a *cantilever beam*, the geometric boundary conditions at the fixed end are

$$v(0, t) = 0, \qquad v'(0, t) = 0$$

and the natural boundary conditions at the free end are

$$v''(l, t) = 0, \qquad v'''(l, t) = 0$$

These conditions imply that

$$\phi(0) = 0, \qquad \phi'(0) = 0$$

$$\phi''(l) = 0, \qquad \phi'''(l) = 0$$

Substituting these conditions into Eq. 104 yields the frequency equation

$$\cos \eta l \cosh \eta l = -1$$

The roots of this frequency equation can be determined numerically. The first six roots are (Timoshenko et al., 1974), $\eta_1 l = 1.875$, $\eta_2 l = 4.694$, $\eta_3 l = 7.855$, $\eta_4 l = 10.996$, $\eta_5 l = 14.137$, and $\eta_6 l = 17.279$. Approximate values of these roots can be calculated using the equation

$$\eta_j l \approx (j - \tfrac{1}{2})\pi$$

The fundamental natural frequencies of the system can be obtained using Eqs. 92 and 101 as

$$\omega_j = \eta_j^2 c = \eta_j^2 \sqrt{\frac{EI_z}{\rho A}}$$

The first six natural frequencies are

$$\omega_1 = 3.51563 \sqrt{\frac{EI_z}{ml^3}}, \qquad \omega_2 = 22.03364 \sqrt{\frac{EI_z}{ml^3}}$$

$$\omega_3 = 61.7010 \sqrt{\frac{EI_z}{ml^3}}, \qquad \omega_4 = 120.9120 \sqrt{\frac{EI_z}{ml^3}}$$

$$\omega_5 = 199.8548 \sqrt{\frac{EI_z}{ml^3}}, \qquad \omega_6 = 298.5638 \sqrt{\frac{EI_z}{ml^3}}$$

In this case it can be verified that the mode shapes are

$$\phi_j(x) = A_{6j}[\sin \eta_j x - \sinh \eta_j x + \bar{D}_j(\cos \eta_j x - \cosh \eta_j x)], \qquad j = 1, 2, \ldots$$

where A_{6j} is an arbitrary constant and

$$\bar{D}_j = \frac{\cos \eta_j l + \cosh \eta_j l}{\sin \eta_j l - \sinh \eta_j l}$$

7.4 ORTHOGONALITY OF THE EIGENFUNCTIONS

In this section, we study in more detail the important property of the orthogonality of the eigenfunctions of the continuous systems. This property can be used to obtain an infinite number of decoupled second-order ordinary differential equations whose solution can be presented in a simple closed form. This development can be used to justify the use of approximate techniques to obtain a finite-dimensional model that represents, to a certain degree of accuracy, the vibration of the continuous systems. Furthermore, the use of the orthogonality of the eigenfunctions leads to the important definitions of the *modal mass*, *modal stiffness*, and *modal force coefficients* for the continuous systems. As will be seen in this section, there are an infinite number of such coefficients since a continuous system has an infinite number of degrees of freedom.

Longitudinal and Torsional Vibration of Rods The partial differential equations that govern the longitudinal and torsional vibration of rods have the same form, and consequently the resulting eigenfunctions are the same for similar end conditions. Therefore, in the following we discuss only the orthogonality of the eigenfunctions of the longitudinal vibration of rods.

It was shown, in Section 1, that the partial differential equation for the longitudinal vibration of rods can be written as (see Eq. 7)

$$\rho A \frac{\partial^2 u}{\partial t^2} = \frac{\partial}{\partial x}\left(EA \frac{\partial u}{\partial x}\right) \tag{7.114}$$

where $u = u(x, t)$ is the longitudinal displacement, ρ and A are, respectively, the mass density and cross-sectional area, and E is the modulus of elasticity. The solution of Eq. 114, which was obtained by using the separation of variables technique, can be expressed as

$$u(x, t) = \phi(x)q(t) \tag{7.115}$$

where $\phi(x)$ is a space-dependent function, and $q(t)$ depends only on time and can be expressed as

$$q(t) = B_1 \sin \omega t + B_2 \cos \omega t \tag{7.116}$$

By using Eqs. 115 and 116, the acceleration $\partial^2 u/\partial t^2$ can be written as

$$\frac{\partial^2 u}{\partial t^2} = -\omega^2 \phi(x) q(t) \tag{7.117}$$

Therefore, Eq. 114 can be written as

$$-\rho A \omega^2 \phi(x) q(t) = (EA\, \phi'(x))' q(t)$$

which leads to

$$-\rho A \omega^2\, \phi(x) = (EA\phi'(x))' \tag{7.118}$$

For the jth eigenfunction ϕ_j, $j = 1, 2, 3, \ldots$, Eq. 118 yields

$$(EA\phi_j'(x))' = -\rho A \omega_j^2 \phi_j(x) \tag{7.119}$$

Multiplying both sides of this equation by $\phi_k(x)$ and integrating over the length, one obtains

$$\int_0^l (EA\phi_j'(x))' \phi_k(x)\, dx = -\omega_j^2 \int_0^l \rho A \phi_j(x) \phi_k(x)\, dx$$

Integrating the integral on the left-hand side of this equation by parts, one obtains

$$EA\phi_j'(x)\phi_k(x)\Big|_0^l - \int_0^l EA\phi_j'(x)\phi_k'(x)\, dx = -\omega_j^2 \int_0^l \rho A \phi_j(x)\phi_k(x)\, dx \tag{7.120}$$

where l is the length of the rod.

For *simple end conditions* such as *free ends* ($\phi_j'(l) = 0$) or *fixed ends* ($\phi_j(l) = 0$), the first term in Eq. 120 is identically zero, and this equation reduces to

$$\int_0^l EA\phi_j'\phi_k'\, dx = \omega_j^2 \int_0^l \rho A \phi_j \phi_k\, dx \tag{7.121}$$

Similarly, for the kth eigenfunction, we have

$$\int_0^l EA\phi_k'\phi_j'\, dx = \omega_k^2 \int_0^l \rho A \phi_k \phi_j\, dx \tag{7.122}$$

Subtracting Eq. 122 from Eq. 121, one obtains

$$(\omega_j^2 - \omega_k^2) \int_0^l \rho A \phi_j \phi_k\, dx = 0$$

Assuming that ω_j and ω_k are distinct eigenvalues, that is, $\omega_j \neq \omega_k$, the preceding equation yields for $j \neq k$

$$\int_0^l \rho A \phi_j \phi_k \, dx = 0$$
$$\int_0^l EA\phi'_j \phi'_k \, dx = 0 \tag{7.123a}$$

and for $j = k$ we have

$$\int_0^l \rho A \phi_j^2 \, dx = m_j$$
$$\int_0^l EA\phi_j'^2 \, dx = k_j \tag{7.123b}$$

The coefficients m_j and k_j, $j = 1, 2, 3, \ldots$, are called, respectively, the *modal mass* and *modal stiffness* coefficients, and they have, respectively, the units of mass and stiffness. It is also clear from Eq. 121 that

$$k_j = \omega_j^2 m_j, \qquad j = 1, 2, 3, \ldots \tag{7.124}$$

That is, the jth natural frequency ω_j is defined by

$$\omega_j^2 = \frac{k_j}{m_j} = \frac{\int_0^l EA\phi_j'^2 \, dx}{\int_0^l \rho A \phi_j^2 \, dx}, \qquad j = 1, 2, 3, \ldots \tag{7.125}$$

For *torsional systems* one can follow a similar procedure to show that the jth natural frequency of the torsional oscillations is given by

$$\omega_j^2 = \frac{k_j}{m_j} = \frac{\int_0^l GI_p \phi_j'^2 \, dx}{\int_0^l \rho I_p \phi_j^2 \, dx}, \qquad j = 1, 2, 3, \ldots \tag{7.126}$$

where G is the modulus of rigidity and I_p is the polar moment of inertia.

Perhaps it is important to emphasize at this point that, if the end conditions are not simple the definitions of the modal mass and stiffness coefficients given by Eq. 123b must be modified, and in this case the general relationship of Eq. 120 must be used to define the modal coefficients.

Transverse Vibration It was shown in Section 3 that the partial differential equation that governs the free transverse vibration of beams is given by

$$\rho A \frac{\partial^2 v}{\partial t^2} = -\frac{\partial^2}{\partial x^2}\left(EI_z \frac{\partial^2 v}{\partial x^2}\right) \tag{7.127}$$

where $v = v(x, t)$ is the transverse displacement, ρ is the mass density, A is the cross-sectional area, E is the modulus of elasticity, and I_z is the moment of inertia of the cross section about the z-axis. The solution of this equation can be written using the separation of variables technique as

$$v = \phi(x)q(t)$$

where $\phi(x)$ and $q(t)$ are, respectively, space- and time-dependent functions. The function $q(t)$ is defined as

$$q(t) = B_1 \sin \omega t + B_2 \cos \omega t$$

Using the preceding two equations, Eq. 127 yields

$$\omega^2 \rho A \phi(x) q(t) = (EI_z \phi''(x))''\, q(t)$$

That is,

$$\omega^2 \rho A \phi(x) = (EI_z \phi''(x))''$$

Therefore, for the jth natural frequency ω_j, one has

$$(EI_z \phi_j'')'' = \omega_j^2 \rho A \phi_j$$

Multiplying this equation by ϕ_k and integrating over the length, we have

$$\int_0^l (EI_z \phi_j'')'' \phi_k \, dx = \omega_j^2 \int_0^l \rho A \phi_j \phi_k \, dx, \qquad j = 1, 2, 3, \ldots \tag{7.128}$$

The integral on the left-hand side of this equation can be integrated by parts to yield

$$\int_0^l (EI_z \phi_j'')'' \phi_k \, dx = (EI_z \phi_j'')' \phi_k \Big|_0^l - EI_z \phi_j'' \phi_k' \Big|_0^l + \int_0^l EI_z \phi_j'' \phi_k'' \, dx$$

Therefore, Eq. 128 can be written as

$$(EI_z \phi_j'')' \phi_k \Big|_0^l - EI_z \phi_j'' \phi_k' \Big|_0^l + \int_0^l EI_z \phi_j'' \phi_k'' \, dx = \omega_j^2 \int_0^l \rho A \phi_j \phi_k \, dx, \qquad j = 1, 2, 3, \ldots \tag{7.129}$$

This is the general expression for the orthogonality condition of the eigenfunctions of the transverse vibration of beams.

One can show that if the beam has *simple end conditions* such as fixed ends, free ends, or simply supported ends, the orthogonality condition of Eq. 129 reduces to

$$\int_0^l EI_z \phi_j'' \phi_k'' \, dx = \omega_j^2 \int_0^l \rho A \phi_j \phi_k \, dx \tag{7.130}$$

Similarly, for the kth natural frequency ω_k, we have

$$\int_0^l EI_z \phi_j'' \phi_k'' \, dx = \omega_k^2 \int_0^l \rho A \phi_j \phi_k \, dx \tag{7.131}$$

Subtracting Eq. 131 from Eq. 130, we obtain the following relatinships for the simple end conditions in the case $j \neq k$:

$$\left.\begin{aligned}\int_0^l \rho A \phi_j \phi_k \, dx &= 0 \\ \int_0^l EI_z \phi_j'' \phi_k'' \, dx &= 0\end{aligned}\right\} \tag{7.132}$$

and for $j = k$

$$\left.\begin{aligned}\int_0^l \rho A \phi_j^2 \, dx &= m_j \\ \int_0^l EI_z \phi_j''^2 \, dx &= k_j\end{aligned}\right\} \tag{7.133}$$

where m_j and k_j are, respectively, the modal mass and modal stiffness coefficients which from Eq. 130 are related by

$$k_j = \omega_j^2 m_j \tag{7.134}$$

or

$$\omega_j^2 = \frac{k_j}{m_j} = \frac{\int_0^l EI_z \phi_j''^2 \, dx}{\int_0^l \rho A \phi_j^2 \, dx} \tag{7.135}$$

If the end conditions are not simple, Eq. 129 can still be used to define the modal mass and stiffness coefficients.

Example 7.5

Find the orthogonality relationships of the mode shapes of the longitudinal vibration of the rod shown in Fig. 10.

Solution. The boundary conditions for this system are

$$u(0, t) = 0$$

$$m \frac{\partial^2 u(l, t)}{\partial t^2} = -ku(l, t) - EA \frac{\partial u(l, t)}{\partial x}$$

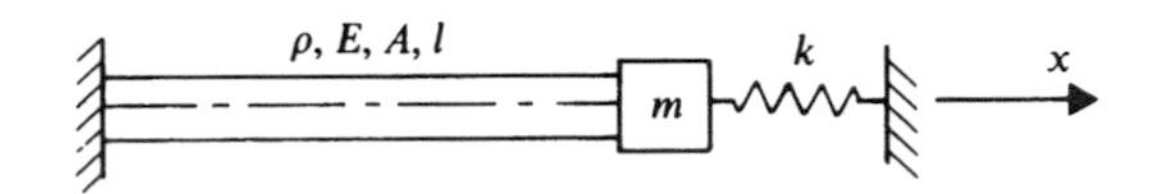

FIG. 7.10. Longitudinal vibration of bars.

These conditions yield

$$\phi(0) = 0$$

and

$$(k - \omega^2 m)\phi(l) = -EA\phi'(l)$$

The general orthogonality relationship of Eq. 120 can be written as

$$EA\phi_j'(l)\phi_k(l) - EA\phi_j'(0)\phi_k(0) - \int_0^l EA\phi_j'(x)\phi_k'(x)\,dx$$

$$= -\omega_j^2 \int_0^l \rho A\phi_j(x)\phi_k(x)\,dx$$

By using the boundary conditions of this example, this orthogonality relationship can be written as

$$-(k - \omega_j^2 m)\phi_j(l)\phi_k(l) - \int_0^l EA\phi_j'(x)\phi_k'(x)\,dx = -\omega_j^2 \int_0^l \rho A\phi_j(x)\phi_k(x)\,dx$$

or

$$k\phi_j(l)\phi_k(l) + \int_0^l EA\phi_j'(x)\phi_k'(x)\,dx = \omega_j^2\left[m\phi_j(l)\phi_k(l) + \int_0^l \rho A\phi_j(x)\phi_k(x)\,dx\right]$$

Similar relationship can be obtained for mode k as

$$k\phi_j(l)\phi_k(l) + \int_0^l EA\phi_j'(x)\phi_k'(x)\,dx = \omega_k^2\left[m\phi_j(l)\phi_k(l) + \int_0^l \rho A\phi_j(x)\phi_k(x)\,dx\right]$$

Subtracting this equation from the one associated with mode j, we obtain the following orthogonality relationships for $j \neq k$:

$$m\phi_j(l)\phi_k(l) + \int_0^l \rho A\phi_j(x)\phi_k(x)\,dx = 0$$

$$k\phi_j(l)\phi_k(l) + \int_0^l EA\phi_j'(x)\phi_k'(x)\,dx = 0$$

and for $j = k$ we have

$$m\phi_j^2(l) + \int_0^l \rho A\phi_j^2(x)\,dx = m_j$$

$$k\phi_j^2(l) + \int_0^l EA\phi_j'^2(x)\,dx = k_j$$

where m_j and k_j are, respectively, the modal mass and stiffness coefficients. The jth natural frequency of the system can be defined as

$$\omega_j^2 = \frac{k_j}{m_j} = \frac{k\phi_j^2(l) + \int_0^l EA\phi_j'^2(x)\,dx}{m\phi_j^2(l) + \int_0^l \rho A\phi_j^2(x)\,dx}$$

Note that if m and k approach zero, the natural frequency ω_j approaches the value obtained by Eq. 125 for the simple end conditions.

7.5 FORCED LONGITUDINAL AND TORSIONAL VIBRATIONS

In this section, we use an analytical approach for developing the differential equations of the forced longitudinal and torsional vibrations of continuous systems. As shown in the preceding sections, the vibrations of continuous systems are governed by partial differential equations expressed in terms of variables that are space and time dependent. In this section, the orthogonality of the eigenfunctions (mode shapes) are used to convert the partial differential equation to an infinite number of uncoupled second-order ordinary differential equations expressed in terms of the modal coordinates. These equations are similar to the equations that govern the vibration of single degree of freedom systems.

Longitudinal Vibration It was shown in Section 1 that the partial differential equation that governs the longitudinal forced vibration of rods is given by

$$\rho A \frac{\partial^2 u}{\partial t^2} = \frac{\partial}{\partial x}\left(EA \frac{\partial u}{\partial x}\right) + F(x, t) \tag{7.136}$$

where ρ, A, and E are, respectively, the mass density, cross-sectional area, and modulus of elasticity, $u = u(x, t)$ is the longitudinal displacement, and $F(x, t)$ is a space- and time-dependent axial forcing function.

Using the technique of the separation of variables, the displacement u can be written as

$$u(x, t) = \sum_{j=1}^{\infty} \phi_j(x) q_j(t) \tag{7.137}$$

where ϕ_j is the jth space-dependent eigenfunction (mode shape) and q_j is the time-dependent modal coordinate. A virtual change in the longitudinal displacement u is an infinitesimal displacement consistent with the boundary conditions (Shabana, 1991) and is

$$\delta u = \sum_{j=1}^{\infty} \phi_j \delta q_j \tag{7.138}$$

Multiplying Eq. 136 by δu and integrating over the length of the beam leads to

$$\int_0^l \rho A \frac{\partial^2 u}{\partial t^2} \delta u \, dx = \int_0^l \frac{\partial}{\partial x}\left(EA \frac{\partial u}{\partial x}\right) \delta u \, dx + \int_0^l F(x, t) \delta u \, dx \tag{7.139}$$

Substituting Eqs. 137 and 138 into Eq. 139 yields

$$\sum_{j=1}^{\infty} \sum_{k=1}^{\infty} \int_0^l [\rho A \phi_k(x) \phi_j(x) \ddot{q}_k - (EA\phi_k'(x))' \phi_j(x) q_k - Q_j] \, dx \, \delta q_j = 0 \tag{7.140}$$

where

$$Q_j = F(x, t) \phi_j(x) \tag{7.141}$$

By using the integration by parts, one has

$$\int_0^l (EA\phi_k'(x))'\phi_j(x)\,dx = EA\phi_k'(x)\phi_j(x)\Big|_0^l - \int_0^l EA\phi_k'(x)\phi_j'(x)\,dx \quad (7.142)$$

Substituting this equation into Eq. 140 yields

$$\sum_{j=1}^{\infty}\left[\sum_{k=1}^{\infty}\int_0^l [\rho A\phi_k(x)\phi_j(x)\ddot{q}_k + EA\phi_k'(x)\phi_j'(x)q_k - Q_j]\,dx - EA\phi_k'(x)\phi_j(x)\Big|_0^l q_k\right]\delta q_j = 0 \quad (7.143)$$

Using the boundary conditions and the orthogonality relationship of the eigenfunctions, one can show in general that

$$\sum_{k=1}^{\infty}\left[\int_0^l (\rho A\phi_k(x)\phi_j(x)\ddot{q}_k + EA\phi_k'(x)\phi_j'(x)q_k)\,dx - EA\phi_k'(x)\phi_j(x)\Big|_0^l q_k\right] = m_j\ddot{q}_j + k_jq_j \quad (7.144)$$

where m_j and k_j are, respectively, the modal mass and stiffness coefficients that depend on the boundary conditions and can be defined using the orthogonality relationships of the eigenfunctions. If Eq. 144 is substituted into Eq. 143, one gets

$$\sum_{j=1}^{\infty}[m_j\ddot{q}_j + k_jq_j - Q_j]\delta q_j = 0 \quad (7.145)$$

Since the virtual changes δq_j are linearly independent, Eq. 145 yields

$$m_j\ddot{q}_j + k_jq_j = Q_j, \qquad j = 1, 2, 3, \ldots \quad (7.146)$$

These equations, which are uncoupled second-order ordinary differential equations, are in the same form as the vibration equations of the single degree of freedom systems. Therefore, their solution can be obtained using Duhamel's integral as

$$q_j = q_{j0}\cos\omega_jt + \frac{\dot{q}_{j0}}{\omega_j}\sin\omega_jt + \frac{1}{m_j\omega_j}\int_0^t Q_j(\tau)\sin\omega_j(t-\tau)\,d\tau, \qquad j = 1, 2, 3, \ldots \quad (7.147)$$

where q_{j0} and $\dot{q}_{j0}$ are the initial modal displacements and velocities and ω_j is the jth natural frequency defined as

$$\omega_j = \sqrt{\frac{k_j}{m_j}}, \qquad j = 1, 2, 3, \ldots \quad (7.148)$$

Having determined q_j using Eq. 147, the longitudinal displacement $u(x, t)$ can be determined by using Eq. 137.

In the case of free vibration, the modal force Q_j is equal to zero and Eq. 146 reduces to

$$m_j\ddot{q}_j + k_j q_j = 0, \qquad j = 1, 2, 3, \ldots \tag{7.149}$$

The solution of these equations is

$$q_j = q_{j0} \cos \omega_j t + \frac{\dot{q}_{j0}}{\omega_j} \sin \omega_j t, \qquad j = 1, 2, 3, \ldots \tag{7.150}$$

Concentrated Loads If the force F is a concentrated load that acts at a point p on the beam, that is, $F = F(t)$, there is no need to carry out integration in order to obtain the modal forces. In this case, the virtual work of this force is given by

$$\begin{aligned} \delta W &= F(t)\delta u(x_p, t) \\ &= F(t) \sum_{j=1}^{\infty} \phi_j(x_p)\delta q_j(t) \\ &= \sum_{j=1}^{\infty} Q_j \delta q_j(t) \end{aligned}$$

where Q_j, $j = 1, 2, 3, \ldots$, is the modal force associated with the jth modal coordinate and defined as

$$Q_j = F(t)\phi_j(x_p) \tag{7.151}$$

in which $\phi_j(x_p)$ is the jth mode shape evaluated at point p on the beam.

Torsional Vibration The equation for the forced torsional vibration, which takes a similar form to the equation of forced longitudinal vibration, was given in Section 2 as

$$\rho I_p \frac{\partial^2 \theta}{\partial t^2} = \frac{\partial}{\partial x}\left(G I_p \frac{\partial \theta}{\partial x}\right) + T_e(x, t) \tag{7.152}$$

where ρ, I_p, and G are, respectively, the mass density, polar moment of inertia, and modulus of rigidity, $\theta = \theta(x, t)$ is the angle of torsional oscillation, and $T_e(x, t)$ is the external torque which is time and space dependent. The solution of Eq. 152 can be expressed using the separation of variables technique as

$$\theta(x, t) = \sum_{j=1}^{\infty} \phi_j(x) q_j(t) \tag{7.153}$$

Following the same procedure as in the case of longitudinal vibration, one can show that the equations of the torsional vibration of the shaft in terms of the modal coordinates are given by

$$m_j\ddot{q}_j + k_j q_j = Q_j, \qquad j = 1, 2, 3, \ldots \tag{7.154}$$

where m_j and k_j are, respectively, the modal mass and stiffness coefficients and

$$Q_j = \int_0^l T_e(x, t)\phi_j(x)\,dx \tag{7.155}$$

The solution of Eq. 154 is defined by Eq. 147, and therefore, the mathematical treatment of the linear torsional oscillations of shafts is the same as the one used for the longitudinal vibration of rods.

Example 7.6

If the rod in Example 5 is subjected to a distributed axial force of the form $F(x, t)$, determine the equations of forced longitudinal vibration of this system.

Solution. The boundary conditions in this example are

$$u(0, t) = 0$$

$$m\frac{\partial^2 u}{\partial t^2} = -ku(l, t) - EA\frac{\partial u(l, t)}{\partial x}$$

which yield

$$\phi(0) = 0$$

$$EA\phi'(l)q(t) = -k\phi(l)q(t) - m\phi(l)\ddot{q}(t)$$

That is,

$$\phi_k(0)\phi_j(0) = 0$$

$$EA\phi_k'(l)\phi_j(l)q_k(t) = -k\phi_k(l)\phi_j(l)q_k(t) - m\phi_k(l)\phi_j(l)\ddot{q}_k(t)$$

The last term on the left-hand side of Eq. 144 can be written as

$$EA\phi_k'(x)\phi_j(x)\Big|_0^l q_k = (EA\phi_k'(l)\phi_j(l) - EA\phi_k'(0)\phi_j(0))q_k$$

which upon using the boundary conditions yields

$$EA\phi_k'(x)\phi_j(x)\Big|_0^l q_k = -k\phi_k(l)\phi_j(l)q_k - m\phi_k(l)\phi_j(l)\ddot{q}_k$$

Therefore, Eq. 144 can be written as

$$\sum_{k=1}^{\infty}\int_0^l (\rho A\phi_k(x)\phi_j(x)\ddot{q}_k + EA\phi_k'(x)\phi_j'(x)q_k)\,dx + k\phi_k(l)\phi_j(l)q_k + m\phi_k(l)\phi_j(l)\ddot{q}_k$$

$$= \sum_{k=1}^{\infty}\left[m\phi_k(l)\phi_j(l) + \int_0^l \rho A\phi_k(x)\phi_j(x)\,dx\right]\ddot{q}_k$$

$$+ \sum_{k=1}^{\infty}\left[k\phi_k(l)\phi_j(l) + \int_0^l EA\phi_k'(x)\phi_j'(x)\,dx\right]q_k$$

Comparing this equation with the orthogonality relationships obtained in Example 5, it is clear that

$$\sum_{k=1}^{\infty}\left[m\phi_k(l)\phi_j(l) + \int_0^l \rho A\phi_k(x)\phi_j(x)\,dx\right]\ddot{q}_k$$

$$+ \sum_{k=1}^{\infty}\left[k\phi_k(l)\phi_j(l) + \int_0^l EA\phi_k'(x)\phi_j'(x)\,dx\right]q_k = m_j\ddot{q}_j + k_jq_j$$

where m_j and k_j are the modal mass and stiffness coefficients defined by

$$m_j = m\phi_j^2(l) + \int_0^l \rho A\phi_j^2(x)\,dx$$

$$k_j = k\phi_k^2(l) + \int_0^l EA\phi_j'^2(x)\,dx, \qquad j = 1, 2, 3, \ldots$$

Therefore, the equations of motion of the forced longitudinal vibration of the rod, expressed in terms of the modal coordinates, are

$$m_j\ddot{q}_j + k_jq_j = Q_j, \qquad j = 1, 2, 3, \ldots$$

7.6 FORCED TRANSVERSE VIBRATIONS

The partial differential equation of the forced transverse vibration of the beams was given by Eq. 89 as

$$\rho A\frac{\partial^2 v}{\partial t^2} + \frac{\partial^2}{\partial x^2}\left(EI_z\frac{\partial^2 v}{\partial x^2}\right) = F(x, t) \tag{7.156}$$

where ρ, A, I_z, and E are, respectively, the mass density, cross-sectional area, second moment of area, and modulus of elasticity, $v(x, t)$ is the transverse displacement, and $F(x, t)$ is the forcing function which may depend on the spatial coordinate x and time t. By using the technique of separation of variables we may write the transverse displacement v as

$$v = \sum_{j=1}^{\infty}\phi_j(x)q_j(t) \tag{7.157}$$

Multiplying both sides of Eq. 156 by the virtual displacement δv (Shabana, 1991) and integrating over the length of the beam we obtain

$$\int_0^l\left[\rho A\frac{\partial^2 v}{\partial t^2}\delta v + \frac{\partial^2}{\partial x^2}\left(EI_z\frac{\partial^2 v}{\partial x^2}\right)\delta v\right]dx = \int_0^l F(x, t)\delta v\,dx$$

which upon using Eq. 157 yields

$$\sum_{j=1}^{\infty}\sum_{k=1}^{\infty}\int_0^l [\rho A\phi_k(x)\phi_j(x)\ddot{q}_k + (EI_z\phi_k''(x))''\phi_j(x)q_k]\delta q_j\,dx$$

$$= \sum_{j=1}^{\infty}\int_0^l F(x, t)\phi_j(x)\delta q_j\,dx \tag{7.158}$$

Integration by parts yields

$$\int_0^l (EI_z\phi_k''(x))''\phi_j(x)\,dx = (EI_z\phi_k'')'\phi_j\Big|_0^l - EI_z\phi_k''\phi_j'\Big|_0^l + \int_0^l EI_z\phi_k''\phi_j''\,dx \quad (7.159)$$

Substituting this equation into Eq. 158 and using the boundary conditions and the orthogonality relationships of the eigenfunctions, one gets

$$\sum_{j=1}^{\infty} [m_j\ddot{q}_j + k_jq_j - Q_j]\delta q_j = 0 \quad (7.160)$$

where m_j and k_j are the modal mass and modal stiffness coefficients that depend on the boundary conditions and can be defined using the orthogonality of the mode shapes, and Q_j is the modal forcing function defined as

$$Q_j = \int_0^l F(x, t)\phi_j(x)\,dx \quad (7.161)$$

Since the modal coordinates $q_j, j = 1, 2, 3, \ldots$, are independent, Eq. 160 yields the following uncoupled second-order ordinary differential equations:

$$m_j\ddot{q}_j + k_jq_j = Q_j, \qquad j = 1, 2, 3, \ldots \quad (7.162)$$

These equations are in a form similar to the one obtained in the preceding section for the two cases of longitudinal and torsional vibrations.

Problems

7.1. Determine the equation of motion, boundary conditions, and frequency equation of the longitudinal vibration of the system shown in Fig. P1, assuming that the rod has a uniform cross-sectional area.

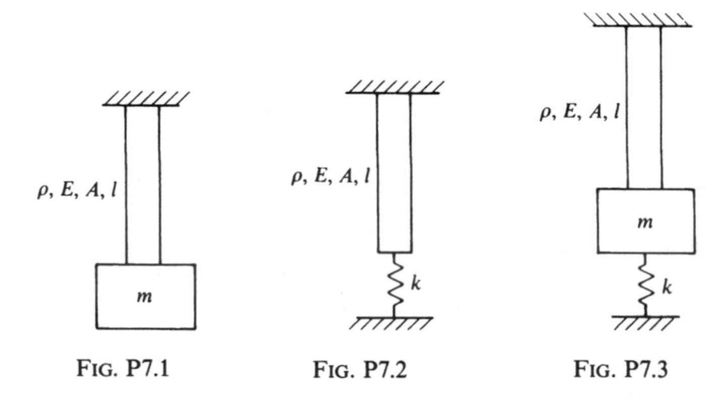

FIG. P7.1 FIG. P7.2 FIG. P7.3

7.2. The system shown in Fig. P2 consists of a uniform rod with a spring attached to its end. Derive the partial differential equation of motion and determine the boundary conditions and frequency equation of the longitudinal vibration.

7.3. Obtain the equation of motion, boundary conditions, and frequency equation of the system shown in Fig. P3.

7.4. Determine the equation of motion, boundary conditions, and frequency equation of longitudinal vibration of a uniform rod with a mass m attached to each end. Check the fundamental frequency by reducing the uniform rod to a spring with end masses.

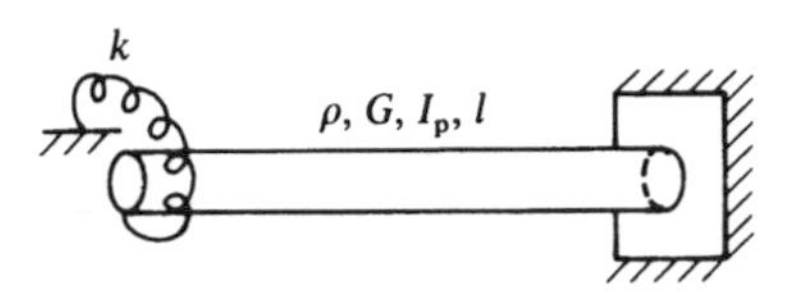

FIG. P7.4

7.5. Determine the equation of motion, boundary conditions, and natural frequencies of a torsional system which consists of a uniform shaft of mass moment of inertia I with a disk having mass moment of inertia I_1 attached to each end of the shaft.

7.6. Derive the equation of motion of the system shown in Fig. P4 which consists of a uniform shaft with a torsional spring of torsional stiffness k attached to its end. Obtain the boundary conditions and the frequency equation.

7.7. Obtain the equation of torsional oscillation of a uniform shaft with a torsional spring of stiffness k attached to each end.

7.8. Determine the equation of transverse vibration, boundary conditions, and frequency equation of a uniform beam of length l clamped at one end and pinned at the other end.

7.9. Obtain the frequency equation of the transverse vibration of the beam shown in Fig. P5.

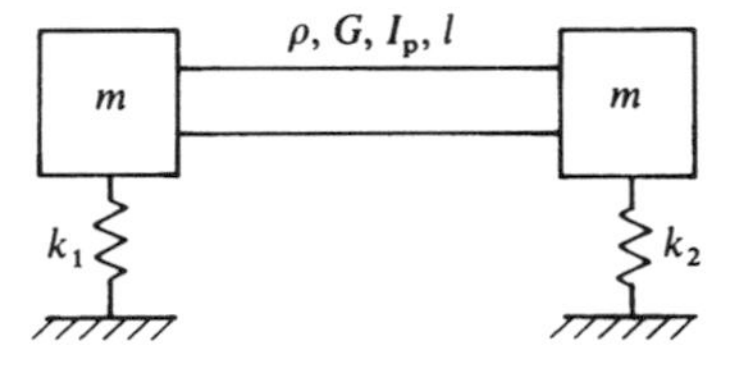

FIG. P7.5

7.10. A uniform rod which is fixed at one end and free at the other end has the following initial conditions:

$$u(x, 0) = A_0 \sin \frac{\pi x}{2l}$$

$$\dot{u}(x, 0) = 0$$

Obtain the general solution of the longitudinal vibration.

7.11. A uniform shaft which is fixed at one end and free at the other end has the following initial conditions:

$$\theta(x, 0) = A_0 \sin \frac{n\pi x}{2l}$$

$$\dot{\theta}(x, 0) = 0$$

where n is a fixed odd number. Obtain the solution of the free longitudinal vibration of this system.

7.12. A uniform rod with one end fixed and the other end free is subjected to a distributed axial load in the form

$$F(x, t) = x \sin 5t$$

Determine the response of the system as the result of application of this forcing function.

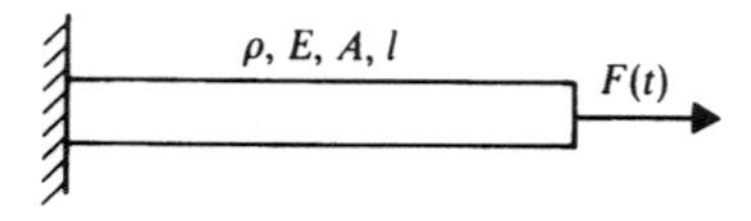

FIG. P7.6

7.13. Determine the response of the system shown in Fig. P6 to a concentrated harmonic forcing function.

References

Atkinson, K.E. (1978) *An Introduction to Numerical Analysis*, Wiley, New York.

Beer, E.P., and Johnston, E.R. (1977) *Vector Mechanics for Engineering; Statics and Dynamics*, third edition, McGraw-Hill, New York.

Carnahan, B., Luther, H.A., and Wilkes, J.O. (1969) *Applied Numerical Methods*, Wiley, New York.

Clough, R.W., and Penzien, J. (1975) *Dynamics of Structures*, McGraw-Hill, New York.

Craig, J.J. (1986) *Introduction to Robotics; Mechanics and Control*, Addison-Wesley, Reading, MA.

Friedland, B. (1986) *Control System Design: An Introduction to State Space Methods*, McGraw-Hill, New York.

Greenberg, M.D. (1978) *Foundations of Applied Mathematics*, Prentice Hall, Englewood Cliffs, NJ.

Hartog, D. (1968) *Mechanical Vibration*, McGraw-Hill, New York.

Hutton, D.V. (1981) *Applied Mechanical Vibrations*, McGraw Hill, New York.

Inman, D.J. (1994) *Engineering Vibration*, Prentice Hall, Englewood Cliffs., NJ.

Kuo, B.C. (1967) *Automatic Control Systems*, second edition, Prentice Hall, Englewood Cliffs, NJ.

Lewis, F.L., Abdallah, C.T., Dawson, D.M. (1993) *Control of Robot Manipulators*, Macmillan, New York.

Meirovitch, L. (1986) *Elements of Vibration Analysis*, McGraw-Hill, New York.

Miller, R.K., and Michel, A.N. (1982) *Ordinary Differential Equations*, Academic Press, New York.

Muvdi, B.B., and McNabb (1984) *Engineering Mechanics of Materials*, Macmillan, New York.

Nayfeh, A.H., and Mook, D.T. (1979) *Nonlinear Oscillations*, John Wiley & Sons, New York.

Paul, R.P. (1981) *Robot Manipulators; Mathematics, Programming, and Control*, MIT Press, Cambridge, MA.

Pipes, L.A., and Harvill, L.R. (1970) *Applied Mathematics for Engineers and Physicists*, McGraw-Hill, New York.

Rao, S.S. (1986) *Mechanical Vibrations*, Addison-Wesley, Reading, MA.

Shabana, A.A. (1991) *Theory of Vibration; Vol. II: Discrete and Continuous Systems*, Springer Verlag, New York.

Shabana, A.A. (1994) *Computational Dynamics*, Wiley, New York.

Shigley, J.E. (1963) *Mechanical Engineering Design*, McGraw-Hill, New York.

Singer, F.L. (1962) *Strength of Materials*, Harper and Row, New York.

Steidel, R.F. (1989) *An Introduction to Mechanical Vibration*, Wiley, New York.

Thomson, W.T. (1988) *Theory of Vibration with Applications*, Prentice Hall, Englewood Cliffs., NJ.

Timoshenko, S., Young, D.H., and Weaver, W. (1974) *Vibration Problems in Engineering*, Wiley, New York.

Wylie, C.R., and Barrett, L.C. (1982) *Advanced Engineering Mathematics*, McGraw-Hill, New York.

Answers to Selected Problems

CHAPTER 1

1.7. $m_1\ddot{x}_1 + (k_1 + k_2)x_1 - k_2x_2 = F_1(t)$

$m_2\ddot{x}_2 + k_2x_2 - k_2x_1 = F_1(t)$

1.8. $m_1\ddot{x}_1 + (k_1 + k_2)x_1 - k_2x_2 = 0$

$m_2\ddot{x}_2 + k_2x_2 - k_2x_1 = 0$

CHAPTER 2

2.1. $x(t) = C_1 \cos 3t + C_2 \sin 3t$

2.3. $x(t) = A_1e^{3t} + A_2e^{-3t}$

2.5. $x(t) = Ce^{(3/4)t} \sin\left(\frac{\sqrt{47}}{4}t + \phi\right)$

2.7. $x(t) = C_1 \cos 3t + C_2 \sin 3t + \frac{3}{5}e^t$

2.9. $x(t) = C \sin\left(\frac{5}{3}\sqrt{3}t + \phi\right) + \left(\frac{1}{37}t^2 - \frac{24}{1369}t + \frac{66}{50653}\right)e^{2t} + \frac{1}{25}t$

2.11. $x(t) = Ce^{-(3/4)t} \sin\left(\frac{\sqrt{47}}{4}t + \phi\right) - \frac{2}{37}\cos 2t + \frac{12}{37}\sin 2t - \frac{9}{34}\cos t + \frac{45}{102}\sin t$

2.15. $x(t) = \frac{5}{8}\sin\left(3t + \frac{\pi}{2}\right) + \frac{3}{8}\cos t$

2.17. $x(t) = 2\cos\left(\frac{4\sqrt{5}}{5}t\right)$

2.19. $x(t) = 2\cos\left(\frac{4}{\sqrt{5}}t\right) + \frac{3\sqrt{5}}{4}\sin\left(\frac{4}{\sqrt{5}}t\right)$

2.21. $x(t) = e^t\left(\frac{1367}{1690}\cos 2t + \frac{1577}{845}\sin 2t\right) + e^{2t}\left(\frac{3}{26}t - \frac{3}{338}\right)\cos 3t$

$+ e^{2t}\left(\frac{t}{13} - \frac{41}{338}\right)\sin 3t + \frac{1}{5}e^{2t}$

CHAPTER 3

3.1. $\omega = 24.495$ rad/s, $\dot{x}_{max} = 0.9798$ m/s

3.3. $x(t) = 0.204\sin(24.495t)$

3.5. $x(t) = 0.03\cos(22.147t)$, $E = 1.104$ N·m

3.7. $k_e = \dfrac{4(k_1a^2 + k_2l^2)}{l^2}$

3.9. $x(t) = 0.311\sin(16.059t)$

3.11. $\omega = \sqrt{\dfrac{k_1l^2/4 + k_2l^2 + mgl/2}{I + ml^2/4}}$

3.13. $\omega = \sqrt{\dfrac{kl^2/4 + mgl + m_rgl/2}{I + m_rl^2/4 + ml^2}}$

3.15. $\omega = \sqrt{\dfrac{kr^2 + G_1J_1/l_1 + G_2J_2/l_2}{I}}$

3.17. $\omega = \sqrt{\dfrac{k_1a^2 + k_2b^2 + mg(b-a)/2}{I + m(b-a)^2/4}}$

3.19. $\theta(t) = 0.0524\cos(95.5t)$

3.21. $\theta(t) = -0.0349\cos(88.09t)$

3.23. $x(t) = 0.055e^{-10t}\sin(43.589t + 1.1406)$

3.25. (a) $x(t) = 0.02236e^{-7.639t} - 0.02236e^{-52.361t}$

(b) $x(t) = 0.05854e^{-7.689t} - 8.541 \times 10^{-3}e^{-52.361t}$

(c) $x(t) = 0.0809e^{-7.639t} - 0.0309e^{-52.361t}$

3.27. $\omega = 35.018$ rad/s

3.29. $\omega = \sqrt{\dfrac{k_1k_3 + k_2k_3}{m(k_1 + k_2 + k_3)}}$

3.31. $ml^2\ddot{\theta} + ca^2\dot{\theta} + (kl^2 + mgl)\theta = 0$

3.33. $\theta(t) = 5te^{-44.94t}$

3.35. (a) $\xi = 0.09$, $c = 19.333$ N·s/m

(b) $\xi = 1$, $c = 214.814$ N·s/m

(c) $\xi = 1.2$, $c = 257.776$ N·s/m

3.41. $\left(I + \dfrac{ml^2}{4}\right)\ddot{\theta} + \dfrac{cl^2}{4}\dot{\theta} + \dfrac{kl^2}{4}\theta = 0$

$\theta(t) = 0.0884e^{-5t}\sin(31.225t + 1.412)$

3.47. $\delta = 0.1223$, $\xi = 0.0195$

3.51. $\xi = 1.072 \times 10^{-3}$

CHAPTER 4

4.1. $x(t) = X\sin(22.361t + \phi) - 0.05\sin 30t$

4.3. $x(t) = 0.1383\sin 25.82t - 0.08572\sin 30t$

4.11. $x(t) = -4.132 \times 10^{-2}e^{-3t}\sin(29.85t - 1.5529) + 9.820 \times 10^{-3}\sin(15t - 0.1343)$

4.15. $x(t) = 0.1174e^{-3.333t}\sin(29.814t + 1.4595) + 0.0667\sin\left(30t - \dfrac{\pi}{2}\right)$

4.19. $\theta(t) = 0.1001e^{-60t}\sin(49.289t + 0.6155) + 0.03675\sin(40t - 0.826) + 0.07844\sin(20t - 0.403)$

4.25. $(F_t)_{\max} = 35.3587$ N

4.27. $x(t) = -0.01568e^{-6.667t}\sin(29.25t - 0.8978) + 0.02557\sin(15t - 0.2881)$

4.31. $\theta(t) = 0.0323e^{-60t}\sin(91.652t + 0.03115) + 0.01\sin(10t - 0.1005)$

4.33. $\theta(t) = 0.0957e^{-23.684t}\sin(50.96t + 0.0327) + 0.0204\sin(10t - 0.1537)$

4.35. $\theta(t) = 0.01587e^{-10t}\sin(126.142t - 1.972) - 5.0\sin(5t - 0.006)$

4.37. $\theta(t) = 0.0376e^{-10t}\sin(76.918t + 0.01) + 0.011\sin(10t - 0.034)$

CHAPTER 5

5.1. $F(t) = \dfrac{F_0}{2} + \sum_{n=2,4,6}^{8}\left(\dfrac{-2F_0}{n\pi}\right)\sin nt$

5.7. $x(t) = A\sin(\omega t + \phi) + \dfrac{F_0}{2k} - \sum_{n=1,3,5}^{\infty}\dfrac{4F_0\omega^2}{(n\pi)^2k(\omega^2 - n^2)}\sin\left(nt + \dfrac{\pi}{2}\right)$

5.11. $\theta(t) = A\sin(\omega t + \phi) + \dfrac{F_0 l}{2k_e} + \sum_{n=2,4,6}^{\infty}\dfrac{(-2F_0 l)}{n\pi k_e(1 - r_n^2)}\sin nt, \qquad k_e = mgl$

5.13. $F(t) = \dfrac{3}{4}F_0 - \sum_{n=1,3,5}^{\infty}\dfrac{2F_0}{(n\pi)^2}\cos\dfrac{2n\pi}{T}t - \sum_{n=1}^{\infty}\dfrac{F_0}{n\pi}\sin\dfrac{2n\pi}{T}t$

5.17. $M(t) = \sum_{n=1,3,5}^{\infty} \left(\frac{-4M_0}{n\pi} \right) \sin \frac{n\pi}{T} t$

5.19. $\theta(t) = Ae^{-\xi\omega t} \sin(\omega_d t + \phi) + \sum_{n=1,3,5}^{\infty} \frac{(-4M_0)}{n\pi k_e \sqrt{(1 - r_n^2)^2 + (2r_n\xi)^2}} \sin\left(\frac{n\pi}{T} t - \psi_n \right)$

$\psi_n = \tan^{-1}\left(\frac{2r_n\xi}{1 - r_n^2} \right)$

CHAPTER 6

6.1. $\omega_1^2 = \frac{(2 + \sqrt{2})k}{m}, \qquad \omega_2^2 = \frac{(2 - \sqrt{2})k}{m}$

6.3. $\mathbf{M} = \begin{bmatrix} m_1 l_1 & m_2 l_1 \\ 0 & m_2 l_2 \end{bmatrix}, \qquad \mathbf{K} = \begin{bmatrix} (m_1 + m_2)g & 0 \\ -m_2 g & m_2 g \end{bmatrix}$

6.5. $x_1(t) = -0.0171 \sin\left(18.48t - \frac{\pi}{2} \right) + 2.927 \times 10^{-3} \sin\left(7.65t + \frac{\pi}{2} \right)$

$x_2(t) = 7.07 \times 10^{-3} \sin\left(18.48t - \frac{\pi}{2} \right) + 7.07 \times 10^{-3} \sin\left(7.65t + \frac{\pi}{2} \right)$

6.9. $\omega_1^2 = 0, \qquad \omega_2^2 = \frac{k(m_1 + m_2)}{m_1 m_2}$

6.11. $\omega_1^2 = 0, \qquad \omega_2^2 = \frac{k\left[\left(I + \frac{ml^2}{4} \right) + m_1 l^2 \right]}{\left(I + \frac{ml^2}{4} \right) m_1}$

6.13. $\begin{bmatrix} m_1 l^2 + I + \frac{ml^2}{4} & 0 \\ 0 & m_2 \end{bmatrix} \begin{bmatrix} \ddot{\theta} \\ \ddot{x}_2 \end{bmatrix} + \begin{bmatrix} kl^2 & -kl \\ -kl & k \end{bmatrix} \begin{bmatrix} \theta \\ x_2 \end{bmatrix} = \begin{bmatrix} 0 \\ 0 \end{bmatrix}$

6.17. $\omega_1^2 = \frac{-b + \sqrt{b^2 - 4ac}}{2a}, \qquad \omega_2^2 = \frac{-b - \sqrt{b^2 - 4ac}}{2a}$

$a = mI, \qquad b = -\left[(k_1 + k_2)I + k_1 ma + \frac{k_2 ml^2}{4} \right]$

$c = k_1 k_2 a(1 + l) + \frac{k_1 k_2 l^2}{4}$

CHAPTER 7

7.1. The frequency equation $\quad \gamma \tan \gamma = \mu$

$\gamma = \frac{\omega l}{c}, \ \mu = \frac{M}{m}, \ c = \sqrt{\frac{E}{\rho}}$

7.3. The frequency equation $\tan\frac{\omega l}{c} = \frac{EA\omega}{c(m\omega^2 - k)}$

7.5. The frequency equation $\sin\frac{\omega l}{c} = 0, \omega_n = \frac{nc\pi}{l}, c = \sqrt{\frac{G}{\rho}}$

7.7. The frequency equation $\sin\frac{\omega l}{c} = 0, \omega_n = \frac{nc\pi}{l}, c = \sqrt{\frac{G}{\rho}}$

7.9. $u(x, t) = A_0 \cos\frac{\pi ct}{2l} \sin\frac{\pi x}{2l}$

7.11. $u(x, t) = A_0 \cos\frac{n\pi ct}{2l} \sin\frac{n\pi x}{2l}$

Index

Mechanical Engineering Series *(continued)*

Laminar Viscous Flow
V.N. Constantinescu

Thermal Contact Conductance
C.V. Madhusudana

Printed in the United States
123209LV00002B/193-195/A

9 780387 945248